AF726140

HOW TO USE ROBOT TO

IMPROVE EFFICIENCY

JOHN LOK

Copyright © John Lok
All Rights Reserved.

This book has been self-published with all reasonable efforts taken to make the material error-free by the author. No part of this book shall be used, reproduced in any manner whatsoever without written permission from the author, except in the case of brief quotations embodied in critical articles and reviews.

The Author of this book is solely responsible and liable for its content including but not limited to the views, representations, descriptions, statements, information, opinions and references ["Content"]. The Content of this book shall not constitute or be construed or deemed to reflect the opinion or expression of the Publisher or Editor. Neither the Publisher nor Editor endorse or approve the Content of this book or guarantee the reliability, accuracy or completeness of the Content published herein and do not make any representations or warranties of any kind, express or implied, including but not limited to the implied warranties of merchantability, fitness for a particular purpose. The Publisher and Editor shall not be liable whatsoever for any errors, omissions, whether such errors or omissions result from negligence, accident, or any other cause or claims for loss or damages of any kind, including without limitation, indirect or consequential loss or damage arising out of use, inability to use, or about the reliability, accuracy or sufficiency of the information contained in this book.

Made with ♥ on the Notion Press Platform
www.notionpress.com

Copyright, June 2021 published
No anyone permit to download or copy any part of this book to sell.

Contents

Foreword

Introduction

Robots had been organizational internal one kind of main technological resources to be used. Some manufacturers had chosen robots to replace human workers to do simple or complex tasks in order to raise efficiency. How to use robots to help organizations to raise efficiencies or improve performance.

Is only robots to replace all human workers in one team , they can raise more efficient ? Can half human workers and half robotic cooperations work together to raise more efficient? Can robots real be major technological resources to future any organizations? In my this book , I shall attempt to explain how and why robotics may become organizational resource to help any organizations to improve efficiency and performance.Readers can have more clear understanding whether robots may be future major organizational resources in order to raise efficiencies.

My view explores the premise that robots serve as critical technological resources capable of augmenting human labor rather than merely replacing it. and argues that the strategic implementation of robotics allows organizations to transition from labor-intensive processes to high-efficiency models, where machines handle repetitive or high-precision tasks, thereby freeing human workers to focus on complex problem-solving and creative oversight.

Preface

Prologue

Table of content

Organizational robotic intangible resource management

- How to implement effective intangible resources management strategy to achieve performance improvement to Amazon e-commerce organization? Why does non intangible resources effective negligence management to Amazon organization, it will cause worse performance to Amazon e-commerce organization any one e-buyer? p.162-190

Chapter 3

Can apply robot to improve consumer behavior to improve performance

Why can (AI) be applied to help businesses to predict consumer behaviors ? p.91-169

How can artificial intelligent tools predict consumer behavior in vehicle market retail industry ?

What is (AI) consumer behavioral prediction tool?

How any why will (AI) tool assist manufactures to attempt to predict consumer behavior before and after consumption occurrence?

What is (AI) deep learning techniques to help farmers to raise food growth

How can apply (AI) technology to help book shops to decide book sale price

How can (AI) provide businesses with better-informed decisions?

How can artificial intelligent tools predict travelling consumer behavior in airline and air agent travelling market ?

Why is (AI) big data gathering tool better
than psychological and survey methods to predict traveler individual travel choice behavior?

How (AI) big data determine future travel behavior from past travel experience and perceptions of risk and safety for the benefits to travel consumers?

How (AI) influences organizational change

Can artificial intelligence development to tourism industry

- Artificial intelligence will bring what benefits to
Influence traveler consumption behavior or raising
Travelling leisure desire?

(AI) raises driven automation industry development

1.1 (AI) - driven automation industry development how to influence work nature change p.241-261

Chapter 4

Competitive influences between artificial
intelligence and human job to improve efficiency
and performance

(AI) journalism, media publishing, digital
communication technology trend p.170-200

Chapter 5

How robots help manufacturers to raise productivities

How Robots bring positive impact to raise organizational

CHAPTER ONE

Robots whether can help organizations to avoid resource waste

To judge whether robots can help any organizations to avoid resources waste behavior. We need to know whether robots can own recyle function. How do robots helpthe environment? Robots can help reduce waste that is implemented by effectively sorting materials that can be recycled and put to use again. They can even help sort waste materials quickly and move efficiently than humans reduce the input power amd cost unsociated with such processes. However, robots also may help business organizations with recycling, instead of natural environment . For example, tobots may help organization facility cut operate resource cost and allowed to take on a second load of recyclables to sort throgh.

How robotic technology help organizations reduce waste?

Advanced software has made it easier to plan out routes that can efficiently guide waste collecting the waste and recycling materials that need to be collected. It also makes collection more fuel efficient and reduces energy usage to any offices, plants, warehouses, business organizations. It seems that robots can help organizations to bring recycling advantages. We should definitely keep on teaching robots to recycle our waste, e.g. clean robotics makes waste management smart with a robotic trash can that automatically separates used and not used materials in workplace environment.

How to use robotic technology to help reduce waste?

There is no financial incentive to reduce waste. One of a few manufacturers of AI (arificial intelligence) powered recycling robots. We can imagine a future where waste-collectig robots will move through air land, and water, cleaning our natural environment, workplace environment conveniently. For example, now robots are being put on duty to help solve the environment pollution, material waste, recycling challenges. AI assisted robotis technology that can work with humans in workplace organizational environment to have robots to do a better job at sorting garbage and reduce the building materials is wasted, sorting trash is a dirty and dangerous job, recycling robots may help human to do thin kind of sorting trash job, even further atomic bombs soon could be delivered over intercontinental distances aboard " pick a back" multi-stage rockets by robotis assistance. So, in our future, robotic can be applied on those avoiding resource waste job aspects, such as degradation and resource depletion , reduce , reuse, recycle and recover, have always been needed to avoid resource waste to future global business organizations. When, AI machine learning and robotics could help us , more efficiently organization, digital tools and new business models.

How green robots are helping with environmental sustainablity?

Robots can help reduce waste that is implemented by effciently sorting materials that can be recycled and put to use again. They can even help sort waste materials quickly and most efficiently than humans reducing the input power and cost associated with such processes.

How robots may help organizations to avoid resource waste?

Waste robotics autonomous recycling technology integrates advanced waste technology to avoid the costly deployment to any organizational users, e.g. the clean sea robot is an autonomous, electric floating aqua-droue that sweeps and collects plastic trash that has collected in coastal areas. It uses a has collected in coastal areas . It uses a combination of computer vision and remote sensing basd on 3D laser scanning technologies . So, clean sea robots can bring sea environmental protection advantages to global seas. Future global sea environment protection organization must need its clean sea tasks helping.

Instead of clean sea rubbish aspect, robots can also help organizations to do digital waste management tasks, digital technologies and their current use in waste management transparent, more economic and more resource-efficient processes, better souring of current important robotis digital waste manaement, technological trends, are robotics,the internet of things, cloud, digital rubbish robotic can help organizations to reduce digital rubbsh easily. then, any organizational digital system can do more efficient tasks daily. Hence, robotics can help e-commerce organizations to do digital resource waste management tasks in order to provide high efficient online purchase services to any countries online buyers in short time.

How robots are applied to reduce waste in hotel organizations?

Reducing waste in hotels in within reach with the help of digital technologies. How are robots used in hotels? Throughout the hotels, robots are deployed to provide information, front desk services, storeage services as well as check in and cleck out services , with technology including voice and facial recognition. An example, of artifical intelligence in the hospitality industry is the use of AI to deliver in-person customer service. The robot is able to provide tourist information to customers who interact with it. Most impressively, it is able to learn from human speech and adapt to individuals. waste management in hotels is important or it is getting increasingly difficult to dispose of waste. Therefore, hotel resource management robots can help hotels to decrease cost of waste disposal as the start with: Using refillable dispensers for soaps, shampoos and conditioners.

Could hotel service robots help the hospitality industry on resource management of robotized hotels, e.g. avoding food waste in hotel robotized hotels, example restaurants robotics technology can help chefs measure, manage and reduce food wase. Morevoer, a restuaurant robots do not get waste and can perform to help chels measure, manage and reduce food waste easily. From a human resource management point of view, in that sense, robotisation may help automation and hotels to improve it service and reduce restaurant food waste cost.

Management science role solves
resources shortages

● What does management resource science mean ?

Although, management science could include the study of all activities of groups that means a management function, it generally include discovering, developing, defining and evaluating the goals of the organization and alternative policies that will lead towards the goals. So, management science is the broad interdisiplianary study of problem solving and decision making in human organizations with strong links to management economics, business , engineering , management consulting fields. Management science helps business to achieve golas using of organizational resources to produce goods and service. It is a contemporary approach to management that is an extension of scientific management that measures the worker to task mix and ratio to raise efficiency.

Thus, the nature of management science is a science because it contains a systematic body of knowledge in the form of general principles. Management principles are important in nature, but they can not be expected to give same results in every suitation. Therefore, management is a social science. If any organizations feel resources are shortages, they need to find that the main factors to influence their resources are often shortage. Management science may be one kind the best management research choice, because it uses various scientific research-based principles, strategies and analytical methods including mathematical modeling, statistics and numerical methods to improve an organization's ability to enact rational and accurate management decisions by arriving at optional or near oprimal solutions to complex decision.

● Why does management science help organizations to avoid/solve resource shortage?

Because management is a core function of every business and a number of theories try to explain how any why resource will be use rapidly and resource shortage causes to any organization. The core function of the management science approach is to compare possible outcomes. In addition, managers in various functions have aware of the potential contribution of analytical modes . So, the managerical science function involves arranging equipment perform functions, such as procurement, production etc. activities.

For drug manufacturing organization or hospital, clinic, medical service organizations exmaple, any kinds of drug must be important resource to provide patients to eat. If they have not enough drugs to be provided to the hospital, clinic, medical service organizational patients to eat. Then, their sicknesses may become serious suddenly. So, managing enough drugs stick is very important. However, management science may help any one these medical organization to implement rationality in strategic decisions: Choice which kinds of drugs purchase number, in order to keep enough drugs supply in drug store room. Understanding in a drug resource pool choice, due to cash available is limited to any one medical organizations.

Management science aims to avoid drug shortage to any one medical organizations. Such as this drug organization case, management science may help them to identify the processes, which kinds of drugs are needed to buy immediately, notice the areas of weaknesses, and realize the future possibilities and needs of the medical organization's patient customers needs. The approach makes the drug utilization of resources easier, since the framework can notice the availability of drug resources and the proper eat of them to be provided to any sickness kinds of patients to eat immediately when the medical organization can have enough drug stocks supply.

How to apply management science to help hospital better prepare for a drug shortage management method may include: Having a plan, the director of pharmacy should consider developing a plan for managing drug shortages, implement structured communications. Ask the right questions,instead of , hospital management also needs to know what is causing drug shortages. IN fact, drug shortages are caused by many factors, including: regulatory issues, and business decisions as well as many other disturbances within the supply chain.

Hence, in management science view point, hospital can attempt to apply mobile App technolgies tool to solve drug shortage problem, because mobile App technologies tool can be one kind of immediate communication tool, it can identify when the kind of drug will shortage and their impact on hospitals, e.g. how many the kind of sickness patient number and automates the entire life cycle of managing any kinds of drug shortage. Mobile App, communication technologic tool can provide clinical information that allows teams to understand and mitigate the ramifications for shorted medications. Any possible drug shortage solution in hoepital, it may be a management strategy that includes clear policies and producers for information. It defines a drug shortage is as a supply issue that affects how the drug store room's daily drug supplies.

Thus, management science may help any medical organizations to solve drug shortage crisis, it may help any organizations to attempt to reduce health care cost, one common solution for mitigating shortages is to keep more drug inventory, but drug inventory number can not be excess, in order to avoid drug out of patients (old drug) to let patients to eat in difficult. Hence, any medical organizations need to implement good pharmaceutical supply chain management strategy in order to solve drug resolve shortage challenge causes.

- How to apply management science technological tool to solve resource shortage to organization challenge?

Management science technological tool predict method may also be applied to water scarcity management aspect for government water supply organization, it may provide the scientific basis to apply optimum water resource management practices in the affected areas, methodological procedure for drought and water scarcity management or applying the economic scarcity potential method to rare earth elements highlights. It can capture resource is a useful approach as is it has an technological tool to avoid scarcity of resources to farm from atmospheric physic to robotic farm water resource shortage technological tool. A key component of this farm management approach is the use of information gathering concerns weather change. It aims to help farmers can predict when ther are enough rain supply to the farm land, they ought grow rice or fruit seed or vegetabe seed to farms. hence, it seems that management science method may be applied to help any organizations to solve natural resource or manufacture resource supply shortage challenge to any organizations.

ON conclusion, management science may be applied to resource shortage aspect, because management science is resource management technique. The two techniques are as follows: resource technique is a technique in which start and finish dates are adjusted based on resource constraints with the goal of balancing demand for resources against available supply. Any organizations may encounter resource shortage challenge. Resource scarcity is essentially about current demand for a resource exceeding available supply. Resource scarcity occurs when demand for a natural resource is greater than the available supply , leading to a decline in the stock of available resource . However, any organizational resource shortage may due to failure management, difficult management technique causes. The common types of resource management may include human resource management natural resource management, project resource management, financial management, infrastructure management , facility management enterprise asset management public asset management at organizational resource management aspects. Hence any organizations must need efficient and effective resource management strategy in order to avoid resource shortage causes.

In organizational studies, resource management is the efficient and one resource management technique choice is resource shortage solution leveling, to keep the enough stock of resources on hand, reducing both excess inventories and shortages. Any kind of resource management technique aims to ensure access to resoures and avoid supply shortages efficiency in their use, facilitate their proper and of life management, hence, effective resource management strategy may help the organization to avoid resource shortge challenge causes easily.

- Reasons to organizations need management resource strategies ?

Why ought organization implement resource managment strategies ? better utilization means a happier and healthier team helping to reduce being overburdened and stressed. Resources are used to their maximum potential, keeping projects on time and on budget. It helps project managers keep to reduce oversights. The reasons that organizations ought to prepare effective managment science to implement the most reasonable and the most efficient and the most effective resource management strategy, because effectively managing resources can help companies more consistently deliver projects and services on time.

This is because better resource management helps to improve insights into resource availability as well as improves timelines projections. Why it is importsant to have enough resources supply and resource management strategies implementation? Resources are important for the growth to one organization, even development for any country, for example, generate energy , one need fossil fuels, and for industrial with the development, we require mineral rsources . On macro-social development aspect, natural resources are getting scarce with the increasing populatin, so it is essential to conserve them.

Why do we need resource management to organizations ? Resource management is the process of pre-planning, scheduling, and allocating the kind of resources to maximize efficiency. A resource is anything that is needed to eecute a task or project, this can be the skill set of employees or the adoption of software. Due to large organizations might be dealing with multiple projects. Effective allocation of resources help project maangers to avoid to use resources inefficiently, e.g. human resources, financial resource, physical resource and information resources. Hence, effective resource management science strategy can benefit the organization team and overall organization.

- How applying right science resource management principle to organizations

Scientific management is an attitude and a philosophy to be accepted and applied of the method of scientific investigation for the solution of the problems of industrial management . They may include experimentati, collection of data, analysis of data and formaulation of certain principles on the basic of such analysis. It aims to seek the most efficiency for plant operations. The main objective is improving economic efficiency, especically labour productivity. However, scientific management can be summarized in four main principles to determine and standardize the one best way of doing a job . A clear division of tasks and responsibilities. High pay for high performing employee, for Toyalor's scientifi managmement principle example, he proposed that by optimizing and simplifying jobs, productivity would increase.

Nowadays, most organizations in the industry make use of scientific management, e.g. hospitals, car and computer , restaurants, among others. The advantages may include : reduction in the cost of production, better quality products, benefits of division of labour , avoidance of disputes between labour and management increased wage, gains to owners/investors.

The main aim of scientific management is to develop all men to their greatest efficiency . The specific objectives aim to enhance production and productivity, decrease cost of production and maximize prosperity both for employer and employees having common interests (not opposite to each other).

● How Amazon e-commerce applies resource management principle to bring avoiding resource waste benefit?

For Amazon e-commerce example, it does not apply scientific management principle, due to it is not one product manaufacturing industry, it is online sale service and foreign delivery service organization. It ignores one of the key of scientific management, its creators genuinely believed that you had to pay higher wages to anyone asked to puch themselves to their physical limits. Under scientific management wages are paid to the workers as per the piece-wage system. Minimum wage is not assured, so every work needs to pay incentive wafe when he can manufacture more piece product, but Amazon believes that pay higher wages to any one can bring incentive productivity , such as it only provide online sale and foreign deli ery service to any one online buyer client, e.g. it can pay higher wage to the warehouse workers, because they need to cooperate with robotics to work hard and the human warehouse workers need to learn how to dominate any one warehouse logistic robotics to know hoe to delive and put goods to the right shelves in order to avoid to put wrong goods to the not right shelves position in warehouse.

So, in Amazon scientific management skill to workers view, its warehouse workers are smart workers, they need to know how to dominate any one warehouse robotics to do the right goods delivery to put on right shelves tasks daily. Their wages ought need to pay higher in order to avoid theig goods wrong delivery to wrong sheleves in careless. SO, Amazon feels that it needs to pay higher salary to the warehouse workers because they need have more smart and robotic control skills. When they and robotic to works together in Amazon warehouses.

So, such as Amazon warehouse workers case, Amazon ecommerce organization can apply scientific management principle to reduce resource waste. I shall explain as below:

Amazon 's warehouses have save diffeent kinds of products to prepare to deliver to different countries buyers, after Amazon had confirmed that it has receive visa card payment from online channel by each online buyer successfully. So, it must have smart warehous workers, they know how to control and dominate any one warehouse robotics to cooperate to send every right shelf position message to every one warehouse robotis to know, when the warehouse robotic receives the right product delivery to the right shelf message from the warehouse warehouse, it will delvier the product to the shelf position carefully. So, avoiding none of any wrong goods putting on the wrong shelves positions occurs easily. When, every one, there has none any wrong goods are putted on wrong sheleves positions occurrence, the spending investigation time to any wrong goods putting on wrong shelves position, it does not need to any one warehouse manager to do every day, So, Amazon does not need to waste time to do any goods putting on wrong warehouse workers number, when it applies many warehouse robotics to assist them to work in order to raise goods delivery efficiency in warehouse, e.g. Amazon warehouse can apply three logistic robotics and one human warehouse worker number to do one goods shelf delivery task. Before, it needs to employ ten human warehouse, logistic workers to reponsible to do one goods shelf delivery task. I assume that the goods shelf can put total 300 pieces of different kinds of products per day. So, Amazon can reduce none warehouse logistic workers number when it increases three logistic robotics to help them to do these 300 goods delivery task per day. Hence, its wage expenditure must decrease. Moreover, logistic robotics do not feel tried , bored, overtime work, these three robotics only follow any one of warehouse worker's message to let them to know whether which kind of product is needed to put on which number of the shelf positiion. Then, these three logistic workers can remember where the product is putted on the shelf position and help any one logisitic worker to get the right product to already deliver to the foreign buyer's home from Amazon's warehouses easily, they must raise efficiency more than only workers , they work in Amazon warehouses.

Hence, Amazon believes higher wage can enourage smart logistic workers can have good performance to dominate how every logistic robot , e.g. avoiding to send wrong message to let any one logistic robotic to put wrong product to the wrong shelf number position. SO, Amazon can apply warehouse scientific management method to avoid warehouse worker individual wrong message delviery to any one logistic robotic occurrence, when they can receive higher wage, and the three logistic robotics and one warehouse worker cooperation relationship is the most suitable workers cooperation number.

Amazon 's warehouse does not need more nine workers to often move in the crowd warehouse space environment to avoid worker accident and wrong goods putting on wrong shelves number position occurrences both. Hence, Amazon can apply scientific management method on warehouse avoiding resource waste aspect, when it decides to apply logistic robotics and workers cooperation in order to avoid wasting time to investigate whether which kinds of goods are put on where the wrong shelves number positions per day tasks occurrence in possible and it can bring delay to deliver goods to the online buyer's home. it is one good example of scientific management avoiding time waste method to Amazon warehouse organization.

Can Robotic Help Warehouses To Avoid Resource Waste On Behavioral Economic View

● Behavioral economy view whether robotic can help
warehouse to avoid time and human resource
waste

IN fact,robts are being used in different types manufacturing to create more efficiency with fewer resource. Robots also reduce errors, to leass waste is produced. Less waste is produced and the robots are able to final and separate the small parts more efficiently than human hands can. For example, on environment recycled aspect, robots can help reduce waste that is incinerated by efficiently sorting materials that can be recycled quickly and more efficiently than humans reducing the input poser and cost control with such processes. So, robots can bring positive affect the environment, because robots use less energy and produce less waste.

As a whole, there are multiple benefits to using robots to fight climate change,e.g. robots can prevent pollution and emissions through careful monitoring optimize the manaufacturing processes to reduce energy consumption. Moreover, robots can help with recycling, the use of robots allows facility operators some new flexibility. Most technologies used in recycling allow to sort materials. The sensing robots (sensoes) allow robots to receive information about a certain measurement of the environment , or internal components. This is essential to robots to perform their tasks, and act upon any changes in the environment to calculate the appropriate response.

- How Amazon warehouse applies logistic robots to help it to waste resource waste

Hence, although robots can take our jobs, because they can help organizations to avoid resource waste, and it can bring negative effect to influence we lose jobs. ON behavorioral economic view, robots can help employers to reduce employees number, but it won't influence organizational overall performance to be worse or inefficiency, such as Amazon warehouse applies logistic robots to assist workers to deliver the right kind of goods to put on every correct shelf number position rapidly every day. Hence, one logistic robot can replace at least 10 store workers to do goods delivery tasks every day, e.g. one store worker needs to spend one minute to find the right kind of good to deliver to prepare to arrange to deliver it to fly to overseas client. Logistic robots only need 10 seconds to find the right kind of goods from the near 200 number shelves in Amazon warehouse as well as they are putting 300 different kinds of goods on these 200 shelves in Amazon warehouse every day.

Because each logistic robot has very good memory. Each logistic robot must remember any kinds of goods , their putting number position on which shelf, e.g. when the worker needs to find the model laptop product from 300 different kinds of products,in Amazon warehouse. They are putting on 200 number shelves number following positions. IN general, human worker will need to spend about one minute to find the model of laptop product from these 200 number shelves in warehouse. For example, when one worker needs to find the brand Apple of one laptop product model: PHZ0123, when the warehouse has total 300 diferent kinds of products are putting on total 200 numbers of different shelves positions. Any one Amazon store worker must need to type this Apple brand laptop" Apple" name and its model number" PHZ0123 on the store computer as well as to search its putting on shelf number position from computer. Then, the Amazon store computer will find this laptop product to find its present putting on the correct shelf number position , e.g. 50 number of the shelf positon, or none stock record of all this model PHZ20123 laptop is sold out. So, Amazon store computer must need time to help this store worker to search this laptop product's putting on shelf number position as well as the worker needs time to walk to the right shelf number position to find this laptop product.

However, logistic robotic does not need to spend time to type this laptop product brand name and model number in order to search where it is putted on the shelf number position. The Amazon store worker only needs to speak this laptop product brand name, e.g. Apple and the kind of product, e.g.laptop and model number, e.g. PHZ0123 and its piece number, e.g. one piece number. Then, the Amazon logistic robot can follow the store worker's sound to find its past this kind of product's shelf number position memory to move to this shelf correct number position and finds it to deliver to the worker immediately. So, if the worker speaks 10 kinds f different products one time, then the logistic robotic can help this worker to find these 10 of different kinds products from their correct shelves numbers rapidly. Hence, it seems that logistic robots can help Amazon store workers to reduce each product search time as well as logistic robots can help workers to do goods delviery tasks. So, in logistic robotic behevioral economic vire, logistic robotics can replace many workers to do product position research and delivery tasks, store workers only need to speak the kind of product name, model, brand name and delivery piece number to let the logistic robotic to know. Then, the logistic robotic can follow the store worke's sending message to find the product's past memmory in order to tell the store worker, whether the product has how many stocks on shelf, or none of stock on shelf and where is putted on . Hence, any one store worker does not need to spend much time to search where any kinds of products shelves number position are. They only need logistic robotics to help them to do any kinds of products deliver to , or 20 or more different kinds between products location and the store worker's location. It means that the store worker only needs to stay on the same location to wait the logistic robotic brings his products comes back after he speaks to let the logistic rotic to know whether which kinds of products and piece number he needs . Then, he checks the logistic robotic's all products where they are correct or not. He may put all of these different kinds of gatherng products to the lorry to prepare to send to airport to fly to another country to deliver to the overseas client's home immediately when he confirms that all goods are his correct.

Hence, such as Amazon warehouse case, logistic robotics can help it to reduce many products searching time tasks and avoiding delivering wrong product to any overseas client's home rick occurence. Also, logistic robotics can reduce store workers number, because logistic robotics can replace 10 to 20 human store workers number absolutely. Otherwise, human store workers may have errors in their product search process, e.g. finding the wrong product from shelf or putting the product to the wrong shelf number position, but logistic robotics can reduce to 0 error to put wrong product on the shelf number position or spends long time to search the kind of product from the shelf number position. So, in logistic robotic behavioral economic view, robotics can help businesses to avoid products putting on wrong shelves number position error risk, reduce store workers number, reduce product shelf number position search economic time.

Can Robotic Help Warehouses To Avoid Resource Waste On Behavioral Economic View

- Behavioral economy view whether robotic can help
warehouse to avoid time and human resource
waste

IN fact,robts are being used in different types manufacturing to create more efficiency with fewer resource. Robots also reduce errors, to leass waste is produced. Less waste is produced and the robots are able to final and separate the small parts more efficiently than human hands can. For example, on environment recycled aspect, robots can help reduce waste that is incinerated by efficiently sorting materials that can be recycled quickly and more efficiently than humans reducing the input poser and cost control with such processes. So, robots can bring positive affect the environment, because robots use less energy and produce less waste.

As a whole, there are multiple benefits to using robots to fight climate change,e.g. robots can prevent pollution and emissions through careful monitoring optimize the manaufacturing processes to reduce energy consumption. Moreover, robots can help with recycling, the use of robots allows facility operators some new flexibility. Most technologies used in recycling allow to sort materials. The sensing robots (sensoes) allow robots to receive information about a certain measurement of the environment , or internal components. This is essential to robots to perform their tasks, and act upon any changes in the environment to calculate the appropriate response.

- How Amazon warehouse applies logistic robots to help
it to waste resource waste

Hence, although robots can take our jobs, because they can help organizations to avoid resource waste, and it can bring negative effect to influence we lose jobs. ON behavorioral economic view, robots can help employers to reduce employees number, but it won't influence organizational overall performance to be worse or inefficiency, such as Amazon warehouse applies logistic robots to assist workers to deliver the right kind of goods to put on every correct shelf number position rapidly every day. Hence, one logistic robot can replace at least 10 store workers to do goods delivery tasks every day, e.g. one store worker needs to spend one minute to find the right kind of good to deliver to prepare to arrange to deliver it to fly to overseas client. Logistic robots only need 10 seconds to find the right kind of goods from the near 200 number shelves in Amazon warehouse as well as they are putting 300 different kinds of goods on these 200 shelves in Amazon warehouse every day.

Because each logistic robot has very good memory. Each logistic robot must remember any kinds of goods , their putting number position on which shelf, e.g. when the worker needs to find the model laptop product from 300 different kinds of products,in Amazon warehouse. They are putting on 200 number shelves number following positions. IN general, human worker will need to spend about one minute to find the model of laptop product from these 200 number shelves in warehouse. For example, when one worker needs to find the brand Apple of one laptop product model: PHZ0123, when the warehouse has total 300 diferent kinds of products are putting on total 200 numbers of different shelves positions. Any one Amazon store worker must need to type this Apple brand laptop" Apple" name and its model number" PHZ0123 on the store computer as well as to search its putting on shelf number position from computer. Then, the Amazon store computer will find this laptop product to find its present putting on the correct shelf number position , e.g. 50 number of the shelf positon, or none stock record of all this model PHZ20123 laptop is sold out. So, Amazon store computer must need time to help this store worker to search this laptop product's putting on shelf number position as well as the worker needs time to walk to the right shelf number position to find this laptop product.

However, logistic robotic does not need to spend time to type this laptop product brand name and model number in order to search where it is putted on the shelf number position. The Amazon store worker only needs to speak this laptop product brand name, e.g. Apple and the kind of product, e.g.laptop and model number, e.g. PHZ0123 and its piece number, e.g. one piece number. Then, the Amazon logistic robot can follow the store worker's sound to find its past this kind of product's shelf number position memory to move to this shelf correct number position and finds it to deliver to the worker immediately. So, if the worker speaks 10 kinds f different products one time, then the logistic robotic can help this worker to find these 10 of different kinds products from their correct shelves numbers rapidly. Hence, it seems that logistic robots can help Amazon store workers to reduce each product search time as well as logistic robots can help workers to do goods delviery tasks. So, in logistic robotic behevioral economic vire, logistic robotics can replace many workers to do product position research and delivery tasks, store workers only need to speak the kind of product name, model, brand name and delivery piece number to let the logistic robotic to know. Then, the logistic robotic can follow the store worke's sending message to find the product's past memmory in order to tell the store worker, whether the product has how many stocks on shelf, or none of stock on shelf and where is putted on . Hence, any one store worker does not need to spend much time to search where any kinds of products shelves number position are. They only need logistic robotics to help them to do any kinds of products deliver to , or 20 or more different kinds between products location and the store worker's location. It means that the store worker only needs to stay on the same location to wait the logistic robotic brings his products comes back after he speaks to let the logistic rotic to know whether which kinds of products and piece number he needs . Then, he checks the logistic robotic's all products where they are correct or not. He may put all of these different kinds of gatherng products to the lorry to prepare to send to airport to fly to another country to deliver to the overseas client's home immediately when he confirms that all goods are his correct.

Hence, such as Amazon warehouse case, logistic robotics can help it to reduce many products searching time tasks and avoiding delivering wrong product to any overseas client's home rick occurence. Also, logistic robotics can reduce store workers number, because logistic robotics can replace 10 to 20 human store workers number absolutely. Otherwise, human store workers may have errors in their product search process, e.g. finding the wrong product from shelf or putting the product to the wrong shelf number position, but logistic robotics can reduce to 0 error to put wrong product on the shelf number position or spends long time to search the kind of product from the shelf number position. So, in logistic robotic behavioral economic view, robotics can help businesses to avoid products putting on wrong shelves number position error risk, reduce store workers number, reduce product shelf number position search economic time.

CHAPTER TWO

Robots whether can help Amazon to avoid resource waste

Organizational robotic intangible resource management

● How to implement effective intangible resources management strategy to achieve performance improvement to Amazon e-commerce organization? Why does non intangible resources effective negligence management to Amazon organization, it will cause worse performance to Amazon e-commerce organization any one e-buyer?

One efficient organization must need have efficient and effective resources management strategy in order to provide enough resources for its organization overall different departments cooperation effectively and efficiently. How to implement effective resources management strategy to achieve performance improvement? Why does ersources shortage to organization, it will cause worse performance? I shall explain the reasons as below:

For Amazon e-commerce organization example, it is one global the most large goods transport delivery service moddleman role between global online customers and their product salespeople. Amazon owns its webstores, so global any one country buyer clicks to its different countries webstores , then he/she can choose any kinds of products to buy from Amazon any one country webstore. However, the product is owned the another seller. So, any products are not owned by from Amazon any one country webstores. However, the product is owned by the another seller. SO, any products are not owned by Amazon. It only provides webstores to let global any one e-buyer to buy the product after he/she has paid visa payment. So, Amazon's role is one middleman. It needs to provide goods transport service to help the product's seller to deliver the product to whose e-buyer individual hoime in the most short time, e.g. when one China e-buyer clicks to Amazon China webstore , after he chooses any brands of computers product from Amazon China webstore, he makes purchase of the brand of comouter decision. Then, he needs to pay visa to Amazon 's China webstore . When Amazon confirms that ie can accept payment by the China e-buyer's visa. Then, Amazon will deliver the product to the China e-buyer's home within one week or longer time, the delivery days time depends on how much delivery fee, the China's e-buyer , he can pay. So, Amazon only can receive commision infomr from the computer brand product seller. Because Amazon can build famous loyalty of rapid goods delivery service provider barnad and its different countries websites can provide above one million different kinds of brand products to let global different conuntries e0buyers to choose in order to make the most fair and the most reasonable purchase price decision from its different Amazon different countries webstores per day. So, Amazon can help its different countries sellers to apply Amazon itself unique different countries' websotes design to attract global many different countries e-buyers to click to Amazon's webstores to find any kinds of products to choose to buy conveniently.

So, such as Amazon e-commerce organization case, if it hopes to attract global different countries e-buyers prefer to click to Amazon itself any one webstore more than other firms themselves webstores . Then, it must need have enough resources (tangible and intangible) both in order to provide raoid goods delviery service and many different kinds of goods choice provision service and reasonable price consumption channel to let any one countrye-buyer feels confidence and safe payment transaction and enough product advertisement, photo, price, function information in order to make final purchase decision from Amazon itself different country webstores more easily.

Hence, the tangible and intangible resources are needed to supuply to Amazon e-commerce organization. They may include: Product photos, price information, product advertisement which are shown to Amazon's different countries webstores , enough online customer service enquiry employees number, enough computers number, enough computers number, different countries offices and warehouses number, computers, internnet technology etc. tangible resource as well as customer service enquiries feedbacks, global rapid goods delivery transport service to any one country e-buyer's home, safe e-payment channel, providing to buy global any one country sellers their products in the most reasonable and the most fair purchase transaction. All ot these issues are any one e-buyer individual purchase feeling to Amazon. SO, they are intangibel (non-tangible) resources to Amazon. It means that if Amazon can let global any one e-buyer feels it's e-purchase service provision can let they feel more satisfactory , then they will choose to buy Amazon webstores any kinds of products more than other sellers their webstores products, because Amazon webstores can provide the kind of product of different brands choice, it aims to compare whether which brand's price is unreasonable too high or which brand's product's quality is worse, or design is not very atrraction. So, Amazon's intangibale resource, such as rapid goods delivery service, online or phone customer enquiry service, webstores' products information whether e-buyers can feel satisfactory, e.g. reasonable price, clear product photo many different kinds of brands product choice. All of these intangible resources to Amazon, they may also influence Amazon's future e-buyer number absolutely.

- How can Amazon raise intangible resource number to be effective?

SO, I explain that Amazon hadboth kinds of resources. They may include tangible and intangible resources. Internet speed, whether it is intangible resources. Internet speed, whether it is rapid or slow to satisfy global e-buyers online purchase speed and non online traffic jam accidents feeling. Amazon webstores any brands of products, photos, price information images whether they are clear to let any one global ebuyer to feel when they click to Amazon any one webstores. All of these internet technology resources to Amazon any one webstore will influence Amazon e-commerce organization's e-buyers number will increase or decrease . For example, if Amazon's China webstore can not let Chinese e-buyers to feel that it can not provide different kinds of brand products photos' clear image to let any one Chinese e-buyer to see the product photo clearly. When one Chinese e-buyer clicks to Amazon Chinese webstore to find any brands of laptop products to prepare to choose one to buy. But when he clicks to Amazon China webstore, he can not feel any one brand of laptop's photo is clear to let he feels. Then, theser unclear laptop photos may influence the Chinese e-buyer forgets his laptop purchase decision from Amazon 's China webstore easily. He may clcik to the another brand of laptop seller webstore to buy the brand of laptop from the laptop seller itself webstore. SO, Amazon's webstore design may be the important intangible resource to influence global any one e-buyer's visiting Amazon's any one webstore times or reducing to do to choose to visit Amazon's webstore behavior , due to they do not click to Amazon any one webstore websites again.

I recommend that Amazon ought consider how to design its any countries webstores in order to attract global e-buyers visiting Amazon itself webstores times number increase. So, global e-buyers' visiting Amazon different countries webstores times which will be Amazon's most important intangible resources to cause its future global e-buyers number. This kind of intangible resources may help Amazon e-commerce goods delivery service organization to raise its competitive effort,, because when one country's ebuyer feels Amazon can provide the most attraction and fair purchase channel from its different countries webstores. Then, the country's e-buyer will talk to his friends, families to prefer to choose Amazon if they have online purchase desire. Because Amazon's any one e-buyer , he .she will persduade his/her friends, families to choose Amazon's e-purchase channel., if he feels it can provide excellent e-purchase method to satisfy his online purchase need. Consequently, Amazon's e-buyers number may be influenced to increase.

ON conclusion, such as Amazon case, how to decide its whether which is the most important tangible and/or intangible resources in order to raise competition effort. It depends on whether how the seller sells its product, in order to concentrate on spending to increase the kind of resources number, such as Amazon e-commerce goods sale and delivery service organization case, internet webstores design intangible image dissatisfactory or satisfactory feeling which may be the most influential global e-buyers number increases or decreases. Moreover, Amazon's internet webstores design intangible satisfactory or dissatisfactory feeling resource may also influence e-buyer before or after e-purchase service requiry feedback feeling, safe visa card payment feeling, fair and reasonable brands of products price information and different brands of product photos image seeing satisfactory or dissatisfactory feeling , they are intangible resource asset to Amazon when global any one ebuyer must need to click to its any one webstore to find its any kinds of product to make purchase decision. So, any organizations must need to spend limit money to support its the most influential intangible resource , instead of tangible resources in order to increase clients number. On conclusion, we can increase organizational resources on these several aspects: Organizational resources are all assets that are available to a firm for use during the production process or service process, such as Amazon ecommerce organization case. The four basic types of organization resources are human, monetary, raw materials may be tangible, but internet technology may be another intangible resource to today any organizations. Improving organizational resources aim to improve efficiency, it means as the ability to accomplish something with the least amout of wasted time, money and effort or performance as well as effectiveness, such as improving internet speed, and none online traffic jam accidents occurrence easily. It means as trhe degree to which something is successful in producing a desired result success. However, we can improve resources by these way, they may include : Review who manages resources within the organization, build an-up-to date knowledge raise and company wide resource pool, manage the resource pool in line with the market, focus on education and talent employees growth, keep the customers in mind, work on quality services or products, learn to use technology , such as Amazon needs to learn how to raise high speed internet service for its webstores, and avoid online traffic jam frequent occurrent to its any one country webstore.

To sum up effective resource management strategy can help any organizations to increase resource number. Resource management is acquiring , allocating and managing the reosurces, such as individuals, and their skills, finances, technology , material , machinery, and natural resources required for a project. Hence effective effective resource management strategy ensures that internal and external resources are used effectively on time and budget, resources may be obtained internally from the host organization or procured from external resources. SO, effective resource management can help organizations to save resources being wastes and finances being spend on the wrong things, a significant cost saving factor. Hence, such as Amazon e-commerce organization, it needs to know how to learn how to apply internet technology to improve its different countries webstores design, shorten goods transport delivery time, gathering more different brands of product prices, data and phoducts photos improving safe visa card transaction secret to increase e-buyer individual e-purchase confidence. When Amazon can concentrate on effective allocate limited resource to achiev ethese objectives. Then, its global e-buyer number may increase significantly.

CHAPTER THREE

Can apply robot to improve consumer behavior to improve performance

Why can (AI) be applied to help businesses to predict consumer behaviors ? How can artificial intelligent tools predict consumer behavior in vehicle market retail industry ?What is (AI) consumer behavioral prediction tool? How any why will (AI) tool assist manufactures to attempt to predict consumer behavior before and after consumption occurrence? First, I shall indicate how to apply (AI) tool to predict vehicle product consumer behavior case example. Nowadays, many vehicle manufacturers hope their vehicles can attract to vehicle buyers to choose to buy their vehicles. However, there are many different brands of vehicles to provide to them to choose, so the vehicle market competition is very serious.

How to judge their different kinds of vehicle price which is reasonable acceptance to attract vehicle buyers to choose to buy the brand of vehicle manufacturers‘ any kinds of vehicles, e.g. fast speed sport style vehicles, comfortable and slow speed common cars, for four passengers common small size or more than four passengers common large car size? How to evaluate the vehicle prices issue is important factor to influence vehicle buyers’ choices. Either if the brand of vehicle price is too high to compare brands, it will influence many vehicle buyers choose to buy other brands‘ vehicles or if the brand of vehicle price is too low, it will influence vehicle buyers feel this brand’s vehicle’s quality is worse to compare to other vehicle brands’ similar vehicle products.

Thus, if the brand of vehicle manufacturers can predict how to design vehicles which can attract many vehicle buyers to choose to buy whose any vehicle products. What are future vehicle buyers‘ favorable vehicle styles? Then, the vehicle manufacturer can concentrate on manufacturing the kind style of vehicle products to sell already. It will reduce its vehicle manufacturing investment risk.

How to apply (AI) tools to predict vehicle buyers’ behavioral consumption model? Whether artificial intelligent tools can predict automotive buyers‘ behavioral consumption model and predict future trend. In fact, automotive brands and dealerships are facing an increasingly competition when attempting to manually gathering the vast quantities of data required to create customer focused programs that increase retention, ultimately new sales and service automotive business. Building a based on that client's intrinsic needs and interests to any kinds of automotive vehicles at any given time. This is especially true in the automotive industry where the time span between purchases is measured in years. Because vehicle buyers would not like often to change their old vehicle to another new one. So, their decisions to buying another new vehicle, the time is usually after one year, even longer time. Hence, it seems any vehicles won't be frequent consumption products to the owned at least one vehicle family consumers (vehicle buyers).

Hence, how to predict vehicle consumers' taste or preferable which styles of vehicle choices issues is very important. If the vehicle manufacturers can not manufacture any attractive vehicles to sell easily in this year. Then, it will lose time, money in this year because it won't know when the owned least one vehicle users or non-owned any vehicle users who will decide to buy one new vehicle or change another new vehicle ensure. The different brand vehicle dealers will possible wait more than one year to attract them to buy their vehicles if their styles are not attractive to compare other brands of vehicle competitors.

However, artificial intelligence and machine learning can help any vehicle manufacturers to find solution to solve patterns in highly to solve patterns in highly complex data-sets that are beyond the capability of a human brain, and then building and automatically acting on the customer insights it generates.

Given the automotive customer need for individualized communications, this technology is positioned to become a critical component of any successful vehicle retailer's domestic or/and overseas vehicle markets. How can vehicle manufacturers and retailers use (AI) to enhance their vehicle marketing campaigns? How will (AI) affect their vehicle sale marketing strategy? What criteria would they use when selecting on (AI) solution?

Vehicle consumers today are able to quickly access different brands of vehicle information, research vehicle products and reviews, negotiate prices and compare one vehicle brand or retailer to another resulting of the brands of vehicle customers. At the same time, the rise of " big -data mining", wearable devices that track user's every move and preference and greater contextualization in advertising and social media has resulted in consumer expectations of individualized. Thus, it seems that (AI) tools can be used to gather " big-data" and then they can make human's mind to analyze how to design kinds of vehicles to satisfy vehicle buyers' needs.

As automotive vehicle marketers can apply (AI) tools to achieve messaging strategies to meet the needs of this new generation of informed vehicle consumers, using data from a variety of sources to move from a variety of sources to move from mass- messaging to more personalized messages aimed at particular vehicle buyer segments, e.g. fast speed sport vehicle buyer segment, slow speed comfortable small size or large size of buyer segment. However, when 90% of vehicle marketers believe having a single vehicle buyer view is important, only 6% have achieved it.

However, one of the main issues vehicle marketers facing is the lack of capacity to efficiently sift through and analyze the massive vehicle buyer amounts of data required to create vehicle buyer individualized vehicle customer experiences easily. This is especially difficult for automotive dealers, the long periods between purchase cycles, and the highly considered nature of the vehicle purchase means that each vehicle dealer needs to not only track a large number of potential vehicle customers for an extremely long period of time, but each of those vehicle customers will generate a huge amount of different kinds of vehicle behavioral consumption data as they research their next vehicle purchase. However, by choosing the right (AI) technological tools and programs , vehicle dealers can solve this big data gathering challenge into a major advantage.

For Forrester vehicle brand example, vehicle consumers have more power over the Forrester vehicle brand's reputation than ever before. Mayne, L. (2014) indicated that Forrester calls this new (AI) tools is the " age of the vehicle customer", a 20 year business cycle in which the most successful vehicle enterprises will reinvent themselves to systematically understand and serve increasingly powerful vehicle consumers. To win in this new age, Forrester declares companies must become vehicle customer obsessed and the only sustainable competitive advantage is knowledge and engagement with customers, such as (AI) gathering data knowledge.

Thus, the biggest challenge vehicle businesses currently face is not the collection of a large quantity of vehicle consumer data, but what to do with that data once they have it. Even at a large vehicle data research firm, the data sets are often too big for a single analyze, or even a team of analysts to sort through and draw conclusion from. However, enter artificial intelligence and machine learning , an efficient technology solution that can continuously find patterns in highly complex data sets that are way beyond the capacity of a human brain and then automatic drive action based on the customer insights is generated.

What is (AI) machine learning tool? Machine learning is a type of (AI) that learns from data and is not explicitly program. Think Amazon, face book. Machine learning serves up relevant content based on an individual vehicle purchase behavior and experiences. More simply, machine learning is a computer program that can learn relationships between data, subject those learnings to errors functions, and then learn from its errors. The program in effect, trains itself.

Lee, T. (2016) explained that "Thus, (AI) tools can learn deep a more advanced branch of machine learning inspired by how our brain's nervous function, has also been found to be especial effective in identifying patterns from data."

When this way sound is complicated from a vehicle dealer perspective, the implementation of a marketing program driven by artificial intelligence can take care of these tasks in an automatic vehicle fashion with little to no manual intervention required from the staff at time vehicle stores.

In practice at a vehicle dealership, the program will continue track vehicle customer behavior online, merging that data with any offline source (like CRM or DMS data) and then analyze this aggregated vehicle buyer data set to predict what vehicle customer may be shopping for and what information they might like to relevance from different kinds style of vehicle design photos.

Artificial intelligence refers to complex in vehicle market, machine learning that posses the same characteristics of human intelligence and that have all our sense, all our reason and think just like human do. Besides, machine learning is the practice of using algorithms to collect and examine data, learn from it, and then make a determination or prediction about something in the world.

The machine is " trained" using large amounts of data and algorithms that give it the ability to learn how to automatically perform a task with increasing accuracy. Otherwise, deep learning is primarily based on artificial neural networks inspired by our understanding of the biology of human's brains.

Deep learning breaks down tasks in ways that enables machines to assist us with increasingly complex tasks, driverless cars, better preventive healthcare and more accurate product recommendation (including vehicle recommendations). So, such as why (AI) technology can be applied to predict how vehicle consumer behavior changes to bring to judge whether vehicle consumer will like what kinds of vehicle styles next year. Then, vehicle manufacturers can gather overall vehicle consumer data to analyze and conclude the more accurate vehicle design direction for next year any new design vehicle manufacturing products.

Thus, (AI) machine learning can help vehicle manufacturers to solve how to design any new vehicle products challenge. A vehicle is both one of the most important and carefully considered purchases the majority of people will ever make in their lifetime. It is also a purchase that tends to be fundamentally tied to a person's identify and view of themselves. As the same time, vehicle consumers changing lifestyles result in changing vehicle needs, e.g. the young sport car enthusiast matures into the family driver.

Automotive dealers need to remember that vehicle customers and prospects are individual human beings with risk, complex and ever-changing lives factors, these factors will influence every vehicle consumer why who feels has vehicle purchase need, and how who choose to buy the first vehicle if who decided to buy the first vehicle.

The (AI) technological customer behavioral prediction tool seems to be the best vehicle salespeople in the world are those that know every one of their vehicle customers. Their likes and dislikes which style of vehicle design, preferences and changing tastes to vehicle choices. The capacity of the human brain, however, limits us from achieving this type of vehicle sales and frequent turnover at vehicle dealerships often results in the further loss of vehicle salespeople along with their vehicle customer relationships and knowledge. In this competitive vehicle environment, machine learning enables platforms to assist the vehicle sales team by tracking the vehicle consumer behaviors of each vehicle customer, learning and memorizing their preferences and predicting their future vehicle purchase needs.

Finally, I recommend that for a vehicle dealerships marketing platform to make their customer engagement efficient and fully-functional, I should be able to: applying (AI) tools to track every vehicle customer behavior across the web, connecting to a society of data sources, CRM, DMS, third-party, web vehicle brands, social email, click etc., aggregating and accurately cross-reference data from a variety of sources, leveraging this data to drive insights on a mass scale, as well as on an individualized basis, driving actions and automatically direct customer engagement via multiple channels based on where each customer is in their individual lifecycle.

How can (AI) provide businesses with better-informed decisions?

I shall explain how (AI) technology can provide businesses with better-informed decisions to drive top-line growth, deliver meaningful experience for customers and smooth their path along the consumer journey. The widely understood definition of (AI) involves the ability of machines or computers to learn human thinking, reasoning and decision-making abilities.

A Narrative science study in 2015 year identified that (AI) was being used primarily in voice recognition, machine learning virtual assistants and decision support. This study also highlighted the many branches of (AI) and that techniques and their definition are used interchangeably. It is possible that (AI) can be used to gather big data , then to analyze to help businesses to predict consumer behaviors. For example, one of the most common techniques is machine learning, where algorithms are used to perform tasks by learning from historical data. Another growth branch of (AI) is natural language procession.

However, during 2017 year, search engines will begin to factor additional behavioral data into prediction of customer behavioral results, such as the user's history of searches and locations and previously captures conservations. Artificial intelligence will use this information to power predictive search results, e.g. predictive future consumer's choice behavioral processing for any kinds of businesses.

Predictive search will improve the quality of search results, and provide new insights into consumers' behavior and the moments which matter to them. Search will give recommendation into tailored how consumer individual choice in consumption process. Several of the largest online platforms already use machine learning to improve predictive consumer behavioral search results.

For example, Google's rank brain technology adds research by understanding the context in which the consumer has entered it. Over time, rank brain will learn further from user behaviors Amazon's DSSTNE (pronouned destiny) learns from shoppers' purchasing habits and consumption behavior to offer better product recommend actions, which Amazon can offer before a consumer has entered anything into the search bar. However, this technology is not independent of human input. For example, Google engineers will periodically retain the rank brain system to improve the models it uses. For another example, in 2016 year , Apple computer revamped its photos app to allow consumers to search for specific items in the phots, they want to find, not just dates and locations. Each photo that an intelligent phone or intelligent pad user takes goes through 11 billion computations, so that photos can understand exactly what is the photography.

It seems that in future, (AI) machine learning will allow search to evolve even further. Search engineers will deliver refined recommendations to their business users and use less human input to predict consumers' needs. For IBM computer example, it indicated 90% of the data that exists today has been created in the last two years. This huge explosion of data gives brands the opportunity to quickly spot and react to the latest trends, fashion and fads among its clients and potential clients. This will allow companies to better engage with younger consumers, who gain influence access to the latest trends, and use the brands. They associate with to help define who they are as individuals. Thus, brands have to identify and make use of them before consumers move on, but the vast quantity of data available makes. This a resource-intensive task. For next example, Lesara, a based online clothes store, uses this machine learning to inform its product decision often gathering information from internal and external sources. When its trends -spotting shoes. Lesara has a range of over 20 styles and sells hundreds of pairs a day. It focus on giving consumers, the very latest trends allow Lesara to develop on average of 50,000 new items each year. It compared to 11,000 old items each year. Thus, (AI) brain seems to human brain to own analytical ability to predict consumer behaviors.

For another example, Lesara is one online clothes store, uses machine learning decisions after gathering information from internal and external sources. One of its most popular products, shoes with LED started life when its trend spotting software flagged up a blogger wearing similar shoes. Now Lesara has a range of over 20 styles and sells hundreds of pairs a day. Its focus on giving consumers the very latest trends allows Lesara to develop an average of 50,000 new items each year, compared to 11,000 for its competitor Lara. it seems (AI) machine learning can help Lesara business to predict what kinds of shoes design or style that shoe consumers will prefer choose to buy in future shoe market trend. Thus, Lesara can predict shoe consumers' taste successfully and it can manufacture many attractive style of shoes. (AI) machine learning can gather global past shoe consumer's shoe shopping experiences, then analyzes to make conclusion to give lesara recommendation successfully. This will make the experience more enjoyable for shoe consumers and allow Lesara to advert whose different new style or design of shoes to deliver them move relevant messages by understanding the context of the experience.

However, (AI) machine learning will have this risk who manufacturers need to concern if they applied this technology to predict consumer behavior. It is on sample consumers' privacy issue, in order to avoid complaint chance occurrence. However, machine learning can tie this data together to identify which f the billions of devices are being used by individual consumers. This helps brands understand how consumer engagement and actions can be attributed to different messages in different contexts and at different time. So, machine learning can help brands to build confidence to promote their products by any advertisement channels. When, this new (AI) machine learning technology can conclude how to design their products to be the most attractive, due to it has more accurate to predict consumer behaviors to compare human themselves prediction judgement effort. It seems that (AI) machine judgement effort is more accurate to compare to human judgment effort.

For example, google is moving away from cookies and using logged in data to track and make to users. It plans to expand the scope of the brand lift tool from online video. Thus, consumers are responded will to shippable context, finding it persuasive and easy to navigate by (AI) machine learning decision. For example, fashion brands can aggregate their You tub videos and blogs into a mobile context marketing experience, such as brand centric context into a personal shopping activity gives the shopper an experience, who are likely to remember and tell their friends about any new style of products design promotion from these internet advertisement channels after (AI) machine learning tools' styles of product design recommendation.

What is (AI) deep learning techniques to help farmers to raise food growth

The (AI) deep-learning technology leads to performance enhancement and generalization of artificial intelligent technology. It influences the global leader in the field of information technology has declared its intention to utilize the deep-learning technology to solve environmental problems, such as climate change. So, it will help agriculture farming businesses can raise any plant food: vegetable, fruit, rice which grow up very easily if farmers can apply (AI) deep-learning technology to solve environment problems to influence their plant food grow. If the whole year seasonal change is very good and it is suitable for any plant food to grow in farming land easily, e.g. rain is enough and soil is enough for any plant food to grow in the farm lands. Then, fruit, rice, vegetable etc. agriculture businesses will have much beneficial attribution to global farmers.

The question is how to use deep-learning technologies in the environmental field to predict the status of pro-environmental consumption. We predicted the pro-environmental consumption index based on Google search query data, using a recurrent neural network (RNN model). To certify the accuracy of the index, we compared the prediction accuracy of the RNN model with that of the ordinary least square and artificial necessary network models. For example, the RNN model predicts the pro-environmental consumption index better than any other model. we expect the RNN model to perform still better in a big data environment because the deep-learning technologies would be increasingly as the volume of data grows. So, deep-learning technologies could be useful in environmental forecasting to prevent damage caused by climate change to influence any rice, vegetable, tomato, potato, fruit etc. different plant food grow in any countries' farming land easily.

For South Korea example, over 800 government agencies spent 2.2 trillion Korea won on eco-products in 2014 year. However, green products are rarely purchased outside these agencies. This phenomenon occurs because there is a gap between consumer attitudes and behavior , that is environmental attitude is a major factor in decision making vis-a-vis the consumption of " green" food and services (Jorea Ministry of Environment, 2015). Therefore, it is necessary to understand those consumer attitude, that will lead to sustainability-conductive behavior and consumption.

Environmental consumption prediction

Recently, many researchers have studied pro-environmental consumption and household indexes as well as suicide rate predictions using messages posted by internet users on Google trend, Tweets etc. channel. Whether can environmental consumption be predicted by (AI) deep-learning technological internet channel? How can impact the pro-environmental consumption attitudes of green policies? Korea scientists estimated pro-environmental attitudes using search query data provided by Google trend and confirmed through regression analysis, that pro-environmental attitude has a positive correlation with the pro-environmental attitude index. They also explained that environment-friendly attitude of residents plan an important role in policy making. In the past, most household consumption indexed were calculated through surveys, but (AI) deep-learning technological tool " big data" have recently gained research attention (Lee et al. 2016).

It seems that (AI) deep-learning technology can help agricultural export countries' farmers , e.g. US, UK, Canada, New Zealand, Australia, Japan, China, India etc. they can predict environmental behavioral consumption to any rice, tomato, potato , fruit, vegetable etc. plant food consumers. The beneficial advantages to them include as below:

(a) Assuming they know their countries' weather, when it has less rain to cause drought or when it has more rain in any seasonal time in the year. They can choose not to grow any kinds of above these plant food to avoid loss.

(b) They can make any kinds of above these plant food price raising after their prediction of these bad seasonal time to cause their plant food shortage supply challenge. Because these plant food consumers' demand number is more, but the supply of these above plant food supply number is less. However, due to they had predicted when the bad seasonal time can not allow them to grow these above plant food before. So, they have enough time to grow many these above plant food number in predictive good seasonal time to prepare to supply to their plant food import countries' plant food consumers to eat. Thus, these predictive environmental consumption plant food export countries can raise their plant food price to sell to them. When, the other non-pre-predictive environmental consumption plant food export countries can not supply any one of those plant food to them to eat, due to the bad climate to cause them can't grow any one of these plant food to export to sell.

Thus, (AI) deep-learning technology can be applied to predict how to raise the plant food supply number in order to raise price to the import plant food countries consumers to eat, due to they feel difficult to buy these plant food to eat in the bad climate seasonal time in whole year.

(c) (AI) deep-learning technology can help climate scientists to find what reasons cause their countries; rain sudden increases or cause their countries' rain sudden decreases. After its gathering data analysis, it can assist climate scientists to find solution methods to attempt to control the rain level can be right falling down level to let agricultural export farmers who can grow their plant food to sell to agricultural import countries in whole year.

(d) The agricultural export countries' farmers can apply (AI) deep-learning technology to help them to choose whether growing which kinds of plant food in that whether climate time to earn more plant food consumption number more easily.

Due to the agricultural countries climate will often change, for example, tomato, potato, rice, fruit etc. plant food can be adapt to grow in more rain time, but vegetable can not be adapt to grow in more rain time. If farmers can apply this technology to predict when it will have move rain or when it will have less rain to fall down in their countries. Then, they can choose to grow which kinds of plant food number more, in the suitable seasonal climate time in order to raise plant food growing number productivities to supply to sell to satisfy any agricultural food import countries' demand effectively.

(e) (AI) deep-learning technology can help agricultural import countries to solve agricultural food shortage challenge in long term. When this technology can be popular to base applied by the agricultural plant food export countries. It will solve global agricultural food shortage challenge. For example, when one agricultural export countries' farmers can popular accept to apply this technology to predict when to grow which kinds of plant food more to rise number productivities to sell. e.g. vegetable, fruit, rice Besides another agricultural export countries' farmers can also accept to apply this technology to predict when to grow plant food, e.g. potato, tomato to raise number productivities to sell. Then, they can concentrate on growing the specific kinds of plant food in order to raise the specific plant food number productivities in every seasonal change time every month. Then, global agricultural plant food supply must be raised, due to these predictive environmental change farmers can know who ought grow which kinds of plant food to sell to raise number productivities.

How can apply (AI) technology to help book shops to decide book sale price

(AI) digital channel can be applied to help businesses to evaluate whether how much the product price is the most attractive to persuade consumers feel it is the most reasonable price to sell. It helps consumers to feel which brands of products which ought change the price to let consumers to choose to buy the brand of product. It can be applied to predict whether how many consumer numbers can be increased or decreased when the brand of product's price is variable. It aims to give opinions to help any brand of product manufacturers or sellers to judge whether which price is the most reasonable to let consumers to accept to choose to buy the brand of product in popular.

Thus, (AI) price measurement technology can be preference to be applied online communication ecommerce and mobile phone internet platform aspect. As businesses can enter their past products prices data and past customer number data into computer or mobile. Then, (AI) price measurement technology can gather these data to analyze these product prices and past customer number to compare their prices variable changing range level to find their price variable difference to measure to make conclusion about every product's price variable changing will influence how many customer number increase or decrease changing to choose to sell their different kinds of products more accurate. Then, (AI) price measurement software will help them to analyze all past price variable changing data to compare whether which price range can let customers to feel it is more reasonable and attractive to influence them to choose to buy the product among different brands of product choice.

Because any product's price is one important factor to influence consumers to choose to buy the product, instead of quality, durability, shape, appearance, color, brand familiarity etc. factors. Any online businesses with a focus on Asia should considerate (AI) customer care, and virtual shopping experience, whereas is Europe and North America still value face-to-face and/or real human interaction over (AI) or virtual worlds.

For example, Amazon publish has applied (AI) price measurement technology to help authors to decide how much every different topic of e-book or paper book price, it can attract the largest number of readers to buy. Any one author only needs to type whose book name to Amazon publish author himself/herself Amazon website. Amazon publish (AI) price measurement learning machine will help them to auto-calculate and judge how much e-book or paper book price is the most attractive and the most reasonable in order to increase reader number to buy their e-books or paper books to read. So, (AI) online price measurement machine will gather past similar book names and past every similar book readers' reading times and the number of readers to give opinions to let every author to judge whether his/her very new e-book or paper book ought charge how much price to the e-book or paper book which can attract many readers to choose to buy. Although, it is not ensure that the e-book or paper book price must let readers to feel it is the most reasonable price to choose to buy in reader's view point. However, it has other factors to influence readers' choice to buy the e-book or paper book, e.g. whether the book content is attractive to public, the author's familiarity, the book's page is enough or not to satisfy readers to read etc. factors. But, instead of all these extra factors to influence readers to choose to buy the book to read. (AI) price measurement learning machine can real give opinions to every author to let them to judge the e-book or paper book different price range whether is too high to influence readers to choose to buy to read or tool low to influence readers feel it is possible poor content book to compare other similar content books. Thus, (AI) price measurement machine can help authors to predict every reader's reading behaviors or reading experience and reading habit from online channel in short time easily.

The author only enter the book name to let Amazon publish price measurement machine to check, it will follow past reader's reading habit and reading experience to judge whether the similar all book topic sale record to judge how much price is the reasonable price to attract many readers to buy the book.

Hence, (AI) can be applied to digital channel to help businesses to predict consumer behavior in the future. In the future, mobile/smartphone, laptop, desktop will be most frequent used ecommerce channels to develop online business. So, (AI) can be also applied to these platforms to gather data to make analysis to help businesses to predict consumer purchase behaviors popularly. Due to , ecommerce is popular to global, so digital online and instore channels can be one good channel to let (AI) learning machine to make platform to gather past every online consumer purchase (buying) experience data to help businesses to build brand personality and having a responsible, positive impact on society.

To apply (AI) learning machine technology to understand customer online purchase behavior, it will raise business e-commerce successful chance: For example, (AI) learning machine can help businesses to gather data to analyze to determine whether short-term or long-term signals in the online consumer behavior that indicate higher purchase intents to let every online business to know. (AI) learning machine can find that online users with long-term purchasing intent tend to save and click through on more content.

However, as online users approach the time of purchase their activity becomes more topically focused and actions shift from saves to searches from online consumption channel. Then, (AI) learning machine will further find that the brand product purchase signals in online behavior can exist weakness before an online purchase is made and can also be traced across different online purchase categories. Finally, (AI) learning machine synthesize these insights in predictive models of online user purchasing intent to the brand of product. Taken together, it's work identifies a set of general principles and signals that can be used to model online user purchasing intent across many online content discovery applications. Thus, (AI) learning machine can help online businesses to gather any online users' click online behaviors data to judge whether there are how many online users will choose to find their online business websites to make final decisions to buy their products from online channels. Then, it will give opinions to help the online businesses to let it to judge whether what are the important website factors will help its online business to attract many online consumers, e.g. designing unattractive website issue, online unattractive product photos issue, unclear website color issue, unclear website advertisement message, contents and words impressions issue, lacking image movement frequent attractive seeing issue etc. different website factors. Thus, online digital channel will be one good choice to apply (AI) learning machine to help businesses to predict consumer behaviors.

Can apply artificial intelligent learning machine " big data" gathering method to predict manufacturers‘ behavioral performance ?

In consumer view point, can they apply (AI) learning machine to predict manufacturers' behavioral performance to judge whether whose products are value to buy. Nowadays, (AI) and big data are reshaping the risk in consumer privacy. For example, consumers want to hide their willingness to pay just as firms want to hide their real marginal cost, and buyers have less favorable information, say a low credit shore, prefer to withhold it just as sellers want to conceal poor product quality. So, it implies that it is possible (AI) learning machine can help customers to gather any manufacturers‘ past sale performance, e.g. how many complaints or appreciation from clients, product quality etc. sale data to let consumers to make judgement whether it is value to buy to compare other competitors. So, it has risk to the poor product quality of manufacturers. Otherwise, it has benefits to the good product quality of manufacturers. It also implies all manufacturers' privacy is not protected or secret when (AI) learning machine is popular to be used to predict manufacturers‘ behaviors by consumers.

Information economists suggest that both buyers and sells have an incentive to hide or reveal private information, and these incentives are crucial for market efficiency. Data technology that reveals consumers type could facilitate a better match between product and consumer type, and data technology that helps buyers to assess product quality could encourage high quality production.

Thus, (AI) big data technology can also assist consumers to gather different manufacturers' data to compare what their advantages and disadvantages of their products are. Then, consumers can make comparison to choose which brand of product is the suitable to whom to buy in these more choice consumption market. (AI) learning machine will gather similar brand their products‘ data to analyze to make conclusion to let consumers know or feel to make final judge to find what advantages or disadvantages of these sample brands of similar products' comparison from internet. On the other hand, it means that manufacturers can gather consumers‘ past purchase behaviors or purchase experience from (AI) big data gathering method to record and analyze to give opinions to let manufacturers to know what reasons or factors influence consumers choose not to buy their products from internet.

(AI) big data gathering consumer behavior prediction method can give these benefits to manufacturers and consumers both, such as: New concerns arise because (AI) technological advance which have enables reducing cost of collecting, storing, processing and using data in mass quantities extend information beyond a single transaction. These advances are often summarized by the big data, it means charge volume of transaction-level data that could identify individual consumers by itself or in combination with the datasets.

The popular (AI) takes big data as in input in order to understand, predict and influence consumer behavior. Modern (AI) is used by legitimate companies, could improve management efficiency motivate innovations and better match demand and supply. But (AI) in the wrong hand, also allows the mass production of fraud and deception. Since , data can be stored, traded and used long after the transaction. Future data use is likely to grow with data processing technology, such as (AI) big data gathering consumer and manufacturer behavioral prediction method from internet channel.

Thus, future (AI) big data learning machine can also help consumers to choose the best brand of manufacturer's products among different brands of manufacturers products choice to compare their past sale performance from internet. They can apply (AI) big data statistic method to gather all different manufacturers' similar products past sale data to compare their advantages and disadvantages to make the best decision to choose to buy which brand of product is the most suitable to them to buy to use. It seems (AI) big data can also help consumers to predict any manufacturers' manufacturing behaviors or manufacturing performance whether they are improving their product quality or are deteriorating their product quality. Thus, (AI) big data tool is also important to help customers to predict future the different brands of manufacturer performance will have improvement in possible.

Thus, I believe that artificial intelligent "big data" gathering method can be suggested to be applied to attempt to predict consumer behavioral changes in global business environment, the reasons are as below:

On the consumer's beneficial hand, Consumers can apply this method to attempt to gather any global manufacturers data to be analyzed by this artificial intelligent learning system. Then, it analyzed all the different brands of specific similar product manufacturer' data to compare what are the range of the best past manufacturing history and sale data to the group of best manufacturers, and what are the range of the better past manufacturing history and sale data, and what are the range of the good past manufacturing history and sale data, and what are the range of the common past manufacturing history and sale data. Finally, the (AI) learning system will compare all the specific similar product, e.g. mobile phone or computer, television, car etc. different kinds of specific products of global manufacturers to conclude the result is such as whether which brands will be the best manufacturers to let the consumer to buy the television or mobile phone or computer or car etc. different kinds of products. It can make more accurate judgement to compare general human's phone or questionnaire surveys investigation method, newspapers, television, radios, internet searches etc. different manufacturing news or data gathering channels to find which brands are the most worth confidence to consumers to choose to buy the specific product in the global consumption market.

On the manufacturers' beneficial hand, manufacturers can apply (AI) data gathering method to predict consumer emotion and buying behavioral changes more accurate. For example, the vehicle manufacturer, it plans to gather data to predict potential driving fast speed sport vehicle consumers' preferences trends in order to make the accurate judgement how to design its sport vehicles to attract many sport vehicle buyers who will choose to buy it's brand of any driving fast speed sport vehicles. It can attempt to apply (AI) intelligent learning system to gather global different brands of sport vehicle data concerns that all past driving fast speed sport vehicle buyer's preference of sport vehicle design. Then, the (AI) intelligent learning system gather global different brands of driving fast speed sport vehicle which had ever been purchased by the different country's driving fast speed sport vehicles consumers. After, it can compare divide the range of similar driving fast speed sport vehicle design and similar price to be different groups. The (AI) intelligent

learning system can attempt to follow the past number of different brands of driving fast speed sport vehicle buyers to calculate how many driving fast speed sport vehicle buyers who choose to buy the brand of driving fast speed sport vehicle as well as it will analyze and make judgement to find whether the cheaper price reason attracts the different countries sport vehicle buyers choose to buy the brand of driving fast speed sport vehicle or the attractive design reason attracts the different countries sport vehicle buyers choose to buy the brand of sport vehicle or fast speed reason attracts the sport vehicle buyers choose to buy the brand of sport vehicle.

For example, although some brands of driving fast speed sport vehicle manufacturers' prices are very high, but they can still attract global many sport vehicle consumers to buy. Whether all sport vehicle's attractive design is the main factor to influence them to buy or whether it's fast speed is the main factor to influence them to buy or whether it's safe confidence it the main factor to influence them to buy or it's familiarity brand is the main factor to influence them to buy. (AI) intelligent learning system will attempt to make judgement and analysis to conclude whether the attractive design factor is the main factor to influence many sport vehicle consumers to choose to buy the brand of sport vehicles.

Otherwise, for another example, although some brands of driving fast speed sport vehicle manufacturer's prices are low, but they can not still attract many global many sport vehicle consumers to buy. Whether all vehicle's unattractive design is the main factor to influence them choose not to buy their fast speed driving sport vehicles or whether the unsafe factor is the main factor to influence them choose not to buy their fast speed driving sport vehicles or whether unfamiliarity brand is the main factor to influence many consumers choose not to buy their fast speeding sport vehicles.

Thus, when (AI) learning system had helped the fast speed sport vehicles manufacturer to gather all different brands of fast speed driving sport vehicle's past sale data and price data, design of different sport vehicle, e.g. color choice, method of style, comfortable chair styles and chair sizes and what kinds of steel material to manufacture the sport vehicles data and driving safe and accident occurrence data and the data concerns what reasons of the past complaint to brand of sport vehicle manufacturer from its sport vehicle buyers. Then, it can make more conclusion to give more accurate opinions whether which brands of fast speed driving sport vehicle manufacturer(s) whose sport vehicle design is the main factor to attract consumers choose to buy its any driving fast speed sport vehicle products really. Thus, it seems that it can make more accurate judgement to compare television survey, questionnaire survey to gather data concerns how to design the fast speed sport vehicle to attract consumers to choose to buy the sport vehicle manufacturer's planning sport vehicle products. I believe that (AI) learning system can help the sport vehicle manufacturer to make more accurate conclusion or judgement how to design its fast speed driving sport vehicles to attract it's consumers more easily.

How to apply (AI) tool to help businesses to bring consumer positive emotion ?

If (AI) tool can be confirmed to apply to predict consumer behavior, then I can conclude that it can be attempted to apply to predict what the factor(s) of the product itself can cause the consumer has positive or negative emotion, so the manufacturer can attempt to avoid the bad factors cause to bring negative emotion to influence the consumer chooses not to buy the product more easily, such as vehicle product case.

Economists refer to the consumption desirability is as " utility" and the product or service consumption decision making is arose

influenced by maximizing utility only. However, they neglect consumer individual immediate emotion change will also influence the consumer individual consumption decision consequently. Expected emotions are those that are anticipated to occur as a result of the outcomes associated

with different possible courses of action. For example, if a potential investor, were deciding whether to purchase a

stock, who might imagine the disappointment who would feel if who ought it and it reduced its price. Otherwise, whose emotion would experience , such as regret if it increased in price, but who does not buy it before the stock rise its price. However, I believe nowadays technology, in the future one day, (AI) tool can be attempted to assist consumer psychology profession or marketing research profession to assist them to find what are the bad factors to influence consumers choose not to buy any manufacturers' products. Then, when the manufacturer
can discover what are the bad factor(S) cause(S) consumers who do not choose to buy their products, then the manufacturer can raise whose product of consumption desirability or " utility" to raise whose product's consumption decision making is influenced by maximizing utility. Hence, (AI) tool will be possible to find what the bad factor(S) to cause consumers do not choose to buy the manufacturer's product in order to raise the product's utility to bring consumer positive emotion to choose to buy its product in possible. SO, (AI) tool will be one consumer psychological emotion prediction tool to assist any manufacturers
to help their products to build positive emotion to any consumers in possible.

The key feature of expected emotions is that they are experienced when the outcomes of a decision materialize, but not at the moment of choice, at the moment of choice, they are only feel about future emotion. Such as consumption case, if the consumer chose to buy the product or consume the service before it's price is increased. Then, the consumer will feel happy and it is worth to purchase or consumer the service as well as the consumer's expected emotion is positive before who decides to buy the product or consume the service, because who believes or feels the product or service's price will be raised in short term, e.g. after one month, one week. Thus, it means that if the consumer does not believe or
feel or predict the product or service's price either it will increase or decrease in short term, whose emotion will be negative, those negative emotion will influence who does not decide to buy the product or service, it is possible that who feel it is not worth to buy the product or consume the service immediately. He She will choose to consume the service or buy the product to wait it's price is decreased later. it seems that the consumer's positive or negative emotion will influence who decides to buy the product or consume the service later or earlier. Thus, it has close relationship between the consumer individual immediate purchase or consumption decision and positive emotion or negative emotion (either expected emotion or immediate emotion influences).

Consequently, if (AI) tool can help any manufacturers
to predict when its product price ought to be increased or decreased in order to attract consumer to choose to buy its product. Then, it can help any manufacturers to build positive expected emotion to attract consumers to choose to buy its product more easily. For example, when the (AI) tool can predict when the consumer expects the product price will fall down, then it can give ideas to the manufacturer to raise up the product price in the month, then it predicts many consumers expect the product price will fall down after six months. So, the product price will not be fall down after six months. So, many consumers will feel disappointment and they will choose to buy the product if the manufacturer
decided to raise the product price after six months. Then, the higher product price will cause many consumers worry

about the product price will continue rise up, so they will prefer to choose to buy the product immediately after six months because they afraid the product price will continue to rise up in the year. Then, I assume that (AI) tool has effort to predict when consumers feel the product will rise up or fall down, then it can give ideas to the manufacturer when to rise up or fall down the product price in order to attract or persuade many consumers choose to buy the product in different period in the year.

Psychologists indicate that immediate emotions, by contrast, are experienced at the moment of choice and fall into one of two categories. Integral emotion, like expected emotions, arise from thinking about the consequences of one's decision, but " integral emotion", unlike expected emotions are experienced at the moment of choice. Such as purchase stock case, the share buyer might experience immediate fear at the thought of the stock's losing value. " Incidental emotions" are also experienced at the moment of choice, such as a consumer predicts the product or service price whether it will be risen up or fallen down. If he/she feels the product or service price will fall down after next month and he/she will choose to buy the product or consume the service. But consequently, after next month, the product or service's price won't fall down absolutely.

Then, he/she will have incidental emotion to influence whom to choose whether he/she ought buy the product or consume the service, due to the product or service price is not still fall down. Otherwise, he/she is fear the product or service will not fall down in short term. Even, it will increase price later. Hence, whose incidental emotion will have possible to influence whom to choose to buy the product or consume the service after one month, if the product or service's price is still not increased absolutely. So, (AI) tool can be attempted to apply to predict when the product price ought need to be raised or fallen down in order to attract consumers to choose to buy the manufacturers' product in different period.

Economists indicate utility an individual consumption with an outcome might arise from a prediction of emotion: For example, a dinner eater might choose a higher utility to an Italian restaurant dinner than a French restaurant dinner because who anticipates being happier at the former, even the former's dinner price is higher than the French restaurant.

So, such as this restaurant dinner case, if one (AI) tool can assist the French restaurant owner to find what factor(S) cause(S) the dinner consumers do not choose to go to its restaurant to eat its food, e.g. high price factor, bad taste factor, bad wait service performance factor, bad cooker's cooking skill factor, poor advertisement promotion factor, poor familiar factor, poor location or poor eating environment etc. different factors. Then, the French restaurant owner can find methods to avoid the bad factor(S) cause(S) many dinner consumers do not choose to go to whose French restaurant to eat dinner more easily.

The question is that whether the positive emotion factor can influence the consumer changes whose mind to choose to consume the more expensive service or buy the more expensive product. To answer this question. it depends on whether the consumer has an imperfect understanding of whose own tastes or the consumer has a perfect understanding of whose own tastes to the product or the service.

It means the consumer will choose to buy the product or consume the service, even it's price is higher than other general similar products or services if who has a perfect understanding of whose own tastes to the product or service. Otherwise, who won't choose to buy the product or consume the service, due to it's price is higher than other general similar products or services if who has an imperfect understanding of whose own tastes to the product or service. So, it seems that the consumer's negative or positive emotion arise will be influenced by whose perfect or imperfect understanding of whose own tastes to the product or service factor.

It concludes that whether how much degree of the consumer's utility to the product or service. It is not the only one important factor to influence the consumer to choose to buy the product or consume the service. Otherwise, the consumer's imperfect or perfect understanding own tastes to the product or service factor will influence the consumer to arise positive or negative emotion to make final purchase or consumption decision immediately. It will be one more consumption influential factor to lead the consumer to make the final consumption decision making immediately. So, future (AI) tool ought to be innovate to own how to judge good taste or bad taste for any food in order to predict food consumers to choose to buy the food manufacturer's any foods more attractively.

(AI) tool technical innovation in cruise tourism
immediate positive emotion influence to
cruise travelling consumers in entertainment
service industry

Can apply (AI) tool to cause positive emotion to cruise tourism consumers? Cruise tourism industry is the most influential emotion industry example to influence cruise travelling consumers' travelling entertainment choice. I shall indicate some evidences how it's innovation will influence cruise travelling consumers' emotion to be changed to positive from negative immediately as well as to prove how the cruise traveler higher utility feeling to the cruise tourism provider is not the main factor to influence whom to choose the cruise provider to consume whose cruise journey service arrangement.

Nowadays, cruising has become one of the fastest growing sectors within tourism, cruise service providers need have themselves unique different entertainment service arrangement to satisfy every cruise travelling consumer individual needs in order to attract every one to choose whose cruise arrangement easily, e.g. meals, activities, entertainment and varied destinations create one-stop holiday shop, reasonable competitive ticket fare. Hence, it seems it is one exciting emotion industry. If the cruise service provider can bring positive emotion to influence many cruise travelling consumers immediately. The, even it change higher service fare to compare other similar cruise service providers. I believe it won't influence them to choose other similar cruise service providers if it can often bring immediate positive emotion to its cruise clients during they are staying in its cruises or during they have left its cruises, but they will often remember or won't forget to enjoy their cruise service provider's happing time forever. Hence, if (AI) tool can be attempted to help cruise entertainment providers to arrange different cruise journeys for varied destinations , to arrange different entertainment facilities, to arrange the different taste food to satisfy different countries age cruise consumers' needs. Then, the (AI) tool will assist the cruise providers to bring positive emotion to let every different countries age cruise consumers to feel satisfactory in order to choose to the cruise providers‘ cruise entertainment service more attractively.

4.1 How can apply (AI) tool to predict cruise service providers bring positive emotion to their clients?

Future, (AI) tool can help any cruise providers to design these kinds of any one entertainment service arrangement to satisfy the cruise provider's customers' needs.

There are different special interests cruising , such as wellness at sea, freighter cruises, river cruises. It has increased the attractiveness of cruising: Romance is for lover cruise traveler target, luxury is for rich cruise traveler target, exotica is for enjoyment exciting feeling traveler target. So, every kind of cruise traveler target will have different kind of cruise entertainment service to satisfy their needs. If the cruise service provider can provide the right and attractive cruise entertainment service to satisfy the specific cruise target. Then, it will bring the positive emotion to the specific cruise target consumers more easily.

Cruise travel was shaped for mass tourism. Prices have been very differently segmented. There are basically four types of markets (Biederman, 2008):

- Contemporary market: On board fun and amenities are playing important role and destinations have secondary importance.
- Premium market: This category is more expensive than the contemporary category and where the destination has same importance as on board amenities.
- Luxury market: It was once dominant type of cruise tourism, but now it has only a small portion of the industry. Generally, it is the most expensive cruise category and usually it takes longer than average cruise days.
- Adventure/exploration: It refers relatively long cruises with special and exotic places where the destination is the main purpose of the trip.
- European cruise travel: Duration takes more five days than worth American travel duration. There is a tendency on European market during the years that duration of travel is getting shorter. This short demand of is explained with the strong demand of customers (Hensen, 2003). Beside this, it is most likely that cruise companies try to convince tourists with short haul travels instead of long term cruise trips for more expenditure.

Thus, I believe that even, the cruise service provider charges higher ticket which won't influence cruise consumers who do not choose its entertainment service on its cruises. If it can arrange the attractive cruise entertainment facilities and destination journey arrangement, staying days arrangement to satisfy different specific cruise target market needs absolutely in order to bring whose emotion to be positive to it's service provision. Then, the cruise service provider will attract many potential cruise clients to choose its cruise service absolutely. Otherwise, if it only bring negative emotion to its cruise clients, it will not attract many potential cruise clients to choose it or loses its old cruise clients, even, its cruise ticket price is needed to decreased in order to raise competitive effort.

In conclusion, I believe that future (AI) tools need to learn how to bring cruise consumers to arise individual immediate or expected positive emotion, this positive emotion consideration is more important to compare to how to reduce cruise ticket price in order to attract cruise clients in global cruise competitive cruise industry.

4.2 Differentiation through the characteristics of cruising route method from (AI) tool route judgement

Future, (AI) tool can attempt to help any cruise entertainment service providers to judge how to design different route to attract different countries age cruise clients‘ choices to satisfy their cruise journey entertainment needs. The determinants of the cruising route's characteristics (functional, social, and emotion) is important factor to influence the cruise service provider's success. Cruising product is no longer selected primarily for the cruising service, but for the content of cruising route. So, the cruising route will influence the cruise consumer individual emotion, because it is the main service need for every cruise consumer.

The approach called the " land sea cruising in product development" is increasingly becoming an area of interest, e.g. determining the direction of the effects of the individual cruising route characteristics on service value's perception , and providing an evaluation model of the route's perception , and indicating significance variables of attraction.

The questions that cruise planners need to know: How does each of the identified determinants affect the overall perceived value of the cruise route? How the overall perceived value of the cruise route affects customer behavior intentions?

Because different routes factor will influence cruise consumer individual emotion changing seriously. It means the ship has become only a tool, when the offered route whose attractiveness highly influences the impression of the guests has become crucial.

Consumer behavior in cruising segment includes all the activities and influences in the selection of the specific cruise route. There activities result in decisions and actions related to a defined price, selection and reselection of cruising company (Cannot, Brink and Brijball, 2006).

4.3 How to apply (AI) tool to arrange cruise route planning have close relationship to influence cruise consumer emotion?

Firstly, use value of cruising routes is based on the subjective experience, and shows how individuals assess the route during, or immediately after sailing. It is affiliated with the benefits that cruising guest realize by choosing a route , and it is subjective because it depends on the individual assessment (photo taken on the route for one guest presents just a family souvenir, and for professional photographers are embodied financial capital).

Secondly, the utilitarian value is also subjective-oriented and is tied on the point where the inner and us ability of cruising routes are compared with the sacrifice of the client (money and time). Finally, the value is considered as the outcome of the comparison of scarifies and personal benefits, which is resulted in essentially utilitarian nature.

Hence, route design is the main value of cruising tourism and it is primarily determined and analyzed from the aspect of observed customers. Otherwise, the cruise is only one tool to be caught for the cruise passengers, whether the cruise can let whom to sleep comfortable , providing what kind of food to them to eat, what kind of entertainment facilities are provided to them to play, these issues are not more important to compare how to design route to bring them to travel to anywhere to enjoy in this cruise journey factor. Because how to design the route factor can bring each cruise passenger to influence them to feel either negative or positive emotion directly. The whole route journey planning is the most influential factor to influence the cruise passengers to feel whether they ought choose it's service again or not in the future.

In economic utility or immediate (expected) emotion aspects, whether which is more influential to excite consumption. To analyze whether it is economic utility or immediate (expected) emotion more influential to excite travelling consumption. It depends on the travelling consumer individual consumption choice is in which situations. For example, if the travelling industry's general travelling consumer individual consumption decision is concentrate on emotion influential aspect, such as cruise entertainment industry, hospital care service industry, theme park entertainment industry, movie watching entertainment industry etc. Above all these industries have same nature, it is service. So, it seems that service industry's main influential factor is immediate (expected) emotion influence, it is not economic utility influence. Otherwise, product sale industry's main influential factor is utility.

(AI) judges travelling consumer utility factor

Travelling consumers' emotion and utility factor will be seem to child's emotion and utility factor. For this toy choice situation example, parent choose to buy one toy to give whose child to play. They usually considerate which kind of toy is attractive to their child whom like to play. In many different kinds of toys choice, if the child likes to choose the kind of toy to play. After the child's parents had purchased the kind of toy to let whose child to play one period time, e.g. six month. Then, when the child feel that who has need to buy another new toy to play, due to he/she feels bored to play this toy. So, it seems that the child feels this toy has less utility or it's utility is decreased. So, he/she expects whose parent can buy another new kind of toy to let whom to play. It also implies that it is not emotion factor to influence the child to feel boredom and unfunny to play this kind of old toy after six months. It is the product's utility factor which can not attract the child to play it any more. So, this old toy's utility is decreased when this child spends six months to play it. This toy's value is only six month utility to this child to play. Otherwise, if this kind of toy is bought by another parent. It is possible that the another child like to play this kind of toy one year or more. So, it's utility to another child is one year or more period. So, product's utility period is difference, it depends on how long time of the user's satisfactory time.

As this toy case, the child's decision will influence whose parent choose which kind of toy to buy to whom to play. Usually toy price is not difference too much. Parent won't consider when the toy price will be increase or will be decreased to influence their emotion to decide not to buy the toy immediately. So, when the child like to play the kind of toy, even the product's price is more than other kind of toys, and the parent feel it is possible that the kind of toy's price will be fallen down later. They will still choose to buy the kind of toy to let their child to play, they won't be influenced not to buy this kind product by later cheap price factor. So, immediate emotion is not the main factor to influence this parent does not choose buy this toy at this moment. Otherwise, utility factor will influence when the parent will buy another new kind of toy to provide to this child to play. If the child enjoy to play it only three months, after he/she will feel bore and he/she will tell whose parent to buy another new kind of toy to let whom play when the fourth month is beginning. So, it implies that if the kind of toy product can have more attractive utility time, then it can attract many parent to choose to buy it among different kind of toys. Thus, when this kind of toy's utility time is longer time. Then, it is possible that it can influence many parents choose to buy it's different style or design of similar kind of toys to let their children to play. In general, when many parents accept to buy this kind of different style or design of similar toys to give their children to play. Due to it's popular long time utility factor, it will influence children like to play it longer time to compare other kind of toys. Consequently, it will influence parents do not need often spend too much money to buy other kinds of toys to give their children to play. So, longer time utility factor to the product can attract many consumers to choose to buy the kind of product to compare lesser time utility factor to the product. Hence, it proves the explanation why utility factor is the main influential factor to influence the consumer choose to buy the product.

(AI) judgement tool of Medical care and utility case

Travelling consumers' emotion and utility also seems to medical care patient's emotion and utility. Medical care is an input in producing health, it is subject to law of diminishing marginal productivity. Health yields utility to the consumer. It is subject to law of diminishing marginal utility. It bring this question: Does either the patient's emotion or the medical care service or medical care product utility which one can influence the patient's hospital choice more?

To answer this question: We need to know medical care is one kind of nursing care service in hospitals or clinics and medical care product is one kind of medical care product sale from merchants, e.g. medicine or medical equipment. So, in medical industry which has different kind of medical care services to provide to patients in hospitals or clinics as well as which has different kind of medical care products sale, e.g. medicine or wheelchairs, heart health measurement equipment etc. different medical care products in medical health industry.

In medical care aspect, it is one kind of any medical care service to patients from hospitals or clinics. So, medical care is an input in producing health service to patients from hospitals or clinics. When the patient is admitting to hospital or clinic, who needs to see doctor and the doctor need to give the right medicine to the patient to eat to ill whose illness. Even, if the doctor feels the patient whom needs to live hospital for one time period. Then, the hospital nurses must need to take care the patient during he/she is living in the hospital period. Consequently, if the patent can be health in short time, e.g. within one week leaving time, then he/she will be shortened time to leave the hospital in next weak. Otherwise, if the patent can not be health in short time, e.g. within on week, then he/she needs to live the hospital more than one week, even, one month, three months or more. So, the staying hospital time will influence the patent's emotion to feel whether the doctor's effort. If he/she needs to live the hospital long time, he/she will bring negative emotion to feel the doctor's medical effort is not good. The doctor's medical effort can not achieve or satisfy whose expected emotion during whose staying hospital time.

Thus, medical care is an input service in producing health, it is subject to law of diminishing marginal productivity. When the patient does not need to live the hospital longer time, the patient will feel more satisfactory to the hospital's doctor and nurses' care effort as well as the patient can give less money to spend the expenditure to live the hospital. So, the law of diminishing marginal health productivity will explain the hospital will shorten time to the staying days of the hospital to the patent as well as the patient's care expenditure will be decreased when he/ she only needs to live to the hospital in short time. Otherwise, the patient needs to live longer time in the hospital, it means that the diminishing marginal health productivity to the hospital, the patient's staying hospital days will be increased and the patient's medical expenditure to the hospital will also be increased. It will bring negative emotion to patient and why this negative emotion factor will influence the patient would choose another hospital to live or find other doctors to see if he/she felt illness in future one day.

In medical care product aspect, health yields utility to the consumer. It is subject to law of diminishing marginal utility. Because patient needs to buy different kind of medical equipment to use or medicine to eat to attempt to cure whose illnesses. So, if the patent can choose the right medicine to eat from the doctor's recommendation or if the patent can choose the right medical equipment to use from the doctor's recommendation. When the patient buy less number of medicine to eat, then he/she can be health or when he/she buy the medical equipment to use, then he/ she can be health. Then, he/she will spend less money to buy medicine to eat or medical equipment to use and he/ she can be health in short time. Then, the medical consumer will feel the medicine or medical equipment has good utility to satisfy whose medical needs. So, good medicine and good medical equipment can only need short time and less money to let the medical patient to be health.

How robotics can excite cruise entertainment customer's immediate (expected) emotion

As cruise entertainment case, every cruise journey must provide fixed stay days on the cruise to let every cruise passenger to play to every cruise journey. So, cruise passenger can not change or extend whose fixed stay day choice in every cruise journey, when they had caught the cruise to go to sea on the day. Is implies that cruise entertainment has none longer time utility factor which can influence each cruise passenger's choice to each different design of cruise journey arrangement.

If the cruise passenger feels very satisfactory and enjoyable to the last time of specific cruise journey arrangement, e.g. five days and four nights New Zealand and Australia cruise journey. Due to this cruise journey can bring positive emotion to let the cruise passenger to let whom feel that he/she can not forget or remember this happy cruise journey forever.

It is possible that this time happy five days and four nights New Zealand and Australia cruise journey will bring positive emotion to influence this cruise passenger to choose to find this cruise service provider to help whom to arrange this same cruise journey or another similar cruise journey again after one month, or three month, or six month or one year or more. Due to this cruse passenger felt this cruise service providers' cruise journey design arrangement can satisfy whose needs and it can achieve whose expected emotion to be positive. So, this cruise entertainment industry must be immediate (expected) emotion influential factor more than time utility factor to influence the cruise consumer's cruise service provider and cruise journey choices.

In conclusion, it is not only utility factor can influence consumption decision. It is emotion factor can also influence consumption decision. It depends on situations whether the consumer is choosing to buy one product or consume one service. If the consumer is choosing to buy one product, how long time of the product's utility factor which will influence the consumer choose to buy which product. If the consumer feels the product can give longer utility time among other similar products, then he/she will have more chance to choose to buy the product. Otherwise, if the consumer feels the product can give lesser utility time among other similar products, then he/she will have less chance to choose to buy the product. If the consumer is choosing to consume one service, emotion factor will influence the consumer choose to find which service provider to consume the same service or similar service. If the service provider can provide excellent service to the consumer, then it will bring positive immediate or expected emotion to whom and it can attract the consumer to choose the service provider again. Otherwise, if the service provider can not provide excellent service to the consumer, then will bring negative immediate or expected emotion to whom and it can not attract the consumer to choose the service provider again.

Why and how (AI) judgement tool can judgement what utility factors are to influence travelling consumer emotion

In emotion and utility both aspects, they include these situations. I shall explain how and why emotion and utility factors can influence travelling consumption behaviors in these different situations as below:

(1) In the first situation is brand factor, the brand image, product quality, product knowledge , attitude and

(2) brand loyalty intangible factor will attract the consumer individual purchase.

For example, luxury

products, e.g. luxury fashion brands of clothing. Brands like Zara from Spain and H&M from Seweden began to produce catwalk-style fashion at low cost offering consumers of luxury fashion alternatives at low prices.

Nowadays, the luxury fashion sector is the fourth largest revenue generator in France, and one of the most remarkable sectors in Italy, Spain , the USA and the potential markets of China, Russia and India. The luxury industry has increased having a huge youth in demand. The luxury consumer have much choice in products, shopping channels and pricing of luxury products. It has possible relationship of age, gender, income and other demographic factors with purchasing intentions to influence the rational and emotional buying behavior regarding luxury fashion products.

(2) The second situation concerns the decision-making of make or female consumers are possible experience an emotional desires and cognitive (reasoning) mind in purchasing choice process. Their emotion includes negative or positive buying emotion and mood management and cognitive process components include cognitive deliberation, planning buying with the exception regard for the future.

University had been using analysis of variances tests, male and female students were found significantly different with respect

affective process components including positive buying emotion, and mood management and cognitive process components include planning buying.

Significant differences were also found between the following product categories: shirts/sweaters, skirts, coats, underwear, accessories, shoes, electronic hardware, computer software, music , CD or DVDs, sports, memorabilia, health /beauty products and magazines/books for pleasure reading. No differences were found in regard to suits/ business wear and entertainments.

The investigate proved that some products will have different emotion influence to cause female or male students whose final consumption decision to buy the kind of product. So, the difference od male and female students will have emotion influence to make purchase decision to buy the product in consuming choice process.

(3) The third situation concerns search advertising factor, e.g. online search to influence consumption behavior. Advertising is possible one method to persuade the consumer to choose to buy the product, even the consumer does not know the product exists. For example, proper cloth, a company based in New York, has a site on the social networking site Facebook.

Whenever the company posts a new photos of its clothes, all its face book " fans" automatically receive the information on their own face book pages. "We want to hear what our customers have to say." It seems online advertisement is a potential promotion method to promote any new attractive products to sell to let publicity to know to buy. Internet is one popular communication tool to be used by youth today. So, when one company can have one website to let any youth to find and enter to the website to discover any new things easily. It will cause many consumption chances to let online potential clients to attempt to choose any products to make purchase decision from online advertising tool easily.

How does the role of advertising influence the purchase decision process? Needs and motivations are the starting points of purchase decisions. In fact, advertising is a communication of photo image, sound image, and word advertisement image channel to persuade consumers to choose to buy the brand of product or consume the brand of service between the merchant and its consumers from television, radio, newspapers, magazine, movie etc. channels.

Why does advertisement influence consumer
choice? For this case example, when one buyer waits until more information is gathered before making a decision. The time, two types of cost are involved. First, there are psychological opportunity costs experienced by consumers who are deprived of the product who need and are consequently in a state of psychological tension.

As time elapses, this psychological tension becomes more frustration. Second, buyers experience costs with the information-gathering efforts. They must invest time and energy to visit several retailers, seek out and read advertisements, or inquire for other opinions about the best product to buy.

These delayed decision costs considerably increase as time elapses. The buyer must seek information until it is felt that a search for additional information will bring about more costs than benefits. So, an advertisement is reaching a potential buyer when who is seeking information will have a greater impact, since the buyer is spending time and effort needed to seek out this information himself and he is less likely to find other competing and advertisements to obtain the additional information.

In general, buyers are generally more responsive to different brand advertisements, when they are seeking information on these brands. This is why the becomes a choice target for the advertiser provided the advertiser can identify and locate them. Thus, a client has interested and is in an information-gathering stage is asked.

Then, the advertiser takes advantage of the consumer's having identified him or herself to send a series of informative and persuasive messages or to send a salesperson who will try to conclude a sale. Thus, advertisement gives a chance to let consumers to gather information to choose the best product to buy or the most excellent service to consume.

What of situations do merchants need advertisements promotion? The short purchase cycle markets are characterized by routine purchase decision processes or by limited problem solving when a new brand is introduced on the market, e.g. coffee, bread , sugar, soft drinks, canned vegetables and household and beauty care products fall into this category. Another irregular purchase cycle markets are characterized by products that are purchased more or less regularly , e.g. cookies, cake mixes, wines, food products. Finally, long or unpredictable purchase cycle markets include all durable products, such as cars, household appliances and furniture (products from which occasions of purchase can't be predicted, which most consumers buy only occasionally). Hence, these kinds of products ought need advertisement promotion specially, due to advertisement can build brand image to let consumers to know. Especially , it is a new brand of product. In conclusion, advertisement will be one good channel to let new product to introduce its brand to let clients to remember in minds heart.

Consequently, future (AI) tool needs to learn how to design different kinds of advertisement to attract consumer attention to the product, needs to learn how to find what the bad factor(S) which cause(S) many consumers do not choose to buy the product, needs to learn when the product price needs to be raised up or fallen down in order to bring consumers' positive emotion to choose to buy the product immediately. SO, if (AI) tool can be invented to own itself effort to design different kinds of methods to predict consumer behavior in order to bring their positive emotion to the product successfully, then the manufacturer can earn positive consumer emotion advantage from the (AI) tool assistance for long term benefit to its product.

Main barriers influence artificial intelligent travelling consumer behavioral prediction

In future, it is possible that these barriers will influence how to apply (AI technology) to predict travelling consumer behavior in success. The barriers may include: Lacking of a (AI) digital data gathering vision and strategy, lacking of efficient workforce readiness, (AI) technology constraints., non reaching (AI) consumer behavioral prediction mature stage, time and money and resource constraints, law and regulations prohibition to develop (AI) consumer behavioral prediction bug data gather technology.

However, the recommendation of solutions to attack the barriers to influence artificial intelligence consumer behavioral prediction not success, it may include gaining employee buy in to participate and develop (AI) consumer behavioral prediction technology, making customer experience to a concern (AI) big data gather questionnaire investigation, providing compensation, training to employees in order to achieve (AI) travelling consumer behavioral big data questionnaire investigation research digital technological goals and strategy, task senior leaders manage any (AI) digital big data gather technology changes, putting policies and (AI) big data gather digital technology in place to support a fully remote, flexible workforce in any (AI) digital big data gather questionnaires research projects, teaching all employees how to code/understand (AI) big data gather consumer behavioral prediction software development, appointing a chief (AI) officer to manage any (AI) big data gather customer behavioral prediction projects and automate everything and encourage customers to attempt experience to self-service and (AI) big data gather questionnaire research to earn beneficial consumption aim after they gave feedback to any (AI) digital questionnaire researches. So, in the future, the (AI) digital big data questionnaire researches can include these industries surveyed, such as automat m financial services, public healthcare, private healthcare, technology, telecoms, insurance, life sciences, manufacturing, media and entertainment , oil and gas, retail and consumer products etc.

Hence, in the future, any of these industries can attempt to apply (AI) digital big data gather technology to predict how and why consumer behaviors will change in order to avoid reducing consumer number threat occurrence.

3.1
(AI) digital data gather technology predicts food consumer behavior's main barriers

It seems food consumer‘ (AI) predictive barriers is similar to travelling consumer. What are the main barriers to food industry? When the food manufacturer applies (AI) big data gather technology to predict food consumer behavior? The barriers include that the food manufacturer / provider needs to decide whether when the right time is applied to the right (AI) digital big data prediction tool channel to find the right food consumers to be chose to full food consumption satisfactory questionnaires, how to gather multi-class food consumption classifiers on real-world food consumers transactional data from the food sale domain consistently to show the critical numbers of different kinds of food items at which the predictive performance most accurate? So, any food manufacturer / provider's advanced in (AI) digital data gather warehousing and management technologies can provide that opportunities for food business to enhance long term relationship with the food providers' clients.

However, food industry's (AI) digital data gather aims to improve food customer product targeting, increase food customer loyalty and food purchase probability to the food supplier. To effective identify, understand and satisfy the needs of their food customers, the food suppliers need to develop the right (AI) digital questionnaire questions and find the right food customers to fill every right questions from every digital questionnaire at the right time through the right channel.

Above of all these, they will be the barriers when one food supplier expects its (AI) digital data gather questionnaires which can conclude the most accurate prediction concerns any kinds of consumer food product choices. So, such as (AI) digital data prediction model, it is needed to incorporate into the food market segmentation, food customer targeting, and food challenging decisions with the goal of maximizing the total food customer lifetime. For example, (AI) big data gather transaction data is reasonable and accurate for building predictive models. Transaction data can be electronically collected and readily made available for data mining in lot quantity at minimum extra costs.

Suggestion to apply (AI) prototypes of food customer profiles method to predict food customer behavioral changes. Prototypes of food customer profiles mean to be extracted from the discovered bins and multi-class classifies models are built using those prototypes. The learned models can than be used to predict the class of food customer profiles (e.g. restaurants, school canteens, supermarkets etc. food suppliers) based on their food purchases. The approach is validated on the case study of a food retail and food service company operating in food and beverages market.

So, a food customer profile, it is a description (AI) data gather tool will record every of food customer using available information, which help in understanding their background and food consumption behavior. (AI) data gather tool can well develop every food customer profile, every food customer data is essential in food market analysis as they aid food suppliers in saving time and money by highlighting the real potential food consumers whose needs are to be met rather a range of individuals.

So, (AI) data gather tool can record every food consumer profile and every can be factual or behavioral food consumption. A factual food customer profile consists of a set of characteristics for (AI) big data gather record, e.g. demographic information , such as food customer name, gender, birth date, when a behavioral food customer profile consists of what the food customer is actually doing and is usually derived from (AI) digital transactional data gather record.

So, (AI) big data gather record's every behavioral food consumer profile can be much stronger predictor of the future food supplier consumption choice actions of a food customer. Furthermore, the food supplier's (AI) all past food consumer information that make up demographically based all past food customer profiles are expensive to acquire when the information for the food suppliers' past every food consumer food consumption behaviors. Moreover, food customer profile can be recorded to make real food purchase every time. So, when the food supplier finds the past food consumer's record from (AI) big data gather tool. Then, it can make more accurate judgement whether past every food consumer has chose to buy its food to eat how many times every year in order to predict whether its every past food consumer will choose to buy its foods how many times next year in possible. If the next year, its every past food consumer's consumption time to the food supplier is less than its current year consumption time. Then, the food supplier can attempt to find whether what factors to cause the past food consumers do not choose to increase food purchase times to the food supplier in current year. The factors may be possible be the food supplier's food prices are raised, food quality or taste is poor, the different kinds of food supply is shortage challenge, the food supplier's consumers lose confidence to buy the food supplier's foods to eat, when (AI) big data gather tool can help the food suppliers to find what the main factors to cause the past food consumer number to be reduced in order to predict how future food consumers' behavioral changes will be influenced from the food supplier's competitors in the global food supply market. Hence, (AI) big data gather tool can help every food supplier to attempt to find what the main factors to case the food supplier's food consumer number to be reduced as well as it can help the food supplier to predict how the food supplier's potential (past not every purchase its any food consumers) food consumers who can be persuaded to choose to buy its foods to eat by learning what the main factors influence.

In conclusion, (AI) big data gather tool can help the food supplier to find what the main factors influence its past food consumers do not choose to buy its food more times or find what the main factors will attract its potential (not ever buying its foods consumes) food consumers to choose to buy the food supplier's foods to eat.

3.2 The challenges of (AI) big data gather shaping the future of retail for consumer industries

The future of retail for consumer industries' (AI) big data gather challenges are similar to future travelling industry's entertainment consumption challenges. Another challenge of (AI) big data gather is that how to shape the consumer behavior to let business owner to feel or know oe predict. It means that how it express it's conclusion or opinion for every consumer behavior after it had gather all big data in any data gather period, e.g. three months, half year or one year consumer shopping model data gather period.

Because every kind of industry, consumers will continue to demand price and quality change , with a wide range of convenient fulfilment options among of different kinds of products or services supply. Overall, the (AI) big data gather procedure gives opinion concerns every time retail experience will become more exciting, simple and convenient, depending on the consumer's ever-changing needs. So, I believe that (AI) big data gather every conclusion or result will be different, due to consumer's price and quality demand will often change to every kind of product or service supply in retail industry. So, how to shape (AI) big data gathering's analytical conclusion or result more clear. I shall recommend organizations need to build great understanding of and a stronger connection to increasingly empowered consumers before they plan and implement how to apply (AI) big data gather tool to predict consumer behavior as below:

Firstly, (AI) is empowered by technology, the consumer is redefining value. The traditional measures of cost, choice and convenience are still relevant, but not control and experience are also important. Globally, consumers have access to more than 2 billion different products choice by a wide range of traditional competitors and dynamic new entrants, all experimenting with new business models and methods of client engagement.

As choice increases, loyalty becomes more difficult familiarity and the consumer becomes more empowered. Businesses will have no choice and constantly innovate and disrupt themselves by meeting new technologies of high standards and expectations of consumers. So, (AI) data gather tool will need to follow different target group of consumers' needs to follow their different kinds of product design or style choice preferable to gather data in order to conclude the different target groups of consumer behavior to give opinion more clear and accurate to let businessmen to understand more clear how its customers' behavioral choice trend in the future half month, even to two years period.

Secondly, businessmen need to adopt changing technologies rapidly. Technology will be the key driver of this retail industry. Industry participants will only success if they have a clear prediction to focus on how to using technology to increase the value added to consumers. They must , however, do so will I realistic assessment of their costs and benefits. Hence, (AI) big data gather technological tools will need to design to help them to gather data efficiently by these ways, such as the internet of things (IOT), artificial intelligence (AI) machine learning, augmented reality (AR)/virtual reality (VR), digital traceability. So, future (AI) big data gather tool are predicted to be most influential customer behavioral positive emotion changing tool for retail , due to their widespread applications , ability to drive efficiencies and impact on labor in order to impact consumer behavior changing effort from negative emotion to positive.

Thirdly, (AI) big data gather tool is an advanced data science of consumer behavior predictive tool. Businesses will have to bring the journey from simply collecting consumer data to using it to scale and systematize enhanced decision making across the entire value chain. When focused on their business goals, industry players should not lose sight of the impact that future capabilities and transformative business models may have on society.

However, (AI) big data gather tool will encounter these challenges when any business plans and implements to apply it to predict consumer behavior in retail industry. The challenges include that as below:

1. The high cost and difficulty of implementing new technologies . The (AI) big data gather tool needs capital and capabilities to be designed to implement to be applied to different retail industry users. so, expensive barriers to innovation, an organization and the skillsets of its people to support a new design of (AI) big data gather tool, highly digital technology may be required.

2. Slow pace of cultural change. Consumers need to adapt or accept (AI) new technology consumption model in the traditional retail industry. The rate of change is outpacing the ability of businesses to keep up. (AI) big data gather tool needs to be designed to adopt in new or evolved business model requires, in most cases, a new level of customer behavioral predictive machine operation will impact to influence any retail businesses' consumer behavioral changes at a minimum, an organization's structure, capabilities, culture and decision making. If the retail business expects to apply (AI) big data gather tool to predict how to change its consumer behaviors and how their consumption behaviors will tend to change in order to achieve to change their positive emotion from negative emotion before they choose to buy its product or consume its service in success.

3.3 Challenge to using (AI) neural networks to predict customer behavior from big data gather tool

The challenge to using (AI) neural networks to predict customer behavior is similar to predict travelling consumers' entertainment need challenge. (AI) big data gather tool will encounter the challenge: How can predict customer behavior be represented as sequential data describing the interactions of the customer with a company or an (AI) data gather system through the time, e.g. these interactions are items that the customer purchase or views ? So, every customer data gather , (AI) needs to spend time to analyze how and why to cause whose consumption behavioral choice. It is too difficult matter or judgement for (AI) learning. So, (AI) needs to spend time to learn how to analyze every customer's shopping behavior or action in order to gather all different consumers' past shopping action information in order to help business owners to predict future its potential customer shopping behavior how to change more clear and accurate prediction.

(AI) big data gather tool needs to learn to know that how to judge every customer interaction likes purchases over time can be represented with sequential data. Sequential data has the main property that the order of the information is important. Many (AI) machine learning models are not suited for sequential data, as they consider each input sample independent from previous ones. Therefore, at the end of the sequence, (AI) big data gather learn machines need to keep in their internal state of every customer purchase data, kind of product or service, price , whole year consumption times form all previous inputs, making them suitable for this type of data.

However, consumer behavior can be represented as sequential data describing the interactions through the time. Examples of these interactions are the items that the user purchases or views. Therefore, the history of interactions can be modeled as sequential data, which has the particular trial that an incorporate a temporal aspect. For example, if a user buys a new mobile phone, who might purchase accessories for this mobile phone in the near future or it the user buys a electronic book or paper book , he might be interested in books by the same author. Therefore, to make accurate predictions is important to model this temporal aspect correctly. To solve this predictive challenge of consumers to buy the product. One count the number of purchased products of a particular category in the last N days, or the number of days since the last purchase.

So, the (AI) big data gather designers can attempt to produce a feature vector which can be fed into a machine learning algorithm such as " logistic regression" will be the main feature and function to any (AI) big data gather machine to learn how to apply this " logistic regression" function or feature to predict any customer behavioral change for any product purchase or service consumption to the (AI) predictive consumer behavioral business users. Every different kinds of product purchases or services consumption will be needed to design " different model of logistic regression" in order to follow the kind of business to predict whose consumer purchase or service consumption behavior to predict more accurate.

3.4
Challenges of artificial intelligence, algorithms technology and machine learning impact to consumption market

The challenges of artificial intelligence, algorithms technology and machine learning impact to consumption market are similar to travelling entertainment consumption market. Markets have played a key role in providing individuals and businesses with the opportunity to gain from trade. If (AI) big data gather tool can predict how to change potential customer behavior in success. The challenges to consumers will face that the overall market consumption model will be dominated by the businessmen only. So, it is not fair or reasonable to consumers, because (AI) big data gather tool has controlled or dominated all consumers' minds and it has predicted how and why every kind of product or service consumer shopping model or consumption behaviors how will change.

It will bring this questions: How can market designers learn the characteristics necessary to set optimal, or at least better, reserve prices after they had gather all data to conclude the analytical results of their consumers behaviors how will change? How can market designers better learn the environments of their markets?

In response to these challenges, artificial intelligence (AI) and machine learning are important tools for market design. For example, retailers and marketplaces , such as eBay, Amazon and many others are mining their vast amounts of data to identity patterns that help them create better shopping experiences for their clients and increase the efficiency of their markets. By having better prediction tools, these and their companies can predict and better manage dynamic consumption market environments. The improved forecasting that (AI) and machine learning algorithms provide help marketplaces and retailers better anticipate consumer demand and producer supply as well as help target products and activities for segmented markets. Another important application of (AI) 's strength in improving forecasting to help markets operate more efficiently is in electricity market example. To operate efficiently, electricity marker makers can attempt to apply (AI) machine learning tool to follow every household family electricity consumers' past electricity consumption record to judge (predict) how it will be every family's forecasting in the year.

An inaccurate forecast in the electricity supply and demand that can dramatically affect electricity market bad supply outcomes causing high variance in electricity charge prices or worse, blackouts. By better predicting every family's electricity demand and supply , electricity market makers can better allocate power generation to the most efficient power sources and maintain a more reasonable electricity stable charge market. Any example is design market, the application of (AI) algorithms to market design are already widespread and diverse.

(AI) algorithms technology , it is a safe that (AI) will play a growing role in the design and implementation of market over a wide range of applications. The challenges are that how (AI) can guarantee accurate to predict when and why and how consumer behavioral changes to any retail industries. In fact, retailers will need to discover the value that (AI) can bring to what benefits to influence their customer behaviors.

In the future, (AI) will bring their benefits to influence travelling entertainment customers to build positive emotions to any travelling entertaonment providers in these aspects as below:

1. Future (AI) big data gather tool will be an area of compute science that deals with giving machines , the ability to seem like they have human intelligence. In short, it is the power of a machine to copy intelligent human behavior. For examaple, machine learning algorithms are being integrated into analytics and customer relationship management platforms to uncover information on how to better serve customers, chat bots have been incorporated into websites to provide immediate service to customers.

2. (AI) adoption continue to rise with chat bots taking the lead. Due to increasing ease of deployment , instant availability and improved quality, chat bots will become more and more common to manage customer service queries and to make intelligent purchase recommendations. Also, retailers can engage this kind of technology to answer continue questions and supplement customer support with chat-based shopping experience. So, (AI) and declines personalized, customized and localized experiences to customers.

(AI) will be applied across the entire retail product and service cycle, firm manufacturing to post-sale customer service interactions. Hence, retailers can use (AI) to its fullest potential will be also to influence purchases in the moment and anticipate future purchases, guiding shoppers towards the right products in a regular and highly personalized manner.

3. (AI) technology can rise the conscious customers. Customers are demanding an increased interest in the ethical practice of the brands they buy from. Todays, customers have a well-developed sense of what is solely intended to drive sales. This has lead to a rise in consumers ho make values based judgements about what to buy and where to shop. These consumers believe their purchase habits have an impact on the world. To win customers, retailers need have good conscious to predict consumers' desire. Future, (AI) data gather technology will be a good consumer behavior predictive tool to predict about for years will now become customer expectations and will have drastically changed the path to purchase. So, (AI) data gather tool is the predictive consumer expectations tool on every interaction, they have these brands.

4. Future (AI) can be impacted to influence consumer behaviors by its potential to free up time, enhance, quality, and enhance personalization. The industries include: Healthcare industry can apply (AI) to support diagnosis by detecting variations in patient data, early identification of potential pandemics, imaging diagnostics; automat industry can apply (AI) to autonomous fleets to ride sharing, semi-autonomous features, such as driver assist, engine monitoring and predictive, autonomous maintenance; financial service industry can apply (AI) to design the suitable personalized financial planning, fraud detection and anti-money laundering and automation of customer operation; transportation and logistics industry can apply (AI) to autonomous trucking and delivery, traffic control and reduced congestion and enhanced security; technology, media and telecommunications industry can apply (AI) to search media, and recommendation, customized content creation and personalized marketing and advertising to attract retailers to promote; retail and consumer industry can apply (AI) to design personalized production, anticipating customer demand, , inventory and delivery management; energy industry can apply (AI) to read and record smart metering , more efficient grid operation and storage and predictive maintenance; manufacturing industry can apply (AI) to enhance monitoring and auto-correction of processes, supply chain and production optimization and on-demand production.

Hence, future (AI) technology will impact consumer technology when any retailers apply it to assist its manufacturing processes or product sale or service provision processes to satisfy consumers' needs, it means that it can help any retailers to influence positive emotion to consumers in their whole sale or consumption or purchase proccesses.

5. (AI) and machine learning technologies make it possible to capture, process, and inter data on a massive scale effectively , then any human being could ever do. For example, Criteo's creative technology " Kinetic design" can apply insights from 1.2 billion monthly impressions to select and optimize individual branded advertisements components according to each shopper's preference and intent. This ensures more personalization and visually inspiring on brand ads. resulting in up to 12% more sales for (AI) technology advertiser clients.

Moreover, advertisers can now engage and inspire shoppers on a more personal level, rendering custom ads. it real-time for every impression. So, designer continues to learn from each design's success to make ads. more and more effective over time. Furthermore, brands are increasingly using paid search on retail sites to draw attention to their products on the crowded online shelf, e.g. Google shopping is a key growth area's more users are engaging with shopping ads. and across the globe. Google shopping has become essential to retailers' marketing strategies, but is a difficult channel to apply its tool to be promoted effectively . Thus, future (AI) and machine -learning technologies can dramatically improve digital commerce performance application to apply (AI) and machine learning to digital consumer. So, future (AI) technology can be applied to digital commerce aspect, it will fall into the categories of pattern recognition, classification, prediction and consumer behavior.

In conclusion, the benefits of using (AI) in digital commerce include: improved efficiency in discovering the relationships between datasets over traditional methods, which require complex modeling and coding, improved accuracy for clearly defined processes that involve a lot of manual processing, ability to deal with a large emotion of data with many attributes, for example: customer behavior data, multichannel and multi-device data , complex product data and fraud detection, more accurate analysis, such as customer segmentation sentiment, analysis and personalization frequent algorithum refreshes, such as several times a day, to capture the changes in customer and market behavior.

Finally, however, a lot of types predictive consumption behavior around (AI), in particulars that driven by vendors claiming their solutions are (AI) , ready and can deliver dramatic improvements over existing technologies. Application leaders for digital commerce can be misled into believing that (AI) can solve all their problems, which is not true for n in-depth discussion of the (AI) consumers and market behavioral predictive tool and machine -learning technologies bot. Thus, (AI) prediction consumer behavioral technology can give beneficial quantitative analysis for forecasting in business and market especially in consumer behavior and in the consumer decision-making process (consumer choice model) more effectively and efficiently.

Is Artificial Intelligent the most effective and accurate travelling consumer behavioral tool?

Is (AI) the best and the most effective and accurate travelling consumer behavioral prediction tool to compare other kinds of consumer behavioral prediction tools? Nowadays, travelling entertainment providers' competitions are serious businessmen often find different kinds of methods to attempt to predict travelling consumer travelling entertainment need of changes. The travelling consumer behavioral predictive methods can include as these below methods, instead of (AI) big data gathering tool.

Firstly, statistics is the popular mathematic method, it applies auto-regression, liner regression, structural equation modelling, logistic regression statistic techniques to be used to predict consumer behaviors. Secondly, it is classification method, it sis a support vector machine to assist businessmen to make consumer behavioral prediction, it also includes decision making tress diagram technique. Thirdly, it is rule mining method, it is algorithm, market base analytic etc. business marketing concept analytical tool, it also includes graph mining technique tool. Next, it is psychological prediction model tool, it is psychology prediction model too, it is a kind of psychological method to predict consumer behaviors. Finally, it is the most updated and potential artificial neural network (ANN) machine tool, it gathered big data, then it will carry on analyzing and applies psychological method to conclude the most accurate and reasonable solutions to give recommendation to businesses to predict when and how and why their consumer behaviors will change. So, it is one owned human mind's machine and owned psychological and analytical efforts to replace humans to make any judgement in order to make the most accurate predictive behavioral changes for consumers, instead of the traditional marketing concept and psychological and mathematic methods to predict consumer behavior, (AI) big data gathering tool will be another new tool.

What are the advantages of (AI) tool to be used to predict consumer behaviors as well as what are the different between it and other traditional consumer behavioral predictive tools? I shall explain as below:

Firstly, as above all case studies are explained to (AI) questionnaire design method benefit, I believe (AI) big data gathering tool can be applied to help human to analyze and design any the suitable valid questions to enquire any kinds of business consumers in order to gather the most meaning and useful opinions to conclude the most accurate consumer behavioral prediction for every questionnaire. So, future (AI)'s analytical effort and decision making effort most be exceed above human's judgement efforts. So, future (AI) can help human to design the most useful and meaning different kinds of valid questionnaire (survey) questions as well as assist humans to analyze and make accurate decision making and conclusions to give opinions to help businessmen to predict when consumer behaviors will change and how their consumption behaviors will change to influence their businesses in order to help them to make any efficient and effective and accurate solutions to avoid consumer number to be decreased and the most important benefit is that it can give opinions to help businessmen to explain why (what the factors) cause their consumer behaviors change suddenly. It will be human's efforts can not achieve to exceed (AI)'s efforts in the future.

Secondly, (AI) can make artificial machine judgement and analytical effort, without human misleading or unfair or unreasonable judgement. So, it can make more fair and reasonable and accurate conclusion to give opinions to predict when, how and why consumer behaviors will change suddenly to the kind of business in customer model building process and evaluating the results of customer relationship management –related investment more accurate.

Furthermore, (AI) big data gathering tool will help businesses to improve the success rate of acquiring customers, increasing sales and establishing competitiveness. (AI) big data gathering tool can give opinions how to build customer loyalty to be positive emotion impact and it can find solutions to avoid every client's negative emotion causes to bring complaints behavior to the businessman's product or service. For example, Telecom industry and aggressive research has been conducted in this by applying various data mining techniques to avoid long distance phone call users' complaints. If gathered any long distance phone call users' past complaint data to record what are their general complaint issues. Then, (AI) tool will analyze all these past complaint issues to conclude and give opinions to let Telecom knows whether which aspects encounter challenge that Telecom needs to improve it's long distance phone call services or functions in order to satisfy Telecom's long distance phone call users' needs for long term. After Telecom attempted to improve its services and/ or functions from (AI) opinions and solution methods, when it fell it's long distance phone call users have positive emotions to satisfy its service performance and function performance. Then, it can prove (AI) tool's opinions and solutions are useful. The consequence is that their complain numbers will be decreased and they won't plan to choose another long distance phone call telephone service company to replace Telecom long distance phone call service more easily.

So, (AI) big data gathering tool can concentrate on finding focus on components of customer relationship management method and datasets more accurate and efficient and effective than human's data gathering and analytical effort. It implies (AI) big data gathering tool has unique more efficient and effective and accurate dataset gathering and analytical and judgement and decision making effort, it is human can not achieve.

Thirdly, (AI) big data gathering tool has much customer loyalty predictive effort. It's effort is more easily subsequently selected, reviewed and classified to compare human's gathering data effort in whole data gathering and analytical process.

In (AI) big data gathering process, (AI) can organize whole big data gathering process and technique more easily in short time. It will include these four steps. The first stage is that customer identification stage, customer identification also known as acquisition has to do with targeting the population , who are most likely to become customer segmentation. So, (AI) can help different kinds of businesses to gather their competitors' consumer purchase behavior data in short time, it is human can not achieve. The second stage is that customer attraction stage, after (AI) maker has been segmented for the business when it has ensured to gather the businessman's global competitors' consumers data. Then it analyze these all data to find solutions / methods to give the best opinions to the organizations how to achieve the direct effort and resources into attracting the target customer segments. The third stage is that customer retention, it can be defined as the activity that an organization undertakes in order to reduce customer defections. TO be successful, customer retention starts with the first contact on organization has with a customer and continues throughout the entire lifetime of a relationship involves loyalty programs, one to one marketing and complaints management. SO, (AI) can consist the business to find the best or the most reasonable , efficient , effective solutions or methods and it will conclude all these solutions to find the most reasonable and useful opinions to achieve to the aim to help the business to reduce customer complain numbers and help the business to build confident loyalty relationship between it and its clients. SO, (AI)'s analytical effort and decision making effort can be more accurate than human's analytical effort and decision making effort. IT can achieve it's consumer behavioral predictive aim more accurate and efficient and effective in the shortest time to compare human.

Fourthly, (AI) big data gathering tool can design more accurate dataset program for questionnaire (survey) to compare human's questionnaire (survey) effort. It means that (AI) can spend less time to research and make judgement what are the most reasonable and meaning questions for different kinds of businesses' needs. This includes data conduction a questionnaire, survey or interview of the individual or environment researched, public data repository: This includes commercially available public data; organizational data; this contains data collected from an organizational database, organizational information system. For example, their website log details etc. It also includes company transactional data, data purchased from a company.

For example, one vehicle sale company expects to research all global vehicle sale companies' past the different kinds of vehicle styles, design sale number data, the different kinds of vehicle style, design sale price data, every country's vehicle consumer number to the vehicle purchase number data to the vehicle company in short time. (AI) big data gathering tool can help the vehicle sale company to gather all any one for these global vehicle sale competitors' past data in the short time. It is human effort, who can not achieve this efficient, effective and accurate data gathering aim for this vehicle sale company. Even, when (AI) had gathered all global it's vehicle competitors' past sale data, (AI) can make more accurate analytical and judgement and decision making effort to design different kinds of questionnaire (survey) questions to prepare to enquire it's different target segmentation vehicle potential clients in order to predict what are their needs to choose to buy any vehicles from the vehicle company. SO, (AI) tool can conclude more accurate conclusions and give the most reasonable and useful opinions to let the vehicle company to know in order to predict what are it's potential vehicle buyer's needs and manufacture the suitable vehicle styles or designs to raise their vehicle purchase desires.

Fifthly, (AI) tool is only one perfect tool for big data gathering in order to achieve accurate results and increased profit. What is (AI) big data gathering mean? The term " big data"gathering describes the accumulation and analytical of vast amounts of information, but big data is much more than a big amount of data. It is also the ability to extract meaning to sort through big volumes of numbers and find the hidden patterns, unexpected correlations and surprising connections that can be used in different industries like medical field, security and protection field or marketing that adopt " big data driven" decision making enjoy significantly greater productivity than those that do not. So, the benefits of (AI) is given to the company by using big data repaid complexity of implementation projects and hence project risks, when accelerating time to value. It is why that human's gathering effort can not replace （ＡＩ) data gathering effort.

All analysing above benefits to (AI) big data benefits to any organizations, it brinfs this question: How can (AI)apply big data gathering and analyzing to predict when and how any why consumer behavior will change suddenly? The purchase decision making process is consumers reducing purchase choice behaviors.

Travelling consumers are being considered pure rational beings (consumer tried only to satisfy self-interest). Hence, due to future (AI) owns human's psychological , analytical , emotional predictive, purchasing decision making effort.

(A I) will be assumed to sees one travelling customer how who will make travelling destination and travelling package decisions. So, after the (AI) gathered all data concerns the find of business's past travelling customer segmentation travelling destination and travelling package design activities, e.g. age, sex,. income level, the different kinds of travelling package sale number, the different kinds of travelling package price variable sale etc. different kinds complex data.

It can make more accurate psychological and analytical effort to predict when the travelling provider's travelling consumer behaviors will change as behaviors will change as well as find what reasons their travelling package choice behaviors will change and how trend of their travelling consumer behaviors will change more accurate. For example, today there are a lot of industries that use big data: healthcare (treatment) becoming personalized and patient centric and predictive analysis are used to prevent diseases for example Angelina Jolie underevent a predictive double mastectomy after learning she had 87% rich to developing breast cancer, sports (by using sensors data are collected from players during a game in order to improve their playing schemes), weather(more than 60 years of global weather analysis are used to predict the risk of future extreme events), logistics (smart tucks and smart species, agriculture (monitoring weather and soil conditions for optimum point of harvesting).

Consequently, due to the evolving consumer demands, and the ever growing digitization, the world is digitally transforming which means the new technologies are needed to be used and driven significant business improvement. So, such as why (AI) tool will be our future main predictive tool to help travelling service providers to predict when, how and why their potential travelling customer package choice behavioral will change. Big data is one of the our channels through digital transformation is made, together with cloud, mobile and networks. The challenges for digital transforming and therefore using A I big data gathering tool as main technology are: digital proficiency, legacy systems, security and jobs becoming absolute.

In the future, big data can use data from text to travelling photo picture , sounds, movies, musics satellite coordinates or any other type of input or output data that type of input or uouput data that came from different influential aspect. It is cloud solutions, bring big data will be for predict insight driven by travelling business stategy, differnet kinds of travelling package design strategies and new travelling consumer relationship, predictive travelling consumer's travelling package design behavioral strategy. Using the right data in the right business decision will mean smart decisions, new opportunitites and utimately a big competitive advantage.Hence (AI) big data gathering tool is different is that (AI) can be one depth in-memeory database function, it can make real-time data analytics that provide meaningful information in short time, it is also the visualization tool , such as SAP Lumira, allow this exploration and understanding of the data, and ultimately supports the decision making process. All above these features, which will be human's data gathering effort who won't exceed (AI) big data gathering effort. Hence, future (AI)big data gathering will be the best choice to assist travelling service businesses to predict travelling consumers' travelling package design behaviors successfully.

Reference

Adrian, P. (2012). Introduction to marketing theory & practice, 3 rd edition, London: Oxford press.

Ajzen, I (1991). The theory of planned behavior. Organizational behavior and human decision processes, 50(2), 179-211. doi: 10.1016/0749.5978 (91) 90020-7.

Alba, Joseph W. and J. Wesley Hutchinson (1987). " Dimensions Of Consumer Expertise", Journal of consumer research, 13 March, 411-454.

Bailey, L., Mokhtarian, P.L. Little, A. (2008). The broader Connection Between Public Transportation, Energy Conservation And Greenhouse Gas Reduction, Report Prepared As Part Of TCRP Project J-11/Tasks Transit Cooperative Research Program, Transportation Research Board Submitted To American Public Transportation Association in
http://www.apta.com/research/into/online/land_use.cfmi, accessed 17 April 2008.

Baucer, R,"Consumer Bhavior As Risk Taking , In Risk Taking And Information handling In Consumer Behavior", D. Coxceds Harvard University Press, Cambridge, Mass 1976.

Biederman, P. (2008). Travel and tourism, Pearson Prentice Hall, New Jersey.

Bogers, R. P., Brug, J. Van Assema, P., & Dagnetie, P.C. (2004) , Explaining fruit and vegetable consumption: The theory of planned behavior and misconception of personal intake level. Appetite, 42,157-166.

Bolton, Ruth N. (1998), " A Dynamic Model Of The Duration Of The Customer's Relationship With A Continuous Service Provider: The Role Of Satisfaction", Marketing Science, 17 (1), 45-65.

B.Shiv and A. Fedorikhin, " Heart And Min In Conflict: The Interplay Of affect And Cognition In Consumer Decision Making", J. Consumer Res., vol. 26, pp. 278-292, Dec. 1999.

Brown, K.W., Ryan, R.M. Reswell , J.D. (2007). Mindfulness: Theoretical Foundatins And Evidence For Its Salutary Effects. Psychological Inquiry, 18, 211-237.

Burke, R.R. : Behavioral effects of digital signage, J. Advertising Res. 49(2), 180-185 (2009).

Cant, M., Brink , A. & Brijall, S., Consumer behavior, Cape Town, South Africa: Juta, 2006.

Conner, M. & Abraham, C. (2001). Conscientiousness and the theory of planned behavior: Toward a more complete model of the antecedents of intention and behavior. Social psychology bulletin, 27, 1547-1561.

Cooper C. Mallon, K, Leadbetter S, Pollack L, Peipins (2005) , cancer internet search activity on a major search engine, United States 2001 to 2003, J Med Internet Res. 7(3): e36.

Cope, R. R. Cope and H. Davis (2008). Disney's virtual Queues: A strategic opportunity to co-brand services ? Journal of Business & economics research, vol. 6 no10, 13-20.

Cornelia, B.F. (1999) Rural development news, the North Central Regional Center For Rural Development vol. no 24 , IOWA.

Couper, M.P. J. Blair and T. Triplet (1999). A Comparison Of Mail And E-mail For a Survey Of Employees In USA Statistical Agencies. Journal Of Official Statistics, 15, 39-56.

David J. Nowak & Gordon M. Melsler (2016) " Air quality effects of urban trees and parks." National recreation and park association, USA.

Data monitor (2008). The proctor and gamble company. Retrieved Nov. 15 2009 from http://www.datamonitor.com/

De Hollander, A. E. M., J.M. Melse, Elebret & P. G.N. Kramers (1999), " An Aggregate public health indicator to represent the impact of multiple environmental exposures" Epidemiology: 606-617.

De Visser, R.O., & McDonnell, E.J. (2013). " Man points": Masculine capital and young men's health. Health psychology, 32(1), 5-14. doi:10. 1037/a0029045.

Dunn, J & A Neumsister (2002). Knowledge management in the Information age. E. business review, Fall , 37-45. Jounral of service, spring 2011, vol. 4, no1, De Grovte (2009).

Dyer, D., F. Dalzell & R. Olegario (2004). Rising tide. Lessons learned from 165 years of brand building at Procter and Gamble. Boston, MA: Havard Business School Press.

Eysenbach G (2006) Infodemiology: Tracking flu- related searches on the web for syndromic surveillance. American Medical Informatics Associaion Annual Symposium Proceedings , Curran Associates, Red Hook, NY, pp. 244-248.

Ettredge M, Gerdes, J. Karuga , G (2005) Using web- based search data to predict macro-economic statistics. Commun ACM 48: 87-92.

Felce, D. and Perry, J. (1995). Quality of life: A contribution to its definition and measurement, vol. 16, no.1 pp: 51-74.

Feldman, Jack M. And John G. Lynch Jr. (1988), "Self-
Generated Validity And Other Effects Of Measurement On Belife, Attitude, Intention And Behavior", Journal of applied psychology, 73(3),421-35.

Fiese, M, Hofmann, W., & Wanke, M (2009). The impulsive consumer. Predicting consumer behavior with implicit reaction time measurement. In M. Wanke (ed.) Social psychology of consumer behavior (pp.335-364). New York, NY: Psychology press.

Fitzsimons, Gavan, J. And Vicki G. Morwitz (1996), " The Effect Of Measuring Intent On Brand-Level Purchase Behavior", Journal of consumer research, 23 (1), 1-11.

Hallerman , D. (2008) video Advertising Online: Spending And Pricing , New York. E-Marketer.

Harriet Griffey. (2010) The art of concentration, enhance focus, Reduce, stress and achieve move. Macmillan publishers ltd,Basinastoke and Oxford, London UK.

Helleman, D. (2008) Video Advertising Online: Spending And Pricing , New York, E-Marketer.

Hensen, C. (2003). Kreuzfahrtourismus.www.christoph- hensen.de/Facharbeit.pdf.

Huang, H.I. (2012). An empirical analysis of the strategic Management of competitive advantage: a case study of higher technical and vocational education in Taiwan (Doctoral dissertation,
Victoria University).

Jamieson, Linda F. And Frank M. Bass (1989), " Adjusting Stated Intention Measures To Predict Trial Purchase Of New Products: A Comparison Of Models And Methods," Journal of marketing research, 26 (August), 336-45.

Korea Ministry Of Environment. Public Organizations spend 2.2 Trillon Korean Won To Purchase green Products in 2014; Ministry Of Environment: Sejoung, Korea, 2015.

Kremers, S.P. J., De Bruijn, G.J., droomers, M., Van Lenthe, F. J., & Brug, J. (2005). Environmental interventions for selected dietary behaviors in adults. In J. Brug & F. J. Van Lenthe (eds.) , Environmental determinants and interventions for physical activity, nutrition and smoking: A review pp. 282-315. Rotterdam: Erasmus Medical Center.

Lee, D.; Kim, M. ; Lee, J. adoption of green electricity policies: Investigating the role of environmental attitudes via big data-driven search-queries. Energy policy 2016. 90, 187-201.

Lee, Terrence, " Tech in Asia-connecting Asia's startup system " Tech. in Asia- connecting Asia's startup ecosystem, N.p.,4 July 2016.

Los Angeles Country Department Of public Health (2016), Country Health Ranking Model, Retrieved From www.countryhealthrankgings.org/our-approach. USA.

Mayne, Lonnie. " Evolve of die in the age of the consumer". Entrepreneur, N.P. , 16 Apr. 2014. web of Oct. 2016.

McGregor, S.L. T., & Goldsmith, E.B. (1998). Expanding our understanding of quality of life, standard of living and well-being. Journal of family and consumer science, 90(2), 2-6, 22.

McMichael, A.J. M. Mckee, J. Shkolnikov and T. Valkanen (2004), " Morality trends and setbacks, global convergence or divergence?", Lancet 363, 1155-1159.

Melse, J.M. & A.E. M. De Hollander (2001). " Human Health And The Environment", background document for the OECD Environmental Outlook, OECD, Paris.

Moschis, George p. & Roy, L. Moore (1979), " Decision making among the young. A socialization perspective " Journal of consumer research , 6 (September).

Mulligan, M. Banerjee, T & Thomas, N. (2008) ,European Paid Content And Activity Forecast, (2008 to 2013), Jupiter Research.

Peter, J., Ryan, M, M, " An Investigation Of Perceived Risk At The Brand Level, " Journal of marketing research, 13 May 1976, pp. 184-188.

Pieters, R., & Wedel, M. (2007). Goal Control Of Visual Attention To Advertising: The Yarbus Implication. Journal Of Consumer Research, 34, 224-233 (August).

Parasuaman, and Leonard L. Berry (1985), " Problems And Strategies In Sevices Marketing", Journal of marketing, 49 (Spring), 33-46.

Priesnitz, W. (2007) Counting Our Food Miles. Natural Life, 1 July.

R.C. Oliver, " When is consumer loyalty?" J.Marketing vol. 63, pp.33-44.1999.

Reggiani, A . (ed). 1998, accessibility, trade and locational behavior, Ashgate publishing ltd, England.

Rushe, D. (2013) " The 10 best paid CEO in America". The Guardian , 22 Oct, (online). Available at: http://www.theguardian.com/business/2013/Oct22/best-paid-chief-executives-america (Accessed: 3 May 2014).

Spiekermann and Wegener (2007), update of selected potential accessibility indicators. Final report, urban and regional research (S&W), RRG spatial planning and geoinformation. ESPON. Available online at http:// <www.espon.eu/mmp/online/website/ contentprojects/947/1297/file_2724/ espon_accessibility_update-2006-fr_070207.pdf>, accessed on 1 July 2009.

Starbucks (2014) Our company available at http:// www. starbucks.com/about- us/company-information (accessed: 3 May 2014).

Shostack, G. Lynn (1984), " Designing Services That Deliver", Harvard Business Review, 62 (January-February), 133-9.

Shostack, G. Lynn (1985), " Planning The Service Encounter ,in the service encounter" , John A. Czepiel, Michael R. Solomon, and Carol F. Suprenant, eds. New York: Lexington Books, 243-54.

Shostack, G. Lynn (1987), " Service Positioning Through, Structural Change", Journal of marketing, 51 (Janurary), 34-43.

Soloman, Michael R. (1985), "Packaging The Service Provider", Service Industries Journal , 5(1), 64-71.

Stevens, C.W. (1980), "K-MartStores Try New Look To Invite More Spending" The Wall Street Journal, Nov. 26, 29-35.

Sullivan, Nicholas P(2007). You can hear me now: How Micro loans and cell phones are connecting the world, San Francisco, CA: John Wilsey & Sans, 2007.

T. Ambler, A. Ioannides, And S. Rose, " Brand s On The Brain : Neuroimages Of Advertising ", Business Strategy rev., vol. 11, 3. pp. 17-30. 2000.

Westbrook, Robert A. (1980), " Intrapersonal affective influences on consumer satisfaction with products, " Journal of consumer research , 7 (June) 49-54.

Wiig, k.(1993). Knowledge management foundations: Thinking About thinking. How people and organizations create, represent and use knowledge vol.1 , of knowledge management series schema press: Arlington, TX.

World Health Organization (2003). Diet, nutrition and the prevention of Chronic diseases report of a joint WHO/ FAO. expert consultation. Geneva: World Health Organization.

Wysocki, B. (1979), " Sight, Smell, Sound: They're all arms in retailer's arsenal" The Wall Street Journal, Nov. 17, 1979. 1-35.

Yale Center For Environmental Law And Policy (2006). Environmental Performance Index. Data available on-line at http://epi.yale.edu

How can artificial intelligent tools predict travelling consumer behavior in airline and air agent travelling market ?

I believe that applying (AI) big data tool to predict vehicle buyer consumption choice behavior, it is similar to predict traveler consumption choice behavior. In this chapter, I shall indicate how to apply (AI) big data gathering tool to predict vehicle buyer consumption choice behavior. Then, I shall its what its similar points to be applied to predict traveler consumption choice behavior.

Nowadays, many vehicle manufacturers hope their vehicles can attract to vehicle buyers to choose to buy their vehicles. However, there are many different brands of vehicles to provide to them to choose, so the vehicle market competition is very serious.

How to judge their different kinds of vehicle price which is reasonable acceptance to attract vehicle buyers to choose to buy the brand of vehicle manufacturers' any kinds of vehicles, e.g. fast speed sport style vehicles, comfortable and slow speed common cars, for four passengers common small size or more than four passengers common large car size?

How to evaluate the vehicle prices issue is important factor to influence vehicle buyers' choices. Either if the brand of vehicle price is too high to compare other brands of similar vehicle price, it will influence many vehicle buyers choose to buy other brands' vehicles or if the brand of vehicle price is too low, it will influence vehicle buyers feel this brand's vehicle machine quality or safe driving level or manufacturing steel material or speed or not comfortable sitting etc. different factors is worse to compare to other vehicle brands' similar vehicle products.

Thus, if the brand of vehicle manufacturers can predict how to design vehicles which can attract many vehicle buyers to choose to buy whose any vehicle products. What are future vehicle buyers' favorable vehicle styles? Then, the vehicle manufacturer can concentrate on manufacturing the kind style of vehicle products to sell already. It will reduce its vehicle manufacturing investment risk.

How to apply (AI) tools to predict vehicle buyers' behavioral consumption model? Whether artificial intelligent tools can predict automotive buyers' behavioral consumption model and predict future vehicle design trend. In fact, automotive brands and dealerships are facing an increasingly competition when attempting to manually gathering the vast quantities of data required to create customer focused programs that increase retention, ultimately new sales and service automotive business.

Building a based on that client's intrinsic needs and interests to any kinds of automotive vehicles at any given time. This is especially true in the automotive industry where the time span between purchases is measured in years. Because vehicle buyers would not like often to change their old vehicle to another new one. So, their decisions to buying another new vehicle, the time is usually after one year, even longer time. Hence, it seems any vehicles won't be frequent consumption products to the owned at least one vehicle family consumers (vehicle buyers). It implies that why vehicle manufacturers ought need to spend time to predict future vehicle buyer design choice for whole year vehicle buyer number growth because they won't often change preferable vehicle design to change another new vehicle more easily.

Hence, how to predict vehicle consumers' taste or preferable which styles of vehicle choices issues is very important. If the vehicle manufacturers can not manufacture any attractive vehicles to sell easily in this year. Then, it will lose time, money in this year because it won't know when the owned least one vehicle users or non-owned any vehicle users who will decide to buy one new vehicle or change another new vehicle ensure. The different brand vehicle dealers will possible wait more than one year to attract them to buy their vehicles if their styles are not attractive to compare other brands of vehicle competitors.

However, artificial intelligence and machine learning can help any vehicle manufacturers to find solution to solve patterns in highly to solve patterns in highly complex data-sets that are beyond the capability of a human brain, and then building and automatically acting on the customer insights it generates.

Given the automotive customer need for individualized communications, this technology is positioned to become a critical component of any successful vehicle retailer's domestic or/and overseas vehicle markets. How can vehicle manufacturers and retailers use (AI) to enhance their vehicle marketing campaigns? How will (AI) affect their vehicle sale marketing strategy? What criteria would they use when selecting on (AI) solution?

Vehicle consumers today are able to quickly access different brands of vehicle information, research vehicle products and reviews, negotiate prices and compare one vehicle brand or retailer to another resulting of the brands of vehicle customers. At the same time, the rise of " big -data mining", wearable devices that track user's every move and preference and greater contextualization in advertising and social media has resulted in consumer expectations of individualized. Thus, it seems that (AI) tools can be used to gather " big-data" and then they can make human's mind to analyze how to design kinds of vehicles to satisfy vehicle buyers' needs.

As automotive vehicle marketers can apply (AI) tools to achieve messaging strategies to meet the needs of this new generation of informed vehicle consumers, using data from a variety of sources to move from a variety of sources to move from mass- messaging to more personalized messages aimed at particular vehicle buyer segments, e.g. fast speed sport vehicle buyer segment, slow speed comfortable small size or large size of buyer segment. However, when 90% of vehicle marketers believe having a single vehicle buyer view is important, only 6% have achieved it.

However, one of the main issues vehicle marketers are facing the lack of capacity to efficiently sift through and analyze the massive vehicle buyer amounts of data required to create vehicle buyer individualized vehicle customer experiences easily. This is especially difficult for automotive dealers, the long periods between purchase cycles, and the highly considered nature of the vehicle purchase means that each vehicle dealer needs to not only track a large number of potential vehicle customers for an extremely long period of time, but each of those vehicle customers will generate a huge amount of different kinds of vehicle behavioral consumption data as they research their next vehicle purchase. However, by choosing the right (AI) technological tools and programs , vehicle dealers can solve this big data gathering challenge into a major advantage.

For Forrester vehicle brand example, vehicle consumers have more power over the Forrester vehicle brand's reputation than ever before. Mayne, L. (2014) indicated that Forrester calls this new (AI) tools is the " age of the vehicle customer", a 20 year business cycle in which the most successful vehicle enterprises will reinvent themselves to systematically understand and serve increasingly powerful vehicle consumers. To win in this new age, Forrester declares companies must become vehicle customer obsessed and the only sustainable competitive advantage is knowledge and engagement with customers, such as (AI) gathering data knowledge.

Thus, the biggest challenge vehicle businesses currently face is not the collection of a large quantity of vehicle consumer data, but what to do with that data once they have it. Even at a large vehicle data research firm, the data sets are often too big for a single analyze, or even a team of analysts to sort through and draw conclusion from. However, enter artificial intelligence and machine learning , an efficient technology solution that can continuously find patterns in highly complex data sets that are way beyond the capacity of a human brain and then automatic drive action based on the customer insights is generated.

What is (AI) machine learning tool? Machine learning is a type of (AI) that learns from data and is not explicitly program. Think Amazon, face book. Machine learning serves up relevant content based on an individual vehicle purchase behavior and experiences. More simply, machine learning is a computer program that can learn relationships between data, subject those learnings to errors functions, and then learn from its errors. The program in effect, trains itself.

Lee, T. (2016) explained that "Thus, (AI) tools can learn deep a more advanced branch of machine learning inspired by how our brain's nervous function, has also been found to be especial effective in identifying patterns from data."

When this way sound is complicated from a vehicle dealer perspective, the implementation of a marketing program driven by artificial intelligence can take care of these tasks in an automatic vehicle fashion with little to no manual intervention required from the staff at time vehicle stores.

In practice at a vehicle dealership, the program will continue track vehicle customer behavior online, merging that data with any offline source (like CRM or DMS data) and then analyze this aggregated vehicle buyer data set to predict what vehicle customer may be shopping for and what information they might like to relevance from different kinds style of vehicle design photos.

1.1

Why does travelling market seem to similar to vehicle market which can apply (AI) learning tool to predict travelling consumer behaviors?

Artificial intelligence refers to complex in vehicle market and travelling entertainment market which is very seem to be applied to predict consumer behaviors.

(AI) machine learning that posses the same characteristics of human intelligence and that have all our sense, all our reason and think just like human vehicle buyer who prefer vehicle purchase choice or travelling consumer who prefer travelling package or travelling destination and airline choice. Besides, machine learning is the practice of using algorithms to collect and examine data, learn from it, and then make a determination or prediction about something in the world.

So, it can be attempted to gather data concerns that travelling consumer past travelling destination choice and air ticket price choice and different travelling package, e.g. high, middle, or low class hotel and foods supply and entertainment places choice in their past travelling journeys.

The machine is " trained" using large amounts of data and algorithms that give it the ability to learn how to automatically perform a task with increasing accuracy. Otherwise, deep learning is primarily based on artificial neural networks inspired by our understanding of the biology of human's brains.

Thus, (AI) big data can gather all these past traveler consumption behavioral choice data to make reference to analyze whether how many travelers will choose to go to the specific travelling destination in any time by the past traveler number record to different travelling destinations, then it can gather the past air ticket sale price to different destinations and past travelling package design to different destinations in order to analyze whether it is the cheap airline ticket price factor or attractive travelling package factor or attractive travelling entertainment etc. in order to predict which factor is the most potential influential factor to they choose to go to the destination to travel in different time within one year. Then, traveler agent or airline can collect these big data to judge how to design their package to attract travelers to go to anywhere to travel or what the main factor influence most of them to choose to visit the destination to travel.

For example, travel agents or airlines can apply "Deep learning" breaks down tasks in ways that enables machines to assist them to predict when travelling consumer choice will be changed and why their travelling choice will change and how their travelling choice will change with increasingly complex tasks.

So, such as why (AI) technology can be applied to predict how travelling consumer behavior changes to bring to judge whether anywhere will be many travelling consumers who will prefer to choose travelling hot destinations next year or next month.

Then, travel agents and airlines can gather overall past travelling consumer data to analyze and conclude the more accurate prediction of different travelling destinations to the number of traveler. Then, they can choose how much air ticket price is more reasonable to charge to the travelling destination or how to design the travelling package which can bring more attractive to the prediction number of different travelling destination travelers in order to achieve to raise the different travelling destination number next year.

Thus, (AI) big data machine learning can help airlines or travel agents to solve how to design any attractive travelling package challenge. A travelling package is both one of the most important and carefully considered travelling entertainment consumption the majority of travelling people will ever make in their lifetime at least one travelling time.

It is also a prediction how travelling package will be designed that tends to be fundamentally tied to a travelling person's travelling destination choice identify and travelling package view of themselves. As the same time, travelling consumers' travelling choice changing lifestyles result in changing travelling destination needs, e.g. the country's young travelers can choose to change non-extreme exciting travelling entertainment package from past extreme exciting travelling entertainment package. Due to personal feeling factor in general. However, I believe that (AI) big data can also be attempted to predict when the country's young travelers will choose to change non-extreme exciting travelling behavior.

It is similar to automotive dealers need to remember that vehicle customers and prospects are individual human beings with risk, complex and ever-changing lives factors, these factors will influence every vehicle consumer why who feels has vehicle purchase need, and how who choose to buy the first vehicle if who decided to buy the first vehicle.

It seems that travelling agents or airlines need to remember that travelling consumers and different features or designs are very traveler beings with risk, complex and ever-travelling package attitude personal changing factors in different travel season, these factor will influence every individual traveler why who feels has travel entertainment need, and how who choose to buy different feature or design travelling package if who decide to travel.

The (AI) big data technological travelling customer behavioral prediction tool seems to be the best travelling behavioral prediction tool in the world are those that know every one of different country's traveler need. Their likes and dislikes which style of travelling package, preferences and travel destination changing tastes to travelling destination choices.

The capacity of the human brain, however, limits us from achieving these different type of travel package sales. In this competitive travelling destination choice entertainment environment, (AI) big data machine learning enables platforms to assist the air ticket and travel package sales team by tracking the travelling consumer behaviors of each travelling customer, learning and memorizing their preferences and predicting their future travelling destination choice and travelling package design needs.

Finally, I recommend that for a travel agent or airline travelling marketing platform to make their travelling customer engagement efficient and fully-functional, I should be able to: applying (AI) tools to track every travelling customer behavior across the web, connecting to a society of data sources, CRM, DMS, third-party, web travelling brands, social traveler email, click etc., aggregating and accurately cross-reference data from a variety of sources, leveraging this data to drive insights on a mass scale, as well as on an individualized basis, driving actions and automatically direct travelling customer engagement via multiple channels based on where each customer is in their travelling individual lifecycle.

1.2
Why is (AI) big data gathering tool better
than psychological and survey methods to predict traveler individual travel choice behavior?

Prediction travel behavioral consumption from psychology and survey methods.

How to predict travel consumption? It is one question to any travel agents concern to use what methods which can predict how many numbers of travelers where who will choose to go to travel more accurately. I think that who can consider how to predict travel behavioral consumption from psychology and survey travel choice prediction method, but it is better to apply (AI) big data gathering method to predict travel consumer's destination choice more accurate. The reason is as below:

The first reason is that traveller individual travel psychological desire is difficult to predict accurate more than (AI) big data gathering method, it is due that the data is past traveler's destination choice and travel package and ticket price actual data from (AI) big data gathering method. Otherwise, survey investigation is only traveler psychological thinking method. It lacks enough past actual traveler data gathering.

The second reason is that on the weakness of traveler individual psychological thinking view of survey investigation. It has evidence to support the relationship between self-identify threat and resistance to change travel behavior to any travelers, controlling for whose past travelling behavior, resistance to change if a psychological phenomenon of long standing interest in many applied branches of psychology.

Past travelling behavior has been acknowledged as a predictor of future action. Such as travelling behavior that is experienced as successful is likely to be repeated and may lead to habitual patterns. Some psychologists differentiate habit between two concepts, such as goal oriented and automatic oriented both. Although repeated past travelling behavior is addition goal oriented and automatic oriented. Further non-deliberative nature of habit may make appeals to judge and to predict future individual traveler's behavior accurately.

However, repeated one traveler will choose the destination to repeat to travel without a necessary constraint of goal orientation and automatic oriented both. So, it seems that psychological factor can influence any individual traveler why and how who choose to decide to repeat to choose the destination to travel.

So, survey investigation is only the traveler's thinking to answer the travel firm. It is not sure that the traveler's past travel experience is real answer. Otherwise, (AI) big data gathering method is computer gathering method which gather past traveler consumption actual data to analyze and conclude future traveler possible repeated travel destination choice and travel package choice more accurate.

The third reason is that on the strength of (AI) big data gathering method computer statistic view to predict future traveller consumer's destination and travel package choice. It is structural equation modeling is an extremely flexible linear-in-parameters multivariate statistical modeling technique. It has been used in modeling travel behavior and values since about 1980 year. It is a software method to handle a large number of variables, as well as unobserved variables specified as linear combinations (weighted averages) of the observed variable.

1.3 Can (AI) big data gather data to predict when climate will change to influence poor travelling behaviours?

(AI) big data tool can predict the flexibility of human travelling behavioral change is at least the result of one such mechanism, our ability to travel mentally in time and entertain potential future. Understanding of the impacts is holidays, particularly those involving travel.

Using focus groups research to explores tourists' awareness of the impacts of travel own climate change, examines the extent to which climate change features in holiday travel decisions and identifies some of the barriers to the adoption of less carbon intensive tourism practices.

The findings suggest many tourists don't consider climate change when planning their holidays. The failure of tourists to engage with the climate change to impact of holidays, combined with significant barriers to behavioral change, presents a considerable challenge in the tourism industry. In the future, computer (AI) big data tool can attempt to predict when the country's climate change to influence travelers to choose to go to the country to travel, e.g. next month or next half year or next year hot travelling destinations.

Tourism is a highly energy intensive industry and has only recently attracted attention as an important contributions to climate change through greenhouse gas emissions. It has been estimated that tourism contributes 5% of global carbon dioxide emissions. There have been a number of potential changes proposed for reducing the impact of air travel on climate change. These include technological changes, market based changes and behavioral changes.

However, the role that climate change plays in the holiday and travel decisions of global tourists. How the global tourists of the impacts travel has on climate change to establish the extent to which climate change, considerations features in holiday travel decision making processes and to investigate the major barriers to global tourists adopting less carbon intensive travel practices.

It will bring this question: Will tourists aware the impacts that their holidays and travel have on climate changes to influence their travelling decision?

When, it comes to understand individual traveler's behavioral change, wide range of conceptual theories have been developed, utilizing various social, psychological, subjective and objective variables in order to model travel consumption behavior. These theories of travel behavioral change operate at a number of different levels, including the individual level, the interpersonal level and community level. Whether pro-environmental behavior can be used to predict travel consumption behavior in a climate change. However, the question of what determines pro-environmental behavior in such a complex one that it can not be visualized through one single framework or diagram.

Despite the potentially high risk scenario for the tourism industry and the global environment, the tourism and climate change ought have close relationship.

However, (AI) big data tool can be applied to find what factors to influence the time of travelers‘ travelling choices. What are the important factors and variables which can limit tourism? e.g. money, time, family problem, extreme hot or cold weather change, air ticket price, journey attraction etc. variable factors.

Mention of holidays and travel were deliberately avoided in the recruitment process, so as not to create a connection factor to influence traveler's individual mind. However, the dismissal of alternative transportation modes can be conceived as either a structural barrier, in the sense that flying is perhaps the only realistic option to reach long-haul holiday destination, or a perceived behavioral control barriers in that an individual perceives flying as the only option open to whom.

The transportation tool factor will be depend to extent on the distance to the destination. This can also be interpreted in a social perspective as an intention with the resources available where much international tourism is structured around flying. To increase the availability of different transportation modes, tourists could choose holiday destination closer to home.

Finally, also how to predict future travel behavioral consumption. I feel that travel agents need to predict whether any country's random daily variation of weather factor is also important to influence travel behavior. e.g. in weather, temperature, rainfall and snowfall with traffic accidents factors will have relationship to cause travel demand.

Some scientists estimate suggest that when warmed temperatures and reduced snowfall are associated with a moderate decline in non-fatal accidents, they are also associated with a significant increase in fatal accidents. Thus increase in fatalities and temperature. Half of the estimated effect of temperature on fatalities is due to changes in the exposure to pedestrians, bicyclists and motorcyclists as temperature increase.

So, if any countries have rainfall, snowfall and low temperature to cause traffic accidents, whether this accident occurrence will influence the travelers who liking climb snow hills, riding bicycle, running sports who will avoid to travel to these countries' bad weather after occurs. So, why I feel that this natural climate factor will also be one serious factor to influence travel behavioral consumption. However, (AI) big data tool can predict more accurate than survey method when climate change to influence the country's climate to be poor, then it can predict when which countries are not popular acceptable to global country consumers‘ travel choice next month.

How can apply (AI) to provide travelling businesses with better-informed decisions ?

I shall explain how (AI) big data gathering technology can provide travelling businesses with better-informed decisions to drive top-line growth, deliver meaningful experience for travelling customers and smooth their path along the travelling consumer journey. The widely understood definition of (AI) involves the ability of machines or computers to learn human thinking, reasoning and decision-making abilities.

So, such as (AI) learning machine system can attempt to learn travelling consumer's travel destination or travel package thinking, judgement of their reasons why they choose to go to the destination to travel or why they choose to buy the travel package and learn how and why they make their past travelling decisions from their past travel big data gathering.

A Narrative science study in 2015 year identified that (AI) was being used primarily in voice recognition, machine learning virtual assistants and decision support. This study also highlighted the many branches of (AI) and that techniques and their definition are used interchangeably. It is possible that (AI) can be used to gather big data , then to analyze to help travel businesses to predict travelling consumer travel destination and travel package choice behaviors. For example, one of the most common techniques is traveler machine learning, where algorithms are used to perform tasks by learning from the airline or travel agent whose past all travelers' travelling destination choice and travel package choice historical data.

However, during 2017 year, search engines will begin to find what additional factors can influence past traveler personal travelling destination and travelling package travelling behavioral data into prediction of future travelling customer behavioral results, such as the online traveler (user's) history of travelling data searches, such as anywhere are the most popular travelling locations or travelling destinations and previously captures conservations.

Artificial intelligence will use this past travelling destinations and travelling package information to power predictive search results, e.g. predictive future travelling consumer's choice behavioral processing for where will be their preferable travelling destination choice and how to design travelling package to satisfy future travelling clients' needs. Predictive search will improve the quality of online travelling search results, and provide new insights into travelling consumers' travelling destination and package behavior and the moments which matter to them. Search will give recommendation into tailored how travelling consumer individual travelling destination choice in travelling decision making process. Several of the largest online platforms already use (AI) travelling machine learning to improve predictive travelling consumer behavioral search results.

For example, Google's rank brain technology adds research by understanding the context in which the travelling consumer has entered it. Over time, rank brain will learn further from user behaviors Amazon's DSSTNE (pronouned destiny) learns from shoppers' purchasing habits and consumption behavior to offer better product recommend actions, which Amazon can offer before a consumer has entered anything into the search bar.

Such as (AI) big data can gather past online travelers' e-ticket purchase transactions to conclude that online traveler's travelling choice habits and online traveler consumption behavior to offer better travelling destinations and travelling package opinions to travel agents or airlines. However, this technology is not independent of human input. For example, Google engineers will periodically retain the rank brain system to improve the models it uses.

For another example, in 2016 year , Apple computer revamped its travelling scene photos app to allow travelling consumers to search for specific travelling destinations in the travelling scene phots, they want to find anywhere travelling destination photos, not just dates and locations. Each travelling photo that an intelligent phone or intelligent pad user takes goes through 11 billion computations, so that travelling scene photos can understand exactly where is the travelling destination photography to let online travelling consumer to feel anywhere they plan to go to the location to travel. So, (AI) learning machine can make online travelling photos more attractive to influence potential travelers choose to the destination to travel after they see the travelling destination scene photos from internet.

It seems that in future, (AI) machine learning will allow online travelling search to evolve even further. Search engineers will deliver refined recommendations to airlines' online traveler e-ticket search users and use less human input to predict travelling consumers' needs from internet channel. For IBM computer example, it indicated 90% of the data that exists today has been created in the last two years.

This huge explosion of past traveler's e-ticket consumption data gives the opportunity to quickly spot and react to the latest trends, fashion and fads among its travelling clients and potential clients. This will allow airline or travel agent companies to better engage with younger travelling consumers, who gain influence access to the latest travelling destination and package trends.

They associate with to help define who they are as individuals. Thus, travelling company brands have to identify and make use of them before travelling consumers move on, but the vast quantity of past e-ticket purchase data available makes from internet channel. This a resource-intensive task. For next example, Lesara, a based online clothes store, uses this machine learning to inform its product decision often gathering information from internal and external sources.

When its trends -spotting shoes. Lesara has a range of over 20 styles and sells hundreds of pairs a day. It focus on giving consumers, the very latest trends allow Lesara to develop on average of 50,000 new items each year. It compared to 11,000 old items each year. Thus, travelling agents or airlines can attempt to apply (AI) big data gathering method to gather all past e-ticket purchase data, concerns where they prefer to choose to go to the destinations to travel and what travelling packages are the most attractive to the travelers to choose to buy. It aims to help them to predict where future travelers will prefer to choose to go to travel or what travelling package they will prefer to choose to buy next year.

For another (AI) big data prediction example, Lesara is one online clothes store, uses machine learning decisions after gathering information from internal and external sources. One of its most popular products, shoes with LED started life when its trend spotting software flagged up a blogger wearing similar shoes. Now Lesara has a range of over 20 styles and sells hundreds of pairs a day. Its focus on giving consumers the very latest trends allows Lesara to develop an average of 50,000 new items each year, compared to 11,000 for its competitor Lara.

It seems (AI) big data gathering machine learning can help Lesara business to predict what kinds of shoes design or style that shoe consumers will prefer choose to buy in future shoe market trend. Thus, Lesara can predict shoe consumers' taste successfully and it can manufacture many attractive style of shoes.

(AI) machine learning can gather global past shoe consumer's shoe shopping experiences, then analyzes to make conclusion to give lesara recommendation successfully. This will make the experience more enjoyable for shoe consumers and allow Lesara to advert whose different new style or design of shoes to deliver them move relevant

messages by understanding the context of the experience.
So, online travel agents or online airline can also attempt to apply (AI) big data gathering method to predict where travelers will prefer to go to travel and how they ought design travelling packages to attract them to choose to buy next year. Hence, (AI) big data gathering technology can conclude how to design traveler agents' travelling package products to be the most attractive to excite many travelers choose to buy their travelling package, due to it has more accurate to predict travelling consumer destination and travelling package choice behaviors to compare human themselves prediction judgement effort, e.g. travelling survey or marketing research, or telephone enquire. It seems that (AI) machine judgement effort is more accurate to compare to human judgment effort in travelling industry.

2.1 Future travel consumption behavior

Can (AI) big data gathering tool predict traveler individual habitual behavior , e.g. renting travel transportation tools ?

Can (AI) big data gathering tool can predict past traveler destination and travelling package choice habit and it can be intended to predict of future traveler behavior to people are creatures of habits judgement of future anywhere travelling destination choice next year or next month or next half year destination prediction ?
Many of human's everyday goal-directed behaviors are performed in a habitual fashion, the transportation made and route one takes to work, one's choice of breakfast. Habits are formed when using the some behavior frequently and a similar consistency in a similar context for the some purpose whether the individual past travel consumption model will be caused a habit to whom. e.g. choosing whom travel agent to buy air ticket or traveling package; choosing the same or similar countries' destinations to go to travel ; choosing the business class or normal (general) class of quality airlines to catch planes.
Does habitual rent traveling car tools use not lead to more resistance to change of travel mode? It has been argued that past behavior is the best predictor of future behavior to travel consumption. If individual traveler's past consumption behavior was always reasoned, then frequency of prior travel consumption behavior should only have an indirect link to the individual traveler's behavior. It seems that renting travel car tools to use is a habit example. So, a strong rent traveling car tools useful habit makes traveling mode choice. People with a strong renting of traveling car tools of habit should have low motivation to attend to gather any information about public transportation in their choice of travelling country for individual or family or friends members during their traveling journeys.

Even when persuasive communication changes the traveler whose attitudes and intention, in the case of individual traveler or family travelers with a strong renting travel car tools habit. It is difficult to change whose travel behaviors to choose to catch public transportation in whose any trips in any countries. However, understanding of travel behavior and the reasons for choosing one mode of transportation over another. The arguments for rent traveling car tools to use, including convenience, speed, comfort and individual freedom and well known.
Increasingly, psychological factors include such as, perceptions, identity, social norms and habit are being used to

understand travel mode choice. Whether how many travel consumers will choose to rent traveling car tools during their trips in any countries. It is difficult to estimate the numbers. As the average level of renting travel car tools of dependence or attitudes to certain travel package policies from travel agents. Instead different people must be treated in different ways because who are motivated in different ways and who are motivated by different travel package policies ways from travel agents.

In conclusion, the factors influence whose traveler's individual traveler destination choice behavior The factors include either who chooses to rent traveling car tools or who chooses to catch public transportation when who individual goes to travel in alone trip or family trip. It include influence mode choice factors, such as social psychology factor and marketing on segmentation factor both to influence whose transportation choice of behavior in whose trip. So, (AI) big data can be attempted to gather past traveler transportation tool choice, rent travelling car tools choice or catching public transportation tools choice to predict where destination can provide what kind of transportation tool to attract many travelers to choose to go to the place to travel.

How (AI) big data determine future travel behavior from past travel experience and perceptions of risk and safety for the benefits to travel consumers?

How (AI) big data determine future travel behavior from past travel experience and perceptions of risk and safety for the benefits to travel consumers? Why does individual traveler avoid certain destination(s) is(are) as relevant to tourist decision making as why who chooses to travel to others?

Perceptions of risk and safety and travel experience are likely to influence travel decisions. If travel agents had efforts to predict future travel behavior to guess whether travelers will feel where is(are) risk and unsafe to cause who does not choose to go to the country to travel. Then, the travel agents will avoid to choose to spend much time to design the different traveling package to attract their potential travel consumers to choose to travel. The reason is because in the case of individual traveler's tourism experience, the traveler whose past disappointment travel experience (psychological risk) will be a serious threat to the traveler's health or life (health, physical or terrorism risk). The past safety or unhealthy risk to the country(countries) will influence the traveler decides to choose not to go to the countries(country) to travel again in the future.

2.2 What is push and pull factors to influence any traveler who chooses where is whose preferable travelling destination ?

How to apply (AI) big data to predict individual traveler's behavioral intention of choosing a travel destination?

Understanding why people travel and what factors influence their behavioral intention of choosing a travel destination is beneficial to tourism planning and marketing. In general, an individual's choice of a travel destination into two forces.

The first force is the push factor that pushes an individual away from home and attempt to develop a general desire to go somewhere, without specifying where that may be.

The other force is the pull factor that pull an individual toward in destination, due to a region-specific or perceived attractiveness of a destination. The respective push and pull factors illustrate that people travel because who are pushed by whose internal motives and pulled by external forced of a destination. However, the decision making process leading to the choice of a travel destination is a very complex process.

For example, a Taiwanese traveler who might either choose new travel destination of Hong Kong or another old travel Asia destinations again or who also might choose any one of Western country, as a new travel destination. The travel agents can predict where who will have intention to choose to travel from whose past behavior and attitude, subjective and perceived behavioral control model. When (AI) big data gather past every country traveler number who chose to go to which countries to travel in order to judge where destinations will be the country travelers' travelling choice destinations in the future.

The factors influence where is the traveler choice, include personal safety, scenic beauty, cultural interest, climate changing, transportation tools, friendliness of local people, price of trip, trip package service in hotels and restaurants, quality and variety of food and shopping facilities and services etc. needs. So, whose factors will influence where is the individual travel's choice. It seems every traveler whose choice of travel process, will include past behavior. e.g. travelling experience, travelling habit, then to choose the best seasoned travelling action to satisfy whose travel needs. This process is the individual traveler's psychological choice process, who must need time to gather information to compare concerning of different travel packages, destination scene, climate change, transportation tools available to the destination, air ticket price etc. these factors, then to judge where is the best right destination to travel in the right time.

Hence, (AI) big data can gather past different countries' climate changing data, transportation tool changing data, destination scene environment changing etc. different data to give opinions to travelling businesses whether any country's these above factors will influence about how many traveler number will be increase or decrease in the future.

2.3 Why can expectation, motivation and attitude factor influence travelling behavior?

Social psychology is concerned with gaining insight into the psychological of socially relevant behaviors and the processes. For instance, on a global level bad influence to global warming, it influences some countries extreme cold or hot bad climate changing occurrence, then it ought influence some travelers' behavioral decision to change their mind to choose some countries to go to travel at the moment which do not occur extreme hot or cold climate (temperature). e.g. above than 40 degree in summer or below than 0 degree in winter. Due to the extreme climate changing environment in the countries, it will cause them to feel uncomfortable to play during their trips. So, the global warming causes to climate changing factor will influence the numbers of travel consumption to be reduced possibly. This is global climate changing environment factor influences to bad or uncomfortable social psychological feeling to global travelers' mind of traveling decision. What is individual traveler expectation, motivation and attitude? Tourism sector includes inbound (domestic) tourism and outbound (overseas) tourism both incomes to any countries. According to recent article, a tourist behavior model has been developed, called the expectation, motivation and attitude (EMA) model (Hsu et al., 2010).

This model focuses on the pre-visit stage of tourists by modeling the behavioral process by incorporating expectation, motivation and attitude. Travel motivation is considered as an essential component of the behavioral process, which has been increasing attention from the travel; industry. The economic approach defines "tourism" is an identifiable nationally important industry. It includes the component activities of transportation, accommodation, recreation, food and related service. So, tourism behavioral consumption is concerned the individual tourist's usual habituate of the industry which responds to whose needs, and of the impacts that both the tourist and the tourism industry have on the socio-cultural, economic and physical environment.

However, travel motivation means how to understand and predict factors that influence travel decision making. According to Backman and others (1995, p.15), motivation is conceptually viewed as " a state of need, a condition that services as a driving force to display different kind of behavior toward certain types of activities, developing preferences, arriving at some expected satisfactory outcome." So, motivation and expectancy which has close relationship to any tourist before who decided to do any tourism of behavior.

Some economists confirmed motivation and expectancy which has relations, such as expectation of visiting an outbound destination has a direct effect on motivation to visit the destination; motivation has a direct effect on attitude toward visiting the destination; expectation of visiting the outbound destination has a direct affection on attitude toward visiting the destination and motivation has a mediating effect on the relationship in between expectation and attitude.

Hence, (AI) big data can gather all the country's climate environment change, transportation tool change, entertainment scene change, hotel price and restaurant price change etc. data to give opinions whether the country will attract how many traveler to choose to go to travel in the year.

What is (AI) deep learning techniques to forecast travelling environment behavioral consumption

Prediction how many travelers will choose to go to the country to travel. It is similar to apply deep-learning technology to predict how to raise the agricultural farming productivity in the agricultural export country.

The (AI) deep-learning technology leads to performance enhancement and generalization of artificial intelligent technology. It influences the global leader in the field of information technology has declared its intention to utilize the deep-learning technology to solve environmental problems, such as climate change.

So, it will help agriculture farming businesses can raise any plant food: vegetable, fruit, rice which grow up very easily if farmers can apply (AI) deep-learning technology to solve environment problems to influence their plant food grow. If the whole year seasonal change is very good and it is suitable for any plant food to grow in farming land easily, e.g. rain is enough and soil is enough for any plant food to grow in the farm lands. Then, fruit, rice, vegetable etc. agriculture businesses will have much beneficial attribution to global farmers.

The question is how to use deep-learning technologies in the environmental field to predict the status of pro-environmental consumption. We predicted the pro-environmental consumption index based on Google search query data, using a recurrent neural network (RNN model). To certify the accuracy of the index, we compared the prediction accuracy of the RNN model with that of the ordinary least square and artificial necessary network models.

For example, the RNN model predicts the pro-environmental consumption index better than any other model. we expect the RNN model to perform still better in a big data environment because the deep-learning technologies would be increasingly as the volume of data grows. So, deep-learning technologies could be useful in environmental forecasting to prevent damage caused by climate change to influence any rice, vegetable, tomato, potato, fruit etc. different plant food grow in any countries' farming land easily.
For South Korea example, over 800 government agencies spent 2.2 trillion Korea won on eco-products in 2014 year. However, green products are rarely purchased outside these agencies. This phenomenon occurs because there is a gap between consumer attitudes and behavior , that is environmental attitude is a major factor in decision making vis-a-vis the consumption of " green" food and services (Jorea Ministry of Environment, 2015).
Therefore, it is necessary to understand those consumer attitude, that will lead to sustainability-conductive behavior and consumption. (AI) Deep learning system can be applied to attempt understand those traveler attitude to environment protection to fly to which country. For example, (AI) deep learning system can attempt to gather data concerns how many Hong Kong people concern air pollution challenge to influence their health, then it can attempt to predict how many Hong Kong travelers do not choose to go China travel, due to the air pollution challenge to influence their health.

Environmental travel consumption prediction

Recently, many researchers have studied pro-environmental consumption and household indexes as well as suicide rate predictions using messages posted by internet users on Google trend, Tweets etc. channel.
Whether can environmental consumption be predicted by (AI) deep-learning technological internet channel to influence how many travelers choose to go to the country to travel?
How can impact the pro-environmental consumption attitudes of green policies to influence how many travelers choose to go to the country to travel?
For example, Korea scientists estimated pro-environmental attitudes using search query data provided by Google trend and confirmed through regression analysis, that pro-environmental attitude has a positive correlation with the pro-environmental attitude index. They also explained that environment-friendly attitude of residents plan an important role in policy making. In the past, most household consumption indexed were calculated through surveys, but (AI) deep-learning technological tool " big data" have recently gained research attention (Lee et al. 2016). So, (AI) deep learning technology can attempt to gather whether how many Korea residents who concern environment pollution to influence their eating green food attitude then to judge whether how many Korea residents hope to leave their country to travel anywhere either high risk environment pollution countries to travel or low risk environment pollution countries to travel in the future.
It seems that (AI) deep-learning technology can help agricultural export countries' farmers , e.g. US, UK, Canada,

New Zealand, Australia, Japan, China, India etc. they can predict environmental behavioral consumption to any rice, tomato, potato , fruit, vegetable etc. plant food consumers. The beneficial advantages to them include as below:

(a) Assuming they know their countries' weather, when it has less rain to cause drought or when it has more rain in any seasonal time in the year. They can choose not to grow any kinds of above these plant food to avoid loss.

(b) They can make any kinds of above these plant food price raising after their prediction of these bad seasonal time to cause their plant food shortage supply challenge. Because these plant food consumers' demand number is more, but the supply of these above plant food supply number is less. However, due to they had predicted when the bad seasonal time can not allow them to grow these above plant food before. So, they have enough time to grow many these above plant food number in predictive good seasonal time to prepare to supply to their plant food import countries' plant food consumers to eat. Thus, these predictive environmental consumption plant food export countries can raise their plant food price to sell to them. When, the other non-pre-predictive environmental consumption plant food export countries can not supply any one of those plant food to them to eat, due to the bad climate to cause them can't grow any one of these plant food to export to sell.

Thus, (AI) deep-learning technology can be applied to predict how to raise the plant food supply number in order to raise price to the import plant food countries consumers to eat, due to they feel difficult to buy these plant food to eat in the bad climate seasonal time in whole year.

(c) (AI) deep-learning technology can help climate scientists to find what reasons cause their countries; rain sudden increases or cause their countries' rain sudden decreases. After its gathering data analysis, it can assist climate scientists to find solution methods to attempt to control the rain level can be right falling down level to let agricultural export farmers who can grow their plant food to sell to agricultural import countries in whole year.

(d) The agricultural export countries' farmers can apply (AI) deep-learning technology to help them to choose whether growing which kinds of plant food in that whether climate time to earn more plant food consumption number more easily.

Due to the agricultural countries climate will often change, for example, tomato, potato, rice, fruit etc. plant food can be adapt to grow in more rain time, but vegetable can not be adapt to grow in more rain time. If farmers can apply this technology to predict when it will have move rain or when it will have less rain to fall down in their countries. Then, they can choose to grow which kinds of plant food number more, in the suitable seasonal climate time in order to raise plant food growing number productivities to supply to sell to satisfy any agricultural food import countries' demand effectively.

(e) (AI) deep-learning technology can help agricultural import countries to solve agricultural food shortage challenge in long term. When this technology can be popular to base applied by the agricultural plant food export countries. It will solve global agricultural food shortage challenge. For example, when one agricultural export countries' farmers can popular accept to apply this technology to predict when to grow which kinds of plant food more to rise number productivities to sell. e.g. vegetable, fruit, rice Besides another agricultural export countries' farmers can also accept to apply this technology to predict when to grow plant food, e.g. potato, tomato to raise number productivities to sell. Then, they can concentrate on growing the specific kinds of plant food in order to raise the specific plant food number productivities in every seasonal change time every month. Then, global agricultural plant food supply must be raised, due to these predictive environmental change farmers can know who ought grow which kinds of plant food to sell to raise number productivities.

Consequently, (AI) deep learning can gather where countries will have high risk environment pollution to influence health food supply. Then, it can give opinions to travelling businesses when these high risk environment pollution countries will encounter the traveler number to be decreased, due to the environment pollution serious challenge will occur.

What methods can predict future travel behavioral consumption ?

How to use qualitative of travel behavioral method to predict future travel consumption from (AI) big data ?

I also suggest to use qualitative of travel behavioral method to predict future travel consumption. Methods such as focus groups interviews and participant observer techniques can be used with quantitative approaches on their own to fill the gaps left by quantitative techniques. These insights have contributed to the development of increasingly sophisticated models to forecast travel behavior and predict changes in behavior in response to change in the transportation system. I shall indicate the weaknesses of human travelling investigation methods as below:

First, survey methods restrict not only the question frame but the answer frame as well, anticipating the important issues and questions and the responses. However, these surveys methods are not well suited to exploratory areas of research where issues remain unidentified and the researched seek to answer the question "why?".

Second, data collection methods using traditional travel diaries or telephone recruitment can under represent certain segments of the population, particularly the older persons with little education, minorities and the poor. Before the survey, focus group for example can be used to identify what socio-demographic variables to include in the survey, how best to structure the diary, even what incentives will be most effective in increasing the response rate.

After the survey, focus, focus groups can be used to build explanations for the survey results to identify the "why" of the results as well as the implications. One Asia Pacific survey research result was made by tourism market investigation before. It indicated the travel in Asia Pacific market in the past, had often been undertaken in large groups through leisure package sold in bulk, or in large organized business groups, future travelers will be in smaller groups or alone, and for a much wider range of reasons.

Significant new traveler segments, such as female business traveler. The small business traveler and the senior traveler, all of which have different aspirations and requirements from the travel experience.

Moreover, Asia tourism market will start to exist behaviors in the adoption of newer technologies, a giving the traveler new ways to manage the travel experience, creating new behaviors. This with provide new opportunities for travel providers. The use of mobile devices, smartphones, tablets etc. and social media are the obvious findings to become an integral part of the travel experience. Thus, quality method can attempt to predict Asia Pacific tourism market development in the future. It is such as (AI) big data gathering tool can give traveler quality opinions to any travelling businesses to make the more accurate where will be the popular travel destination choice next month or next half year or next year.

However, improving the predictive power of travel behavior models and to increase understanding travel behavior which lies in the use of panel data(repeated measures from the same individuals). Whereas, cross-sectional data only reveal inter-individual differences at one moment in time, panel data can reveal intra-individual changes over time. In effect, panel data are generally better suited to understand and predict (changes in) travel behavior. However, a substantial proportion was also observed to transition between very different activity/travel patterns over time, indicating that from one year to the next, many people renegotiated their activity/travel patterns.

How to apply advanced traveler information systems (ATIS) to predict future travelling behavior?

Nowadays, information can impact on traveler behavior and network performance. For example, when steadily growing levels of vehicle ownership and vehicle miles traveled information has been identified as a potential strategy towards man aging travel demand, optimizing transportation networks and better utilizing available capacity. Toward, this goal to predict further tourist behavioral consumption. Many countries, government tourism development institutes has applied advanced traveler information systems (ATIS) which travel behavior models and high-fidelity network performance models made increasingly feasible through the rapid advances in computer power. Crucial components of this problem domain are the modeling of individual tourist drivers' response to travel information and the development accurate guidance of relevance to real would trip makers. So, this advanced traveler information systems (ATIS) can assist the tourist who like to rent travelling car tools to travel in any countries own free traveler information systems service conveniently. Also, this travel information system can be intended to assist travelers to make better travel choices. e.g. this system can improve the decision making of individual traveler rather than improvements of network performance overall. So, we need to understand how tourists make their travel plans. Also, understanding decision process that lead to booking of the trip is equally important, as it allows of a potential behavior.

How can online tourism sale channel influence traveling consumption of behavior?

Nowadays, internet is popular, it seems that booking air ticket behavior of using internet is predicted to influence overall tourism air tickets payment method. Tourism industry has grown in the previous several decades. Despite its global impact, questions related to better understanding of tourists and whose habits. Using online travel air ticket booking benefits include booking electronic air tickets can be made from entering any electronic travel agents websites in the short time and electronic travel ticket payers do not need leave home, who can pay visa card to pre booking any electronic travel ticket from online channel conveniently.

How can analyze activity based travel demand ?

Nowadays, human are concerning the traffic congestion and air quality deterioration, the supply oriented focus of transportation planning has expanded to include how to manage travel demand within the available transportation supply. Consequently, there has been an increasing interest in travel demand management strategies, such as congestion pricing that attempts to change aggregate travel demand. The prediction aggregate level, long term travel demand to understanding disaggregate level (i.e. individual levels) behavioral responses to short term demand policies, such as ride sharing incentives, congestion pricing and employer based demand management schemes, alternate work schedules, telecommuting limitation of travel agent traditionally work nature shall influence oriented trip based travel modelling passenger travel demand indirectly.

Finally, online travel purchase will be popular to influence the number of travel behavioral consumption nowadays. Any travel package products can be sold from websites to attract travelers to choose to pre-book air ticket for any trips conveniently. In the past ten years, the internet has become the predominant carrier of all types of information and transactions. Regarding travel decisions, internet has also become an important sales channels for the travel industry, because it is associated with comparably lower distribution and sales costs, but also because it adapts to high supply and demand dynamics in this industry. Consequently, the travel and tourism industry tries to increase the internet sale specific share of sales volumes. So, internet sale channel has changed travel consumption behavioral pattern and characteristics and travel experience. For example, Switzerland has one of the highest population-to-computer ratio in Europe. It is also one of the most highly internet penetrated countries in terms of use of the WWW on a day-to-day basis, with more than 75 percent of the population older than 14 years using the WWW daily (ICT, 2005).

The reason of booking online tourism may include: convenience, fast transaction, finding traveling package choice easily, more airline seats available. So, online booking tourism will influence the traditional tourism agents visiting of sales and air tickets and travelling package numbers to be decreased. Finally, the online booking tourism market shares will be expanded to more than traditional tourism agents visits sale market in the future one day. So, the travel agents who still use the traditional tourism visiting sale channel which ought raise whose features to compare to differ to online tourism sale channel if these traditional tourism agents want to keep competitive ability in tourism industry for long term.

What is actively based patterns of urban population of travel behavioral prediction method?

Actively based patterns of urban population. It is a method of motivational framework means in which societal constraints and inherent individual motivations interact to shape activity participation patterns. It can be used to predict one city or urban the numbers of travel demand in the year. It has two elements: First, capability constraints refer to constraints are imposed by biological needs, such as eating and sleeping and/or resources, such as income, availability of cars etc. to undertake the urban or city's family activities in the year. Second, coupling constraints define where, when and the duration of planning activities that are to be pursued with other individuals. So, this method needs to gather information (data) to get the relationship between activities, travel and spending work time and space time to evaluate whether there are how many families who have real needs to spend time to go to travel in the year.

What is trip based versus activity based approaches?

What is trip based versus activity based approaches? The fundamental difference between the trip-based and activity based approaches is that the former approach directly focuses on trips without explicit recognition of the motivation or reason for the trips and travel. The activity based approach , on the other hand, views travel as a demand derived from the need to pursue travel activities. So, it is better understand the individual or family behavior basis for individual or family travelling decision regarding participation in travelling activities in certain places or cities or countries at given times and hence the resulting travel needs. This behavioral basis includes all the factors that influence the why, how, when and where of performed activities and resulting individuals and household, the cultural/social norms of the community and the travel surrounding environment.

Another difference between the two approaches is in the way travel is represented. The trip based approach represents travel as a collection of trips. Each trip is considered as independent of other trips, without considering the inter-relationship in the choice attributes , such as time, destination and mode of different trips. As tours are chains of trips beginning and ending at a same location , say home or work. The tour based representation helps maintain the consistency across and capture the interdependency and consistency of the modeled choice attributed among the trips of the same tour.

In addition to the tour based representation of travel, the activity based approach focuses on sequences or patterns of activity participation and travel behavior, using the whole day or longer periods of time is the unit of analysis. Such as approach can address travel demand management issues through an examination of how people modify their activity participation, for example, will individuals substitute more out-of-home activities for in home activities in the evening of who arrived early form work due-to a work schedule change?

The major difference between trip based and the activity based approaches is in the way, the time dimension of activities and travel is considered. In the trip based approach, time is reduced to being simply a cost making a trip and a day's viewed as a combination, defined peak and off peak time periods. On the other hand, activity based approach views individuals' activity travel patterns are a result of their time use decisions with a continuous time domain. As individuals have 24 hours in a day or multiples of 24 hours for longer periods of time and decide how to use that travel among or allocate that time to activities and travel and with who, subject to their socio-demographic, transportation system and other and scheduling of trips. So, determining the impact of travel demand management policies on time use behavior is an important step to assessing the impact of such policies on individual travel behavior. The final major difference between this two approaches relates to the level of aggregation. In the trip based approach, most aspect of travel, e.g. number of trips etc. are analyzed at an aggregate level.

Consequently, trip based methods accommodate the effect of socio-demographic attributes of households and individuals in a very limited fashion, which limits the activity of the method to evaluate travel impacts of long term socio-demographic characteristics of the individuals who actually make the activity travel choices and the travel service characteristics of the surrounding environment. So, the activity based models are better equipped to forecast the longer term changes in travel demand in response composition and the travel environment of urban areas. Also, using activity based models, the impact of policies can be assessed by predicting individual level behavioral responses instead of employing trip based statistical averages that are aggregated over defined demographic segments.

Can apply (AI) big data gathering method predict senior age will be main travelling target?

In the past, Germany government had established tourism survey analysis to analyze survey data in order to arrive at reliable conclusions on future trends in travel behavior. To aim to find how demographic change will influence the tourism market and how the industry can adapt to those changes. The travel analysis provided data on tourism consumer behavior, including attitudes, motives and intentions. Since, 1970 year, it is based on a random sample, representative for the population in private households aged 14 years or older. Then, a continuous high scientific standard combined with a national and international users makes the travel analysis a useful tool and reliable source for tourism industry and policy decisions. It aimed to gather statistical data. e.g. on the age structure and on demographic trends, quantitative and qualitative analysis with time series data from the travel analysis. It shows e.g. not only the future volume , quite different from today's seniors, or how who will travel of family holidays will change, e.g. single parents of low, but grandparents of growing significance for tourism.

Demographic change is said to be one of the important drivers for new trends in consumer traveling change behavior in most European countries (e.g. Lind 2001). Because the growing number of senior citizens in the European Union and other industrialized countries, such as the USA and Japan, looks to become one of the major marketing challenges for the tourism industry. United Nations statistics predict that the share of people being 60 age or older will grow dramatically in the coming future, and is expected to rise from 10 percent of the world population in 2000 year to more than 20 percent in 2050 year (United Nations Population Division, 2001). From its statistic, some data showed that travel propensity increased throughout life until the age of about 50 years of age and was then kept stable until very late in life 75 age. The most important results is that the travel propensity when getting older is not going down between 65 and 75 age of course, the overall development of this variable is influenced by a lot of other factors which are responsible for quite a variation over time. It is now possible to suggest that the general pattern of travel propensity is one of the key indicators for holiday life cycle travel behavior, includes three stages. The growth stage tends to increase from early adult hood until 45 age old or when reaching some 80%. The next stage is stabilization from the ages of around 50 age, until 75 age old, starting with a lower increase. Finally, the decrease stage is a slight decrease occurs once people reach the more advanced age of 75 age to 85 age old (Lohmann & Danielsson 2001).

So, it seems Germany government tourism prediction to future travelers' behavior indicated these findings, such as on how future senior generations will travel, who had used survey data to examine the patterns of travel behavior of a generation getting older and applied the findings to draw conclusions on the future. Also, it predicted that on the future of family trips, family segmentation will be the travel behavior patterns in the future. These findings together with the statistical data on demographic change allowed for a better understanding of the coming tends in family holidays. It's aim developed in consumer behavior related to demographic change and predicted what will happen future of tourism one had to consider other influences and drivers as well, for example, trends on the supply side. e.g. low cost airlines or in travelling consumption behavior in general whether how the past may provide a key to predict travel patterns of senior citizens to the future.

Given the projected growth of the senior citizens market, designing specific marketing strategies to meet the prospective needs of elderly tourists will become increasingly important. It has been an implicit assumption that it will be a close relationship between the travel behavior of today's senior citizens and the those of future ones. The growing number of senior citizens in the world. e.g. China, Hong Kong, Japan, USA etc. countries. Global senior citizen tourism market will be based solely on demographic predictions about the future of the population's age structure. However, many of these seniors won't only live longer but will be fitter and more active until later in life. Many of the will also have plenty in life. Many of them will also have plenty of time and money to spend on travel. So, will these new seniors behave like today's senior citizens? Will they adopt the same travel behavior as the previous generation or become a new market of oldies for the leisure and tourism industry? However, to determine the actual number of senior citizens who will be travelling and to sought to evaluate and specify certain difficult to predict the actual numbers of senior citizen to any country. However, they can be based on the implicit assumption that there is a close relationship between the travel behavior of past, present and future seniors. But is this a valid assumption? As the revise- analysis travel analysis survey, which was conducted in Germany every year, offered some interesting data possibilities. It was designed to monitor the holiday travel behavior, opinions and attitudes of Germans and has been carried out since 1970 year, questions in the questionnaire. Data are based on face to face interviews, with a representative sample of more than 7,500 respondents, the interviews being carried out in January each year. All results refer to the average for the defined generated, which ranges generally over ten years. The group of people then at the age of 60 to 69 age is described. This corresponds to the same generation ten years ago, when they had an age of 50 to 59 age. When this methodological approach is not necessarily very sophisticated, it does have the important advantages of being cost effective.

IS (AI) big data gathering method a better psychological method to compare human marketing research method predict travel behavioral consumption?

On the psychological view point, I think individual traveler's character will have those kind of personal characteristics. First, simplicity searchers value above everything ease not transparency in their travel planning and holiday making, and are willing to avoid having to go through extensive research. Second, cultural purists use their travel as an opportunity to immerse themselves in an unfamiliar looking to break themselves entirely from their home lives and engage. Sincerely with a different way of living. Third, social capital seekers understand that to be well travelled is a personal quality, and their choices are shaped by their desire to take maximum of social reward from their travel. They will exploit the potential of digital media to enrich and inform their experiences, and structure their adventures always keeping in mind they are being watched by online audiences. Finally, reward hunters seek a return on the investment who make in their busy , high-achieving lives. Linked in part to the growing trend of wellness, including both physical and mental self-improvement who seek truly extraordinary and often indulgent or luxurious' must have experiences.

Why needs to know the personal character of individual traveler's characteristics? Because if travel agents could feel which kinds of individual traveler's character, then who can predict which kind of travel package to design to them more easily. For example, how to determine future travel behavior from past travel experience and perceptions of risk and safety? We need to concern that the influences of past international travel experience, types of risk associated with international travel and the overall degree of safety feeling during international travel on individual's travelling experiences likelihood of travelling to various geographic regions on their next international vacation trip or avoidance of those regions, due to perceived risk. Because individual traveler's experience of safety risk degree to the countries, it will influence who chooses to go to the countries/country to travel again.

Why travelers avoid certain destinations are as relevant decision making as why who choose to go to the country(countries) to travel. Perceptions of risk and safety and travel experiences are likely to influence travel decisions; efforts to predict future travel behavior can benefit to individual tourist's decision making.

As Weber & Bottorn (1989) defined risky decision is as "choices among alternatives that can be described by probability distributions over possible outcomes" (p.114). Some psychologists judge subjective perceptions of physical reality, i.e. image of a particular tourist destination, whereas value judgement refers to the way individual rank destinations according to whose attributes. i.e. attractiveness, safety, risk etc. factors to form on overall image. So, if the individual traveler had unhappy and worried and unsafe experiences to go to where the place(country) to travel during whose vacation time before. Then, this negative travel experience will influence who is afraid to go to the place (country) to travel again. Risk of place, country, destination or region means the danger is relatively high to the place, i.e. increasing in airplane accidents, crime or terrorist activity targeting citizens of potential traveler's nationality or the probability of occurrence is great , i.e. recent occurrences involving travel regions/destinations under consideration or effective actions to control consequences exist. i.e. selecting safe regions and destinations, taking extra precautions when traveling to risky destinations. These risk factors will influence the individual traveler who chooses to cancel travel plan to go to the country again.

Another interesting research, how to predict behavioral intention of choosing a travel destination, which has focus of tourism research for years, but the complex decision making process leading to the choice of a travel destination has not been well researched. The planned behavior model using its core constructs, attitude, subjective norm and perceived behavioral control, with the addition of the past behavioral variable on behavioral intention of choosing a

travel destination.
Understanding why people travel and what factors influence their behavioral intention of choosing a travel destination is beneficial to tourism planning and marketing. Understanding travel motivation is the push and pull model. The idea of the push and pull model is the decomposition of an individual's choice of a travel destination into two forces. The first force is the push factor that pushes an individual away home and attempts to develop a general desire to go somewhere else, without specifying where that may be. The second force is the pull factor, that pulls on individual toward a destination, due to a region specific travel location or perceived attractiveness of a destination. The respective push and pull factors illustrate that people travel because who are pushed by their internal motives and pulled by external forces of a destination. Nevertheless, how push and pull factors guide people's attitude and how these attributes lead to behavioral intentions of choosing a travel destination have rarely been investigated. The decision making process leading to the choice of a travel destination is a very complex process. The planned behavior model is as a research framework to predict the behavioral intention of choosing a travel destination. The model based on the three constructs of attitude, subjective norm, and perceived behavioral control (Fishbein & Ajzen, 1975).

In conclusion, the factors can influence travelers who decide to choose to travel the country, which include personal safety was perceived to the highest motivation factors among the important factors which include, scenic beauty, cultural interests, friendliness of local people, price of trip, services in hotels and restaurants, quality and variety of food and shopping facilities and services. The factors include both push and pull. Push factors include knowledge, prestige, and enhancement of human relationship etc., whereas, the most significant pull factors include high technologic image, expenditure and accessibility etc. For example, Japanese travelers visiting Hong Kong. Push factors are such as exploration dream fulfillment and pull factors are such as benefits sought, attractions and good climate city. It will be the factor of future travel patterns and motivations of sub-cultural and ethic groups for Japanese choice to go to Hong Kong travelling.

How can apply (AI) digital channel (big data gathering method) predict travelling consumer behaviors?

(AI) big data digital channel can be applied to help travelling businesses to evaluate whether how much the e-ticket price and travelling package price is the most attractive or reasonable to persuade travelling consumers feel it is the most reasonable price to choose to buy the airline's e-tickets or the travel agent's travelling package product from internet channel . It helps travelling consumers to feel which airlines or travelling agents which ought change their e-ticket and/or travelling package price to let travelling consumers to choose to buy the airline e-ticket or the travelling agent's travelling package products from internet channel. It can be applied to predict whether how many travelling consumer numbers can be increased or decreased when the airline e-ticket price is variable or the travelling agent travelling package price is variable . It aims to give opinions to help any online airlines or travelling agents to judge whether which e-ticket or travelling package price is the most reasonable to let travelling consumers to accept to choose to buy which airline's e-tickets or traveling agent's package products more attractive.
Thus, (AI) e-ticket or e-travelling package price measurement technology can be preference to be applied online

communication ecommerce and mobile phone internet platform aspect. As traveling businesses can enter their past e-ticket or travelling package prices data and past travelling customer number data into computer or mobile. Then, (AI) price measurement technology can gather these data to analyze these e-ticket or travelling package product prices and past travelling customer number to compare their e-ticket and/or travelling package prices variable changing range level to find their e-ticket and /or travelling package price variable difference to measure to make conclusion about every travelling package or/and e-ticket product's price variable changing will influence how many travelling customer number increase or decrease changing to choose to sell their different kinds of travelling package or e-ticket products more accurate. Then, (AI) price measurement software will help them to analyze all past e-ticket and/or travelling package price variable changing data to compare whether which e-ticket and/or travelling package price range can let travelling customers to feel it is more reasonable and attractive to influence them to choose to buy their e-ticket or travelling package product among different airlines and travel agent choices. Because any e-ticket or travelling package product's price is one important factor to influence travelling consumers to choose to buy the airline's e-tickets or travelling agent's travelling package products.

For example, Amazon publish has applied (AI) price measurement technology to help authors to decide how much every different topic of e-book or paper book price, it can attract the largest number of readers to buy. Any one author only needs to type whose book name to Amazon publish author himself/herself Amazon website. Amazon publish (AI) price measurement learning machine will help them to auto-calculate and judge how much e-book or paper book price is the most attractive and the most reasonable in order to increase reader number to buy their e-books or paper books to read. So, (AI) online price measurement machine will gather past similar book names and past every similar book readers‘ reading times and the number of readers to give opinions to let every author to judge whether his/her very new e-book or paper book ought charge how much price to the e-book or paper book which can attract many readers to choose to buy. Although, it is not ensure that the e-book or paper book price must let readers to feel it is the most reasonable price to choose to buy in reader's view point. However, it has other factors to influence readers' choice to buy the e-book or paper book, e.g. whether the book content is attractive to public, the author's familiarity, the book's page is enough or not to satisfy readers to read etc. factors. But, instead of all these extra factors to influence readers to choose to buy the book to read. (AI) price measurement learning machine can real give opinions to every author to let them to judge the e-book or paper book different price range whether is too high to influence readers to choose to buy to read or tool low to influence readers feel it is possible poor content book to compare other similar content books. Thus, (AI) price measurement machine can help authors to predict every reader's reading behaviors or reading experience and reading habit from online channel in short time easily. The author only enter the book name to let Amazon publish price measurement machine to check, it will follow past reader's reading habit and reading experience to judge whether the similar all book topic sale record to judge how much price is the reasonable price to attract many readers to buy the book.

Hence, (AI) can be applied to digital channel to help travelling businesses to predict travelling consumer behavior in the future. In the future, mobile/smartphone, laptop, desktop will be most frequent used ecommerce channels

to develop online business. So, (AI) can be also applied to these platforms to gather data to make analysis to help travelling businesses to predict travelling consumer purchase behaviors popularly. Due to , ecommerce is popular to global, so digital online and instore channels can be one good channel to let (AI) learning machine to make platform to gather past every online travelling consumer purchase (buying) experience data to help travelling businesses to build airline or travelling agent brand personality and having a responsible, positive impact on society.
To apply (AI) learning machine technology to understand travelling customer online purchase behavior, it will raise business e-commerce successful chance: For example, (AI) learning machine can help travelling businesses to gather data to analyze to determine whether short-term or long-term signals in the online travelling consumer behavior that indicate higher purchase intents to let every online travelling business to know. (AI) learning machine can find that online users with long-term purchasing intent tend to save and click through on more content.
However, as online travelling users approach the time of purchase their activity becomes more topically focused and actions shift from saves to searches from online travelling consumption channel. Then, (AI) learning machine will further find that the brand airline and/or travelling agent purchase signals in online travelling consumption behavior can exist weakness before an online travelling purchase is made and can also be traced across different online travelling purchase categories. Finally, (AI) learning machine synthesize these insights in predictive models of online travelling user purchasing intent to the brand of airline or/and travelling agent travelling package product. Taken together, it's work identifies a set of general principles and signals that can be used to model online travelling user e-ticket and/or travelling package purchasing intent across many online content discovery applications. Thus, (AI) learning machine can help online travelling businesses to gather any online travelling users' click online travelling behaviors data to judge whether there are how many online travelling users will choose to find their online travelling business websites to make final decisions to buy their travelling package or/and e-ticket products from online channels. Then, it will give opinions to help the online travelling businesses to let it to judge whether what are the important website factors will help its online travelling business to attract many online travelling consumers, e.g. designing unattractive travelling website issue, online unattractive scene photos issue, unclear website travelling photo color issue, unclear website travelling advertisement message, contents and words impressions issue, lacking image movement frequent attractive seeing issue etc. different website factors. Thus, online digital channel will be one good choice to apply (AI) learning machine to help travelling businesses to predict travelling consumer behaviors.

Thus, (AI) big data technology can also assist travelling consumers to gather different manufacturers' data to compare what their advantages and disadvantages of their travelling package products are. Then, travelling consumers can make comparison to choose which airline or travelling agent is the suitable to whom to buy e-ticket or pre-booking travelling package in online travelling consumption market.

.

Thus, I believe that artificial intelligent "big data" gathering method can be suggested to be applied to attempt to predict travelling consumer behavioral changes in global online travelling business environment, the reasons are as below:

On the travelling consumer's beneficial hand, travelling consumers can apply this (AI) big data gathering method to attempt to gather any global airline e-tickets and/or travelling agent's package product data to be analyzed by this artificial intelligent learning system to compare human general marketing research method, e.g. survey, questionnaire, marketing plan etc. different human judgement methods to predict traveler consumption behavioral change model. Then, it analyzed all the different data to compare what are the range of the most reasonable e-ticket and/or travelling package online purchase history and sale in order to make more accurate prediction to future traveler change traveling consumption behavioral model in next month, or next half year or next year short term period traveling consumption change prediction.

Reference

Backman and others "motivation is conceptually viewed as " a state of need, a condition that services as a driving force to display different kind of behavior toward certain types of activities, developing preferences, arriving at some expected satisfactory outcome.", 1995, p.15.

Fishbein & Ajzen, "The model based on the three constructs of attitude, subjective norm, and perceived behavioral control". 1975.

Hsu et al. "A tourist behavior model has been developed, called the expectation, motivation and attitude " (EMA) model ,2010.

ICT,WWW . "Switzerland has one of the highest population-to-computer ratio in Europe." Switzerland, 2005.

Jorea Ministry of Environment, " For South Korea environmental attitude is a major factor in decision making vis-a-vis the consumption of " green" food and services", Korea, 2015.

Korea Ministry Of Environment. Public Organizations spend 2.2 Trillon Korean Won To
Purchase green Products in 2014; Ministry Of Environment: Sejoung, Korea, 2015.

Lind , Lohmann & Danielsson , United Nations Population Division, "Demographic change is said to be one of the important drivers for new trends in consumer traveling change behavior in most European countries". 2001.

Mayne, Lonnie. " Evolve of die in the age of the consumer". Entrepreneur, N.P. , 16 Apr. 2014. web of Oct. 2016.

Lee, D.; Kim, M. ; Lee, J. adoption of green electricity policies: Investigating the role of environmental attitudes via big data-driven search-queries. Energy policy 2016. 90, 187-201.

Lee, Terrence, " Tech in Asia-connecting Asia's startup system " Tech. in Asia- connecting Asia's startup ecosystem, N.p.,4 July 2016.

Weber & Bottorn "risky decision is as choices among alternatives that can be described by probability distributions over possible outcomes" , 1989, p.114

Bibliography

Adrian, P. (2012). Introduction to marketing theory & practice,
3 rd edition, London: Oxford press.

Ajzen, I (1991). The theory of planned behavior. Organizational behavior and human decision processes, 50(2), 179-211. doi: 10.1016/0749.5978 (91) 90020-7.

Alba, Joseph W. and J. Wesley Hutchinson (1987). " Dimensions Of Consumer Expertise", Journal of consumer research, 13 March, 411-454.

Bailey, L., Mokhtarian, P.L. Little, A. (2008). The broader Connection Between Public Transportation, Energy Conservation And Greenhouse Gas Reduction, Report Prepared As Part Of TCRP Project J-11/Tasks Transit Cooperative Research Program, Transportation Research Board Submitted To American Public Transportation Association in
http://www.apta.com/research/into/online/land_use.cfmi, accessed 17 April 2008.

Baucer, R,"Consumer Bhavior As Risk Taking , In Risk Taking And Information handling In Consumer Behavior", D. Coxceds Harvard University Press, Cambridge, Mass 1976.

Biederman, P. (2008). Travel and tourism, Pearson Prentice Hall, New Jersey.

Bogers, R. P., Brug, J. Van Assema, P., & Dagnetie, P.C.
(2004) , Explaining fruit and vegetable consumption: The theory of planned behavior and misconception of personal intake level. Appetite, 42,157-166.

Bolton, Ruth N. (1998), " A Dynamic Model Of The Duration Of The Customer's Relationship With A Continuous Service Provider: The Role Of Satisfaction", Marketing Science, 17 (1), 45-65.

B.Shiv and A. Fedorikhin, " Heart And Min In Conflict: The Interplay Of affect And Cognition In Consumer Decision Making", J. Consumer Res., vol. 26, pp. 278-292, Dec. 1999.

Brown, K.W., Ryan, R.M. Reswell , J.D. (2007). Mindfulness: Theoretical Foundatins And Evidence For Its Salutary Effects. Psychological Inquiry, 18, 211-237.

Burke, R.R. : Behavioral effects of digital signage, J. Advertising Res. 49(2), 180-185 (2009).

Cant, M., Brink , A. & Brijall, S., Consumer behavior, Cape Town, South Africa: Juta, 2006.

Conner, M. & Abraham, C. (2001). Conscientiousness and the theory of planned behavior: Toward a more complete model of the antecedents of intention and behavior. Social psychology bulletin, 27, 1547-1561.

Cooper C. Mallon, K, Leadbetter S, Pollack L, Peipins (2005) , cancer internet search activity on a major search engine, United States 2001 to 2003, J Med Internet Res. 7(3): e36.

Cope, R. R. Cope and H. Davis (2008). Disney's virtual Queues: A strategic opportunity to co-brand services ? Journal of Business & economics research, vol. 6 no10, 13-20.

Cornelia, B.F. (1999) Rural development news, the North Central Regional Center For Rural Development vol. no 24 , IOWA.

Couper, M.P. J. Blair and T. Triplet (1999). A Comparison Of Mail And E-mail For a Survey Of Employees In USA Statistical Agencies. Journal Of Official Statistics, 15, 39-56.

David J. Nowak & Gordon M. Melsler (2016) " Air quality effects of urban trees and parks." National recreation and park association, USA.

Data monitor (2008). The proctor and gamble company. Retrieved Nov. 15 2009 from http://www.datamonitor.com/

De Hollander, A. E. M., J.M. Melse, Elebret & P. G.N. Kramers (1999), " An Aggregate public health indicator to represent the impact of multiple environmental exposures" Epidemiology: 606-617.

De Visser, R.O., & McDonnell, E.J. (2013). " Man points": Masculine capital and young men's health. Health psychology, 32(1), 5-14. doi:10. 1037/a0029045.

Dunn, J & A Neumsister (2002). Knowledge management in the Information age. E. business review, Fall , 37-45. Jounral of service, spring 2011, vol. 4, no1, De Grovte (2009).

Dyer, D., F. Dalzell & R. Olegario (2004). Rising tide. Lessons learned from 165 years of brand building at Procter and Gamble. Boston, MA: Havard Business School Press.

Eysenbach G (2006) Infodemiology: Tracking flu- related searches on the web for syndromic surveillance. American Medical Informatics Associaion Annual Symposium Proceedings , Curran Associates, Red Hook, NY, pp. 244-248.

Ettredge M, Gerdes, J. Karuga , G (2005) Using web- based search data to predict macro-economic statistics. Commun ACM 48: 87-92.

Felce, D. and Perry, J. (1995). Quality of life: A contribution to its definition and measurement, vol. 16, no.1 pp: 51-74.

Feldman, Jack M. And John G. Lynch Jr. (1988), "Self-
Generated Validity And Other Effects Of Measurement On Belife, Attitude, Intention And Behavior", Journal of applied psychology, 73(3),421-35.

Fiese, M, Hofmann, W., & Wanke, M (2009). The impulsive consumer. Predicting consumer behavior with implicit reaction time measurement. In M. Wanke (ed.) Social psychology of consumer behavior (pp.335-364). New York, NY: Psychology press.

Fitzsimons, Gavan, J. And Vicki G. Morwitz (1996), " The Effect Of Measuring Intent On Brand-Level Purchase Behavior", Journal of consumer research, 23 (1), 1-11.

Hallerman , D. (2008) video Advertising Online: Spending And Pricing , New York. E-Marketer.

Harriet Griffey. (2010) The art of concentration, enhance focus, Reduce, stress and achieve move. Macmillan publishers ltd,Basinastoke and Oxford, London UK.

Helleman, D. (2008) Video Advertising Online: Spending And Pricing , New York, E-Marketer.

Hensen, C. (2003). Kreuzfahrtourismus.www.christoph- hensen.de/Facharbeit.pdf.

Huang, H.I. (2012). An empirical analysis of the strategic Management of competitive advantage: a case study of higher technical and vocational education in Taiwan (Doctoral dissertation, Victoria University).

Jamieson, Linda F. And Frank M. Bass (1989), " Adjusting Stated Intention Measures To Predict Trial Purchase Of New Products: A Comparison Of Models And Methods," Journal of marketing research, 26 (August), 336-45.

Korea Ministry Of Environment. Public Organizations spend 2.2 Trillon Korean Won To Purchase green Products in 2014; Ministry Of Environment: Sejoung, Korea, 2015.

Kremers, S.P. J., De Bruijn, G.J., droomers, M., Van Lenthe, F. J., & Brug, J. (2005). Environmental interventions for selected dietary behaviors in adults. In J. Brug & F. J. Van Lenthe (eds.) , Environmental determinants and interventions for physical activity, nutrition and smoking: A review pp. 282-315. Rotterdam: Erasmus Medical Center.

Lee, D.; Kim, M. ; Lee, J. adoption of green electricity policies: Investigating the role of environmental attitudes via big data-driven search-queries. Energy policy 2016. 90, 187-201.

Lee, Terrence, " Tech in Asia-connecting Asia's startup system " Tech. in Asia- connecting Asia's startup ecosystem, N.p.,4 July 2016.

Los Angeles Country Department Of public Health (2016), Country Health Ranking Model, Retrieved From www.countryhealthrankgings.org/our-approach. USA.

Mayne, Lonnie. " Evolve of die in the age of the consumer". Entrepreneur, N.P. , 16 Apr. 2014. web of Oct. 2016.

McGregor, S.L. T., & Goldsmith, E.B. (1998). Expanding our understanding of quality of life, standard of living and well-being. Journal of family and consumer science, 90(2), 2-6, 22.

McMichael, A.J. M. Mckee, J. Shkolnikov and T. Valkanen (2004), " Morality trends and setbacks, global convergence or divergence?", Lancet 363, 1155-1159.

Melse, J.M. & A.E. M. De Hollander (2001). " Human Health And The Environment", background document for the OECD Environmental Outlook, OECD, Paris.

Moschis, George p. & Roy, L. Moore (1979), " Decision making among the young. A socialization perspective " Journal of consumer research , 6 (September).

Mulligan, M. Banerjee, T & Thomas, N. (2008) ,European Paid Content And Activity Forecast, (2008 to 2013), Jupiter Research.

Peter, J., Ryan, M, M, " An Investigation Of Perceived Risk At The Brand Level, " Journal of marketing research, 13 May 1976, pp. 184-188.

Pieters, R., & Wedel, M. (2007). Goal Control Of Visual Attention To Advertising: The Yarbus Implication. Journal Of Consumer Research, 34, 224-233 (August).

Parasuaman, and Leonard L. Berry (1985), " Problems And Strategies In Sevices Marketing", Journal of marketing, 49 (Spring), 33-46.

Priesnitz, W. (2007) Counting Our Food Miles. Natural Life, 1 July.

R.C. Oliver, " When is consumer loyalty?" J.Marketing vol. 63, pp.33-44.1999.

Reggiani, A . (ed). 1998, accessibility, trade and locational behavior, Ashgate publishing ltd, England.

Rushe, D. (2013) " The 10 best paid CEO in America". The Guardian , 22 Oct, (online). Available at: http://www.theguardian.com/business/2013/Oct22/best-paid-chief-executives-america (Accessed: 3 May 2014).

Spiekermann and Wegener (2007), update of selected potential accessibility indicators. Final report, urban and regional research (S&W), RRG spatial planning and geoinformation. ESPON. Available online at http:// <www.espon.eu/mmp/online/website/ contentprojects/947/1297/file_2724/ espon_accessibility_update-2006-fr_070207.pdf>, accessed on 1 July 2009.

Starbucks (2014) Our company available at http:// www. starbucks.com/about- us/company-information (accessed: 3 May 2014).

Shostack, G. Lynn (1984), " Designing Services That Deliver", Harvard Business Review, 62 (January-February), 133-9.

Shostack, G. Lynn (1985), " Planning The Service Encounter ,in the service encounter" , John A. Czepiel, Michael R. Solomon, and Carol F. Suprenant, eds. New York: Lexington Books, 243-54.

Shostack, G. Lynn (1987), " Service Positioning Through, Structural Change", Journal of marketing, 51 (Janurary), 34-43.

Soloman, Michael R. (1985), "Packaging The Service Provider", Service Industries Journal , 5(1), 64-71.

Stevens, C.W. (1980), "K-MartStores Try New Look To Invite More Spending" The Wall Street Journal, Nov. 26, 29-35.

Sullivan, Nicholas P(2007). You can hear me now: How Micro loans and cell phones are connecting the world, San Francisco, CA: John Wilsey & Sans, 2007.

T. Ambler, A. Ioannides, And S. Rose, " Brand s On The Brain : Neuroimages Of Advertising ", Business Strategy rev., vol. 11, 3. pp. 17-30. 2000.

Westbrook, Robert A. (1980), " Intrapersonal affective influences on consumer satisfaction with products, " Journal of consumer research , 7 (June) 49-54.

Wiig, k.(1993). Knowledge management foundations: Thinking About thinking. How people and organizations create, represent and use knowledge vol.1 , of knowledge management series schema press: Arlington, TX.

World Health Organization (2003). Diet, nutrition and the prevention of Chronic diseases report of a joint WHO/ FAO. expert consultation. Geneva: World Health Organization.

Wysocki, B. (1979), " Sight, Smell, Sound: They're all arms in retailer's arsenal" The Wall Street Journal, Nov. 17, 1979. 1-35.

Yale Center For Environmental Law And Policy (2006). Environmental Performance Index. Data available on-line at http://epi.yale.edu

Can artificial intelligence development to tourism industry

● Artificial intelligence will bring what benefits to
Influence traveler consumption behavior or raising
Travelling leisure desire?

Artificial intelligence will bring what benefits to influence traveler consumption behaviors or raising travelling leisure desire? It would be unreasonable not to analyze how artificial intelligence is affecting, such a big part of global economy. Especially, how it leads travel media to develop. We need to answer those questions , before we hope predict how artificial intelligence can bring advantages to influence travelers to raise travelers travelling leisure desires in long time. Which tends will shape the travel industry? What will customers demand in their journeys? Which business models will prevail ? What role of artificial intelligence to play in future tourism industry? What new opportunities and business models do the new technology block chain and artificial intelligence offer for the travel industry? How is artificial intelligence future impacting the travel industry ? The utilization of artificial intelligence will how raise tourism demand?

Future artificial intelligence tourism development may help travel leisure benefits on these several aspects. They may include geographical elements choice, the tourist generating regions choice, the tourist destination regional choices and the transit route regions choice.

It is one exciting tourism service arrangement for travelers when artificial intelligence is participated to any travel agents' travelling leisure provision services. Artificial intelligence can help any travelling customers to arrange whole journey of a tourists, in these stages . In the before booking stage, (AI) journey service arrangement may include these from searching and being inspired stage to discovering and planning stage and booking stages. During booking, (AI) journey service arrangement may include booking and defining and improving stages. Finally, in the alter booking stage, (AI) journey arrangement service may include experiencing and reflecting and share , it is the tourist's leisure feeling after he/she had experienced the whole travel journey service arrangement. So, the (AI) 's whole tourism journey arrangement whether it is suitable to the tourist to let him/her to feel satisfactory ; or more satisfactory ; or the most satisfactory enjoyment feeling to his/her journey. It depends on how the (AI) judges the whole tourism journey service arrangement, e.g. reasonable price and comfortable accommodation choice for the business traveler of leisure travel in order to satisfy his/her accommodation choice. Because artificial intelligence is one complex computer machine, it ought may help any tourists to search accommodation information to search the most reasonable price and the most comfortable accommodation choice from social media to compare the tourist himself/herself accommodation search activity in order to get the most accurate accommodation choice and in order to satisfy the tourist's living need in his/her whole journey. So, future artificial intelligence is applied to tourism leisure service arrangement aspect. (AI) is this a goal oriented execution for any one tourist before he/she makes the final accommodation choice, tourism destination choice. It needs to help the tourist to execute his/her whole journey process, no matter how difficult this process may seem for humans.

● How (AI)helps tourism industry to reduce cost ?

In the future, robotics or artificial intelligence systems dominate the tourism industry, they may include chat bots, travel assistants, and service robots to any travel agents service providers. Their roles may include these several aspects to any travel agents service providers. They may include searching and being inspired, discovering , planning and booking , refining and improving, how excites the tourist experiencing to the journey service arrangement, reflecting and how raising his/her tourism leisure enjoyment feeling during the whole tourism journey from customer intelligence platform service.

However, future travel agent intermediaries as well as inventory providers will start to focus on their quality of offering and journey arrangement service. Thus, they invest comparable less in operational excellence, e.g. concentration of digital systems at the search and booking phase and robots at the experience phase of the customer journey is almost self-explanatory, journey searching, planning and booking via internet is nowadays the norm. This, these are no hybrid systems necessary. However, during the experience phase, in which the traveler leaves the realms of the digital world, physical interaction, which can only be delivered by robots, regains importance. Since char bots , travel assistants as well as service robots are all representatives of systems used at the travelling customer interface. However, supposing when the travelling agent decides to install (AI) systems, can not be explained by the trend of a rebound on customer journey experience. However, during the expert discussions, it was revealed that most of the tourism industry does not consider technology a core competency and often lack the capacities to self-develop and install such (A I) systems.

However, it may also seem that AI can impact automatization tasks to tourism industry, chat bots or service robots are typical representatives in which artificial intelligence is used to automate a task previously done by humans. This can be called the automatization effect. Nevertheless, automatization through (AI) systems should not be equated with other forms of technology utilization which are called automatization too. Nowadays, the automatization through self-serving terminals, whether at airports through self-check terminals or at hotels through self check in systems is quite common in the tourism service industry. However, in travel agents service industry sector, the (AI) systems process itself is not automated, but rather changed from the airport check in / out organizations to the travel agent journey arrangement customer service organizations. Artificial intelligence on the other side truly automates these process. So that neither a human at the airport or travel agent journey service arrangement organizations nor the travelling customer much do it.

A recent example are (AI) supermarkets , such as the " Amazon Go" store in which can artificial intelligence system, which can recognize persons and objects, tracks people and the objects which they put in their shopping-bad and automatically bills their purchases on the customers bank account（Ｓｔａｒｋ，２０１７）.

Another similar system is imaginable at hotels and airports in which a face-recognition system recognizes the arrival of guests and automatically checks them in. JetBlue is already testing such a system a Boston airport（Ｅｎｔｉｓ，２０１７）.

However, how tourism organizations apply (AI) system can be employed to increase its competitive advantages, it means that either be achieved through differentiation, meaning delivery supervisor benefits to the consumer, or lower cost. If the robotic can help any tourists to choose the cheapest and the most comfortable apartment to let the tourist to live in his journey, such as recommender systems in the booking stage generate further benefits for consumers. The effect, however, is often related to automatization cost deduction, such as(AI)system can help the tourist whose the whole journey apartment cost reduction. Thus, one could assume that (AI)systems which automate tasks are mainly implemented to reduce apartments living cost for the tourist's whole journey. For example, chat bots can undoubtedly be used to replace sales and service staff at the tourist interface.

However, many companies use chat bots to expand their sales and service offerings instead, without reducing the respective staff. AirFrance-ＫＣＭ uses its chat bots to sell its products and inform its customers about flight schedules via new channels, such as the face book messenger. This service is available 24 hours a day and seven days a week. This is mainly because chat bots are pure digital systems and can this practically scaled –up indefinitely (Kelly , 2016).

A simple example is that many hotels utilize robots as a marketing tool to attract guests than as a true mean to save costs, it nevertheless shows the feasibility of such as model. A simpler use case would be the utilization of a vacuum cleaning robot to automate the task of cleaning the floor. It would replace a share of the existing cleaning personal and this gave costs. However, also (AI) systems which improve an existing task or just enable one, can and are used for cost savings. For instance, the British AirlineＥａｓｙｊｅｔｕｓｅｓ（ＡＩ）systems to better predict the demand of food and beverage on its flights. It reduced the inventory costs for the airline and help it to sustain a cost advantage (Emarkerer , 2017).

Reference

Emarketer 2017, " How EasyJet uses artificial intelligence to improve operations-eMarketer" Emarketer. Accessed Dec. 10

http://www.emarketer, com/article/how-easyjet-uses.
Artificial-intelligence-improve-operations/1014558.
Entris, Laura: 2 0 1 7 " D e l t a , J e t b l u e f l i g h t s t o t e s t f a c i a l recognition scaual Fortune." Fortune.com accessed Dec. 2019. http://fortune.com/2017/06/01/jetblue-delta-boarding-checkin/
Stark, Hareld. 2017 " Amazon go, A cashierless convenience store
Now available near you' Forbes.com accessed Dec. 9
http://www, forbes. Com/sites/haroldstark/2017/11/20
introducing-amazon-go-a-cashierless-convenience-
store-available-near you/20efofo952f9.

Hence, future robotic may be applied to apartment service industry in hotels, instead of travel agents or airport services. For example, vending machines are found in public locations are doing the work of a hospitality workers and ATM machines have largely replaced the human labor of the bank teller. The design of robot-friendly hospitality facilities may bring cost benefit of adoption of robots for travel, tourism and hospitality companies. They are service robots.

In tourism service industry, customer attitudes can be influences in the mind of the tourist and is considered as an important element to influence the hotel, travel agent, airport services success. So service robotic participation will influence service robotic can bring positive attitude to their customers, such as providing excellent service for hotel book room service, air ticket check in /out service for airport travelers or apartment choice service for tourist journey arrangement choice services.

Thus, (AI)service will influence the country's overall hotel room booking, travel leisure and airport check in arrival service to let oversea tourists feel this country's tourism service can provide the most excellent service arrangement to compare other countries' travelling services. When, service robotic can be participated to the country's overall tourism related services, e.g. service robots will be faster than human employees, robots will deal with calculations better than human employees, robotic will provide more accurate information than human employees, robotic will be friendlier than human employees, robots will be more polite than human employees, robots will be able to understand guest's level of satisfaction, robot can understand more clear to any questions or orders to compare human employees.

Robotic can do special requests, they work not only in a programmed frame. Moreover, when any hotel livers can be influenced to have positive attitude towards the potential use of robotic in hotels, such as being served by robotic will be a memorable experience, being served by robots will be a pleasurable experience, being served by robots will be an existing experience, preference towards the appearance of the robots, preferences towards . To compare the human employees-robots performance ratio in a hotel, if they feel robotic performance can let them to feel better service to compare human employees, then the tourists will choose the hotel to live again, when they come back to the country to travel. So, it seems that future hotels ought may choose service robots to help them to serve their clients, if they can train robotics to let hotel guests to feel more satisfactory service feeling. Their hotel service tasks may include that the reception, multi-lingual robots is a talking dinosaur , responsible for greeting, checking in and assisting guests. At the cloakroom, a robotic aim stores luggage , and porter robots carry them to the rooms, replacing a friendly , human receptionist with a robotic dinosaur may appear questionable for hospitality aficionados, but the concept may appear to be successful.

In travel bots application aspect, it may include these service robots, customer-service bots is usually incorporated for example, in the provider's website and their functions are limited to answering basic questions and assisting the user with navigating through the home page. Facebook chatbots is more interactive than customer-service bots, allowing a possibility to enter search and booking –related data using another interface, e.g. Expedia's or skyscanner's facebook, messager bots as well as the travel AIbots is such applications still rely on instance messaging to interface with the customer , but also utilize algorithms and access to information to make recommendations, e.g. a virtual travel agent uses calendar and email information to produce personalized recommendations.

In conclusion, all of the above service robots will be entertained for purposes in tourism mainly due to their efficiency and reliability asnd providing excellent service performance to compare human service employees in possible , when a number of the sbove mentional technologies are already a reality in tourism, they ought not appear counter-intuitive and unsuitable for service-encounters.

I believe automation in tourism service can bring new enjoyment exciting feeling to future dream-inspiring hospitality of the tourism -sector. We also need to believe tourists will be willing to adopt interact with service robots during their holidays and what is their perceived optimal trade off between efficiency and humanity.

(AI) raises driven automation industry development

1.1 (AI) - driven automation industry development how to influence work nature change

(AI) -driven automation industry will create wealth and expand economy growth to any countries, but it will be accompanied by changed in the skills that workers need to learn. One of main ways that technology increases productivity is by decreasing the number of labor hours needed to create a unit of output. It implies (AI) technology will influence low educated and low skillful labor number to be decreased (reduction employment number).

In contrast, technological change tended to work in a different direction throughout the nowadays. The advance of computer and the internet raised the relative productivity of higher skilled workers. So, routine-intensive occupations that focused on predictable tasks disappearance, such as switch board, operators, filming checkers, travel agents and assembling line workers etc. were particularly replaced by new technologies.

However, today, it may be challenging to predict exactly which jobs will be most immediately affected by (AI) driven-automation. The reason is because (AI) is not a single technology, but rather a collection of technologies that are felt unevenly through the economy to influence job changing both negatively and positively. In positively view point, (AI) driven-automation will make many workers more productive and increase demand for certain skills. Consequently, new jobs are likely to be directly create in areas , such as the development and supervision of (AI) as well as indirectly created in a range of areas throughout the economy as higher incomes lead to expanded demand. Otherwise, in negatively view point, many traditional human needed (demand) skillful jobs will be

threatened by automation are highly concentrated among lower-paid, lower-skilled and less -educated workers. It means automation will cause pressure on demand for this group, pressure and employment, if (AI) can replace the low skilled and less educated workers' jobs. Thus, (AI) will have negative influence to impact on the labor market.
(AI) capabilities will enable automation of some tasks that have long required human labor. Why can (AI) replace some simple human jobs? For example, advances in robotics are expanding machines' abilities to interact with and sharp the physical world. Combined , (AI) and robotics will give rise to smarter machines that can perform more sophisticated functions than ever before and brings more advantages that humans have exercised. This will permit automation of many tasks now performed by human workers and could change the shape of the labor market and human activity.

1.2 How (AI) influences labor market

Today, it may be challenging to predict exactly which jobs will be most immediately affected by (AI)-driven automation. Because (AI) is not a single technology, but rather a collection of technologies that are applied to specific tasks.
Some specific predictions are possible based on the current (AI) technology. For example, driving jobs and house cleaning jobs, bank counter service jobs, telephone enquiry service operators. Restaurant cooking jobs, simple accounting record service jobs etc. that require relatively less education to perform. Advancements in computer vision and related technologies have made the feasibility of fully appear more likely, potentially displacing some workers in driving-dominant professions. Seemingly similar robot, for which the operational tasks is less specific of navigating to a specific destination when following a set of given rules and preserving safety.
In the future, the effects of (AI) on the labor market in the decade ahead will continue the trend toward skill-biased change that computerization and communication innovations have driven in recent decades. Thus, some human driving occupation will be disappeared or replaced by (AI) automation driven. For example, bus drivers, light truck or delivery services drivers, heavy and tractor-trailer truck drivers, school drivers, tax drivers, travel bus drivers.
However, (AI) technology could enable some workers to focus time on other job responsibilities, boosting their productivity, and actually raised wage growth among those still holding the reshaped jobs. For example, salespeople, who currently spend a considerable amount of time driving could find themselves able to do other work when a car drives them from place to place, or inspectors and appraisers could fill out paperwork, when their car drives itself. This (AI) -driven technology should make these workers more productive, with (AI) -driven technology serving as a complement, not a substitute. New jobs will also likely be created, both in existing occupations cheaper transportation costs with lower prices and increase demand for products and all the related occupations, such as service and fulfillment, and in new occupations not currently foreseeable.
What kind of jobs will be created by (AI) technology? Predicting future job growth is extremely difficult, due to it depends on technologies or substitute for existing today as well as they may complement or substitute for existing human skills and jobs. However, (AI) will also lead to substantial indirect job creation to the degree it

raises productivity and wages, it may also lead to higher consumption that would support additional jobs from high-end draft production to restaurant and retail. The future(AI) " augmented intelligence", the technology's role is as assisting and expanding the productivity of individuals rather than replacing human work. Thus, based on the biased-technical change framework, demand for labor will likely increase the most in the areas where humans complement (AI) automation technologies. For example, (AI) technology , such as IBM's Watson may improve early detection of some cancers or other illnesses, but a human healthcare professional is needed to work with patients to understand and translate patients' symptoms, inform patients of treatment options, and guide patients through treatment plans. Shipping companies may also partner workers who pick up and deliver products over the last feet with (AI) enabled autonomous vehicles that move workers efficiently from site to site. In such cases, (AI) augments what a human is able to do and allows individuals to either be move effective in their specially task or to operate on a larger scale. Thus, it seems (AI) technology will also create new jobs, raise productivities and workers' efficiencies.

Redefining management in the workforce of artificial intelligence

2.1 Change management

In the future, due to artificial intelligence influences to some kind of human jobs nature. So, the kind of human jobs of management methods will also need to change to adapt the artificial intelligence technology input to their organizations. It will cause challenges for every executive and manager if who won't have effort to manage their teams how to apply artificial intelligence technology to work efficiently and easily. For example, division of labor will change among humans and machines will increase. Thus, companies will have to adapt their training performance and talent strategies how to emphasize on work that how to make human judgment and skills and experimentation. Thus, (IA)'s greatest impact will be on administrative coordination and control tasks, such as scheduling , resource allocation.

In fact, mangers will encounter this challenges: How to apply human experience and expertise to judge critical business decisions and practices when the information available is insufficient to suggest a successful course of action? Due to this kind of work will require new skills and mindsets. I shall indicate these change management methods to adapt (AI) technology. Such as: administration and routine tasks, scheduling , allocation of resources and reporting will fall within the intelligence machines, responsibilities that have long been reserved for humans. For example, a typical store manager or a lead nurse at a nursing home most constantly arrange shift schedules, accounting for staff members' absences owing to illness, vacation time or sudden departures.

Thus, the managers need to learn how to arrange new division of labor within the organizations after (AI) technology had been implemented to the organization. Artificial intelligence is currently influencing into once considered exclusive to humans: assessing and acting on human emotions and personality traits. The influences to managers need to change their strategies to adapt (AI) technology implements include such as below:

Firstly, managers need to spend the bulk of their time on coordination and control tasks from intelligent system implements. Their time spending on these major three aspects from impact of intelligent system: coordinate and control, solve problems and collaborate and people and community , strategy and innovation three aspects. Thus (AI) will influence managers need to change their judgment method to teach whose teams how to adapt the (AI) system operations in any organizations.

Secondly, (AI) will influence top, middle and low level management needs to change to adapt the (AI) technology operations to any owned (AI) technology organizations in the future. Intelligent machines must be trained in context. Just like humans , on-the-job training is a requirement for such machines because they typically arrive with only very general capabilities. To get the most from (AI), managers at all levels must participate in the instructional experience and in the learning process and provides managers' familiarity with such systems on these aspects, e.g. How the system works and generate advice, how the system has a proven track record , how the system provides convincing explanations , how the system can make simple rule- based decisions.

Thirdly, managers need to learn how to make judgment more accurate (AI) systems assistance. Although (AI) will invariably take on more routine work and even augment human decision-making, it won't judgment work, the application of human experience and expertise to critical business decisions when the information available is insufficient to suggest a successful course of action or reliable enough to suggest an obvious course of action. For a sense of the nature of judgment work, consider big data marketing and sales analytics. Such analytics often provide insights that can inform promotional campaigns, including predicting which promotions will generate desired sales brand further into the future, marketing executives need use judgment, combining analytics with their own and others' insight and experience.

The application of experience and expertise to critical business decisions and practice represents the real value of human judgment. But, when artificial intelligent machines are implemented to any organizations to assist the low, middle and top level management to make any business judgment. These forms of judgment work that managers can gather data interpretation, idea development more absolute from (AI) machine assistance. Thus, why these level management executives need to learn how to apply (AI) machines to help them to make any business judgment more accurate.

2.2 How (AI) influences organizational change

Consequently creative and social intelligence will be in even greater demand as (AI) makes in management and the workforce. This development will represent a long term trend in labor markets , one characterized by intensifying demand and reward for social skills with a growing desire for creative capabilities, managers will seek to fashion of ideas and hypotheses from inside and outside of the enterprise to shape solutions to their most pressing business problems. Thus, (AI) will influence overall organizational team members who have chance to participate any decision to make more accurate business judgment.

Many managers mistakenly view judgment work as only an individual discipline, failing to appreciate that it can also involve decide interpersonal and organizational practices. In more complex settings, judgment is typically a collective outcome of individuals' and teams' diverse perspectives, insights and experiences. And often , the resulting choices are better informed than decisions that an individual would have arrived at on his or her own.

Thus, when any organizations apply (AI) technology to assist managers to gather data and ideas to make any judgment. In these cases, organizations can create the conditions for effective collective judgment by establishing structures , such as " shadow advisory boards" that prompt managers and employees to source and synthesize multiple perspectives. Thus, a traditional organization (firm) might freshen its thinking is t put together a shadow advisory board, comprised of young, digital people who can apply (AI) machine assistance to make judgment work more accurate whether related to people development, problem-solving or strategizing and innovating for considerable degrees of creative and social intelligence.

Thus, on the one hand, (AI) technology machine augmentation and automation can give these advantages to human (organization managers) , e.g. developing people and community, solving problems and collaborating, coordinating and controlling work, shaping strategy and leading innovation. Besides, on the other hand, the next generation managers need have these individual attitude to treat intelligent machines to be as colleagues.

When, judgment is a human skill, intelligent machines can accelerate human learning that supports it, assisting in data -driven simulations, scenarios and search and discovery activities. Focuses on judgment work, some decisions require insight beyond what data can tell them. This is the sweet sport for human judgment, the application of experience and expertise to critical business decisions and practices. Thus, managers will also need to find ways to learn how to use digital (AI) technologies to tap into the knowledge and judgment of partners, customer external stakeholders and role models in other industries after the (AI) machine had been implemented to the organization.

Future works change:Automation, employment
and productivity

3.1 How (AI) influences employment

Human future " micro to macro" industry trends will be affected business strategy and public policy by (AI) technology. In the future (AI) technology will influence those six themes: productivity and growth, natural resources, labor markets, the evolution of global financial markets, the economic impact of technology and innovation and urbanization. However, (AI) technology will bring economic benefits of tackling gender inequality, a new global competition, Chinese innovation and digital globalization.

Nowadays, advances in robotics artificial intelligence, and machine learning are in a new age of automation, as machines match or outperform human performance in a development to any countries. For example, automation of activities can enable businesses to improve performance by reducing errors and improving quality and speed, and in some cases achieving outcomes that go beyond human capabilities. For example, some research indicated automation could raise productivity growth globally by 0.8 to 1.4 % annually; more than 2,000 work activities across 800 occupations. When less than 5% of all occupations can be automated using demonstrated technologies about 60% of all occupations have at least 30% of constituent activities that could be automated. Many occupations will change that will be automated away: Activities most susceptible to automation involve physical activities, in highly structured and predictable environments, as well as the collection and processing of data. They are most prevalent in manufacturing , accommodation and food service and retail trade and include some middle-skill jobs. For example, such as natural language processing is a key factor. Beyond technical feasibility, the cost of technology competition with labor including skills and supply and demand dynamics, performance benefits including and beyond labor cost savings, and social and regulatory acceptance will be affected by (AI) automation technology. Thus, (AI) automation will impact to influence global employment in those aspects as below:

Firstly, assuming that people are displaced by automation will find other employment. The anticipated shift in the activities in the labor force is of a similar order as the long-term shift away from agriculture and decreases in manufacturing share of employment. Both of manufacturing and agriculture industries which would be accompanied by the creation of new types of work not foreseen at the time.

Secondly, for business, the performance benefits of automation are relatively clear. Thus, the businessmen have opportunities for their micro economies to benefits from the productivity growth potential and macro economies to benefit to encourage continued progress and innovation , investment and market incentives. At the same time, employers must innovate policies to help workers and institutions adapt to the impact on employment.

This will likely include rethinking education and training, income support and safety nets , as well as support for those dislocated, when employees need to leave themselves homes to move to other cities to learn new (AI) automation works. Thus, individuals in the workplace will need to engage move comprehensively with machines as part of their everyday activities, and acquire new skills that will be in demand in the new automation age. Consequently , the scale of shifts in the labor force over many decades that automation technologies can be a similar order to the long -term technology -enables shifts in the developed countries' workforces away from agriculture in the 21 th century. Those shifts did not result in long-term mass unemployment because they were accompanied by the creation of new types of work not foreseen at the time. However, human will still be needed in the workforce when the total productivity gains are caused by (AI) technology.

3.2 What occupations will be influenced by (AI) technology.

In the future, scientists predict that these occupations will be influenced by (AI) technology mostly. They include : retail salespeople, food and beverage service workers, language or translation teachers, health practitioners. Since these work activities have a more relevant occupations are made up of a range of activities with different potential for (AI) automation . For example, a retail salesperson will spend more time interacting with customers, stocking shelves , or ringing up sales. Each of these activities is distinct and requires different capabilities to perform successfully.

Thus, these job activities have similar simple control characteristics. Simple activities include greet customers, answer questions about products and services, clean and maintain work areas, demonstrate product feature process sales and transactions. All these activities can have similar simple activities in order to (AI) machines can be learn how to do these activities from (AI) technology . For example, the capability perception includes sensory perception, cognitive capabilities, such as retrieving automation, recognizing known patterns(supervised learning), logical reasoning problem solving.

Thus, (AI) machine is such human, which has feeling and emotion, such as social and emotional sensing, judgement reasoning methods, natural language understanding and physical capabilities, such as mobility , navigation, gross motor skill, fine motor skills. It seems that the future, (AI) human invents machines which will have these human characteristics to do human similar behavioral job duties more easily and efficiently. It implies these above human occupations will be replaced by (AI) human invention machines in the future. Due to (AI) creation, it is possible to cause unemployment number of these above workers will increase because (AI) machines can do their similar job behavioral activities.

Consequently, employers won't need to employ many of these skillful labor. Otherwise, they can buy less number (AI) machines to attempt to do whose job activities more easily and efficiently. So, it seems (AI) machines will have more high work performance to replace these occupation workers' work performance. Finally, these occupation worker unemployment number will only increase when the (AI) machines had been invented to achieve to do their work behavioral activities absolutely success in the future.

3.3 Whether (A) technology machine labor will replace human worker more or assist human worker more

There is no single agreed definition of a robot how outcome of a task that is completed without human intervention. When some definitions require the task to be completed by a physical machine moves and respond to its environment, other definitions use the term robot in connection with tasks completed by software , without physical embodiment.

However, to answer the question : Whether (AI) technology machine labor will replace human worker more or assist human worker more. I shall indicate some examples to let readers to judge whether (AI) technology can create new jobs or reduce old jobs.

Firstly, I shall explain what (AI) function is. (AI) is a service robot that performs useful tasks for humans or equipment excluding industrial automation application . Thus, the classification of a robot into industrial robot or service robot is done according to its intended application. It is also a personal service robot or a service robot for personal used for a non commercial task, usually by lay persons . Examples are domestic servant robot, and pet exercising robot. It is also a professional service robot or a service robot for professional used for a commercial task, usually operated by a properly trained operator. Examples, are cleaning robot for public places, delivery robot in offices or hospitals, fire-fighting robot, rehabilitation robot and surgery robot in hospitals. Thus, these functions will be future (AI) application to our daily life necessaries or business necessaries.

However, some authors agree (AI) will bring negative outcomes of automation, due to raise competiveness, reduce human job nature. Otherwise, other authors argue (AI) will bring positive outcomes of automation, due to raise productivities, job creation, assist humans work.

On the positive outcome hand, robots can increase productivity . This is particularly important for small-to medium sized businesses both are in developed and developing countries economies. It also enables large companies to increase their competitiveness through faster product development and delivery. Increased use of robot is also enabling companies in high cost countries to re shore, or bring back to their domestic base parts of the supply chain that will have previously outsourced to sources of cheaper labor. Currently , the greater threat to employment is not a automation, but an inability to remain competitive. Automation has led overall to an increase in labor demand and positive impact on wages. The reason is that the middle-income/middle-skilled jobs have reduced as a proportion of overall contribution to employment and earnings leading to fears of increasing income inequality, the skills range within the middle income bracket is large. Thus, robots are driving an increase in demand for workers at the higher -skilled and with a positive impact on wages. This issue is how to enable middle-income earners in the lower-income range to unskilled or retain. Finally, the (AI) positive impact supporter who argue the future will be robots and humans can work together.

However, on the negative outcome hand, robots can substitute labor activities, but don't replace jobs. They believe that less than 10% of jobs are fully automatable. Increasingly , robots are used to complement and augment labor activities, the net impact on jobs and the quality of work is positive. Automation can provide the opportunity for humans to focus on higher-skilled, higher-quality and higher-paid tasks. Robots can improve productivity when they are applied to tasks that which perform more efficiently and to a higher and more consistent level of quality than humans. For example, increased productivity is enabling some firms, such as Whirlpool, Caterpillar and Ford Motors company in the US restructure their supply chains, bringing back parts of the manufacturing process to the country of origin. Thus, productivity gains due to robotics and automation are important not just at the company level, but also for build industry and nation competitiveness.

I suppose that productivity can be raised. What are the impacts of robots on employment? Firstly, the main focus of development has been on personal entertainment, which does not drive worker productivity (manufacturing production). When the internet (information and communication technology (ICT)) innovation. This is borne and by findings that manufacturing productivity, which has been driven by innovations in automation rather than consumer technologies, has government strongly than productivity in the services sectors of the economy in most nature economies. It seems (AI) automation will create many jobs in internet communication entertainment game industry. For example, many young people like to use internet to play any electronic games from computer or mobile at home or outside home conveniently. Thus, (AI) automation will increase demand to be invented to any new entertainment game from internet channel. It will need to employ many (AI) entertainment game inventors to create many automation entertainment games. Thus, (AI) automation in internet entertainment game industry will need human (AI) entertainment game inventors to invent the knowledge-based capital of (AI) automation entertainment games. The (AI) entertainment game inventors will need own research and development skills, form specific skills, organizational know-how skills, databased knowledge, design and various forms of intellectual property to do these (AI) automation entertainment game invention occupations in the future.

International Federation Of Robotics(2016) indicated that China will be as a major robotics manufacturer and user of robots, benefiting from jobs created by robot manufacturing and productivity gains from robot use. Chins had sold of robots to any one single market every year since 2017 year. The Chinese government has included a focus on robotics in its 10 year strategy. In order to achieve its target of a robot density of 150 units per 10, 000 workers by 2020 year. Thus, Chinese companies will have to install around 650,000 new industrial robots between 2016 to 2020 year, 2.5 times more than installed globally in 2015 year.

Hence, China (AI) manufacturing industry will need to employ many workers . It implies (AI) manufacturing industry will create many new occupations in China. Also, ministry of economy, trade and industry (2015) also showed that Japan currently has the largest stock of industrial robots in operations, primarily in the automation industry. Driven by a rapidly aging population and low productivity rates, the Japanese government has sights on a 20-fold increase in the use of robots in the non-manufacturing sector and a three-fold growth rate of labor productivity in the service sector both by 2020 year. Thus, it also implies Japan will need many robots to be provide to service industry. Due to robots will provide to serve any businessmen's clients. Thus, it is possible that the service workers won't be dismissed as well as it is depended on the serving job nature to decide whether Japan's service workers can still serve to their employer when the service (AI) robots are applied to whose employers.

Consequently, it seems that (AI) can create employment, Ministry of economy, trade and industry (2015) showed that such as China will develop the major (AI) automation manufacturing industry. The (AI) employers will need to employ many workers to manufacture any these different kinds of (AI) robots to satisfy China or overseas individual or business buyers needs. But, (AI) can also cause unemployment to the low skillful service workers. Such as if Japan some service businesses choose to buy any (AI) service robots to replace their service staffs to serve their clients. It is possible that the service staffs will be dismissed, due to (AI) robots can do such as their same service job duties to achieve better service performance.

Thus, today, it is increasingly common for people to use robots in various situations at home and in retail stores, hotels and hospitals these service industries. Robots are classified into server types based on their functionality (service and utility robots or those designed to communicate with humans) and appearance (humanoid robots or mechanical robots). The type of robot, to which each country allocated particular importance in the advance of robotics, reflects the sense of values and preferences of its population. Thus, if the country has high population needs to use robots, then they will influence either more new jobs creation or more old job loss in the country's (AI) manufacturing or (AI) service industries both. For example, Japan respondents often associate the term " robot " with humanoid robots that can communicate with human and they have a high level of familiarity with robot. The US has the highest level of robot utilization at home and in retail stores with its people being the most enthusiastic about the future use of robots. Germany shows a strong tendency to consider robots for industrial purposes and its people feel strong effort to the presence of robots in their households.

In conclusion, to judge whether how (AI) will influence the country's employment to be better or worse. It will depend on the country home buyers (users) or business buyers (users) how to use (AI) for their daily needs. If the country , such as US retail stores need to use (AI) , it will have possible to reduce some or many retail service workers. Even, if the country , such as Japan has many home users need to use (AI) , it will not influence the employment market. Otherwise, it will raise (AI) salespeople numbers. Even, if the country, such as Germany and China will have many (AI) manufacturers, then it will create many (AI) manufacturing occupations for these (AI) manufactory workers.

Consequently, (AI) robots manufacturing and service needs will have positive or negative impact to any country's employment. It will depend on the (AI) service provision and service workers' job nature as well as the manufacturing workers of (AI) knowledge level to decide their employment chance in their country's employment market.

- Could work activities in China be automated

making in the nation with the world's largest automation potential?

Can (AI) technology influence China economy? Could China workers be affected and jobs made up of routine work activities and predictable? Will programmable tasks be particularly impact to China employment market ? When impact on labor market is likely to be gradual at the aggregate level, it can be sudden and dramatic at the level of specific work activities, rending some job obsolete fairly. Overall (AI) technology will raise digital skills when reducing demand for medium incomer inequality for China workers. It seems (AI) technology's effect on productivity could be crucial to China's future economic growth as the population ages are increasing.

In China, some biggest technological companies driving significant investments in research and development. Moreover, China is one of the leading global (AI) technology development county. However, China will need to focus on building its innovation capacity. For example, United States and United Kingdom are currently producing more influential (AI) technological research. However, if China planed to achieve (AI) technology success, it's traditional industries will need to develop technical know-how –to and overcoming implementation costs prepare to develop (AI) . When (AI) technology is introduced into China society, China government needs to raise concerning ethical, legal, technological security etc. business questions. Also, surrounding issues include privacy, discrimination, legal liability and regulation. It aims to encourage overseas investors to choose to invest (AI) technological industry to raise GDP growth and manufacturing industries income growth for long term in China.

If China encouraged overseas (AI) technology investment in its country. It is possible to influence China employment market to be changed. Because (AI) technology will impact to influence China people daily life. Due to (AI) technology is introduced to China society, many rich people will prefer to spend to buy any high (AI) technological products for entertainment or learning or machine man driving etc. daily necessity activities. Then it will raise GDP growth and will raise (AI) manufacturers or related-(AI) technological manufacturers profit. It is beneficial to China because it can become one high knowledgeable and (AI) technological economical society.

But it will bring bad influences to raise unemployment chance for the low skillful labor. In labor economy aspect influence , how (AI) technology can influence China low skillful labor unemployment ratio raising. The raising low skill labor unemployment reason is because China low skillful human labors are argued or are replaced by (AI) technology creating new challenges to introduce to influence China society of simply human manufacturing job nature to be changed to be high (AI) technology manufacturing job nature in any China factories. Moreover, when (AI) technology introduction to China, it will cause other related social challenges in China. The varied (AI) related challenges, including the difficulty of creating safe and reliable hardware for sensing and affecting (transportation and education), the challenges of gaining public trust, a low resource comities and public safety and security, the challenges of overcoming fears or marginalizing humans in China employment and workplace and the risk of diminishing interpersonal trust because the low skillful labors won't believe any China employers will give chance to employ them , due to (AI) technology will replace their skills and man manufacturing of productivity is much less to compare to (AI) technology manufacturing method.

- How does (AI) technology influence
the future of employment change?

Are future nature of jobs changed to computerization from (AI) technology? Where are the probability of computing occupations from (AI) technology influence? What is expected impacts of future computing on labor market from (AI) technology influence? John Maynard Keynes's frequently cited prediction of widespread technological unemployment " du to our discovery of means of economic the use of labor outrunning the pace of which we can find new used of labor" (Keynes, 1933, p.3).

In the future, (AI) technology will impact some nature of occupations to change computing. This chance will also influence some countries‘ economic change. For example, some factory human labors hand routine manufacturing tasks will be changed to computerization of routine manufacturing tasks by (AI) technological machine men hand manufacturing method. it will cause a structured shift in the labor market, with workers reallocating their labor supply from middle-income manufacturing to low-income service occupations.

Arguably, this is because the manual tasks of service occupations are less computerization, as who require a higher degree of flexibility and physical adaptability. So, (AI) technology will influence the human hand labor skillful occupation nature of task cheaper , such as vehicle manufacturing , ship manufacturing, computer manufacturing, steel manufacturing, television, radio etc. home electronic products of heavy machine industry change. Due to (AI) technology machine man will be proper to be used to manufacturing these electronic products when the (AI) technology innovation can develop to the mature stage. Then, any countries manufacturers will choose to use (AI) technology machine man, instead of human hand production.

Supposing the future prices of computing are fallen, seriously, problem solving skills are becoming relatively productive, explaining the substantial employment growth in manufacturing occupations, involving cognitive tasks where skilled labor has a comparative advantage, as well as the increase education needs for (AI) technology computing of machine man subject study.

Prediction of education needs for (AI) technology student numbers will increase, due to manufacturing industry needs many (AI) technology students in future employment market. Another (AI) technology influence if the future (AI) technological innovation, e.g. machine man manufacturing or machine man service industries will both increase demand, then with more sophistic software technologies will be disrupted labor markets by marketing workers redundant.

For publishing industry, what is striking about the case in paper book publishing industry will be unpopular? Due to the electronic book publishing industry will be popular, e.g. Amazon publish . (AI) technology can influence paper book manufacturing method which is replaced by machine man electronic book manufacturing method as well as it will cause the computerization is no longer confined to routine manufacturing tasks. Due to (AI) machine man manufacturing technology will be proper to be used to manufacture any products in short time efficiently and effectively , e.g. electronic book products. In the future, if it is fact to occur this case, such as (AI) technological machine man manufacturing method will be adopted (applied) to manufacture electronic books or any products in possible. (AI) technology will cause many manufacturing workers are unemployed. It is beneficial to employers, who can reduce to spend much wages expenditure to employ manufacturing workers, but it will cause many manufacturing workers loss jobs and reduce income to support whose families lives. It will cause social challenges, e.g. increasing stealing crimes if the manufacturing workers had not other skills to find other jobs to do easily. So, manufacturers need to concern over technological unemployment which will be hardly future phenomenon if who decided to dismiss all manufacturing workers, due to (AI) technology machine men replace to them.

If (AI) technology can be innovated to produce any kinds of machine man to serve any service or manufacturing industries successfully. Then, it will bring these questions: Can future that workers be influenced to be automation employment and productivity by (AI) technology influence? Does it impact to influence the (AI) technology countries' productivity and growth and natural resources development and labor markets and evolution of global financial markets and economic impact of technology and innovation and urbanization etc. issues? How will automation transform the workplace? What will be the implication for employment? What is likely to be its impact both on productivity in the global economy and on employment?

In fact, automatic of activities can enable businesses to improve performance by reducing errors chance and improving quality and speed, and same cases achieving outcomes that go beyond human capabilities. Some economists indicate (AI) technology would give a needed boost to economic growth and prosperity have of the working age population in many countries. Based on the scenario modeling, they estimate automation could raise productivity growth globally by 0.8 to 1.4 % annually. They also indicated that almost half the activities people are almost $1.6 trillion in wages to do in the global economy have the potential to be automated adapting current demonstrates technology, according to their analysis of more than 2,000 work activities across 800 occupations. When less than 5% of all occupations can be automated entirely using demonstrated technology, about 60% of all occupations have at least 30% of worker made activities, that would be automated. More occupation will change to be automated. They also indicated for business performance benefits of automation are relatively clear, but the issues are more complicated by policy making to attract foreign investors. Beyond technical feasibility, the cost of technology, competition labor will include skills and supply and demand dynamics, performance benefits and beyond labor cost savings and social and regulatory acceptance will affect the automation. Their predictions suggest that half of today work activities could be automated by 2055 year, but this could happen 10 to 20 years earlier or latter depending on the various factors in addition to their wider economic condition.

Some scientists suggest (AI) technology is finally starting to deliver real-life business benefits. Computer power is growing significantly , algorithms are becoming more sophisticated and perhaps most important of all, the world is generating vast quantities of the fuel that powers (AI) technology data billions of gigabytes of it every day. Also, online firms are digital natives, such as Google online search service company is investing on (AI) technology. For new though most of the news if coming from the suppliers of (AI) technologies. And many new users are only in the experimental phase. Few products are on the market or are likely to arrive these soon to drive immediate and widespread adoption. As a result, analysts believe (AI) technology's potential will give true economic benefit in the future. (AI) industry will introduce to suppliers and users to raise economic potential of (AI) technology.

In the future, (AI) technology systems can solve business problems. Some scientists categorized those into five technology systems that are key areas of (AI) technology development: robotics and autonomous vehicles, computer vision language virtual agents and machine learning , which is based on algorithms that learn from data without replying on rules-based programming in order to draw conclusions or direct an action.

Such as computer vision and language includes natural language processing, analytics, speech recognition technology, some are about learning from information, such as about machine learning and others are related to acting on information, such as robotics, autonomous vehicles and virtual agents, which are computer programs that can converse with humans. Machine learning and a subfield called deep learning are artificial intelligence applications.

- Can artificial intelligence impact
global economy growth?

Artificial intelligence (AI) is a term first defined in 1956 year. It is a branch of computer science that aims to create intelligent machines that work and react like humans. In contrast today, 60 years later, (AI) is characterized by a number of applications, including computers playing games against humans and understanding human languages, virtual personal assistants, and robotics which involve computers seeing , hearing and reacting to sensory stimuli. In the future, technologists predict for (AI) technology ranging from (AI) being used as a tool to aid relatively simple processes for robots with human like mental capabilities, who expect (AI) technology can emulate human performance by learning, coming to mind its own conclusions, understanding complex content, engaging in dialog with people, enhancing human cognitive performance or replacing humans in executing both routine and non-routine tasks. In existing industry, (AI) technology is used , such as targeted advertising and virtual used personal assistant as well as the (AI) technology that my exist in the future, such as robots with human vehicle processing capabilities.

The range of (AI) technology's progress in the future will determine the economic impact future of (AI) technology on the global economy with more limited advances and applications (i.e. weak (AI) only) corresponding to more limited economic impacts and more substantial progress, i.e. strong (AI) technology is corresponding to more significant economic impact.

(AI) technology learning that automates analytical model, including predicting cause-and-effect relationship from biological data, identifying new drugs, self-driving cars and protecting against fraud etc. functions. Also (AI) learning can improve natural language processing that allows computers to continue to better analyze, understand and generate language to interface with human using the natural human language, virtual personal assistant, helps users by providing scheduling appointment, reminds organizing personal finance and finding providers of various services, machine vision allows (AI) machine man to identify object, scenes and activities in detect pedestrians and bicyclists. We expect the economic effects of (AI) technology to include both direct GDP growth from sectors that develop or manufacture (AI) technology and indirect GDP growth through increased productivity in existing sectors that employ some from of (AI) technology. If (AI) producing sectors could grow, then it could lead to increase revenues and employment of (AI) technological professionals within these existing firms as well as the potential creation of entirely new economic activities to any countries' societies productivity improvement in existing sectors could be realized through faster and move efficient processes and decision making as well as increased (AI) technological knowledge and access to information available in societies easily.

In the future, if (AI) technology is an increasingly critical component of more products, it will become an integral part of necessary products of many people's lives. The extent of (AI)'s economy effort is also likely to vary from region to region, thought variation may be more dependent on the predominate economic activity of a region and the (AI) ability can influence economic activity, rather then the economic or developmental status of the regions. (AI) technology can move accessibility and can use source development to do international business between one country and another country.

So (AI) technology has the potential to give benefits to different income chooses and to bring significant gains to both developed and developing countries. For agricultural technology, (AI) has the potential to optimize food production around the world by analyzing agricultural regions and identifying what is necessary to improve crop yield. In total, (AI) technology gives greater economic impact to any countries agricultural regions if which implemented (AI) technology to grow crop , fruit etc. food production in the farms.

Investment in (AI) technology is such as capital investment to any countries' public or private enterprises. So, it will have large economic impact to the future . If the (AI) technology is reasonable invested to the different needs aspect by the public or private enterprises in the country. Then, it will have good economic impact to the country in the future. However, when (AI) technology is likely to affect both the productivity and employment components of economic growth in many sectors. Significant public debate has focused on projections of (AI)'s effect on the labor force. However, for instance, some researchers have argued that the rise of (AI) technology and automation will led to significant unemployment as capital is substituted for the low skillful labor. So, they point to the concern that the increasing sophistication of (AI) technology may balance skilled and semi-skilled workers and the reduce the size of the middle class. However, this is not a new argument, due to (AI) technology negatively affecting the labor force and leading to mass unemployment. Because the (AI) technology is the substitution of machinery for human labor. Although, employment in certain industries, has been reduced in the past due to technological advancement. For long term, the labor market has adapted to the introduction of new technology, giving rise to new jobs in new areas.

(AI) technology may also be accomplished without a reduction to total employment in the long-term to some Asia countries, such as Hong Kong and Japan. Because Hong Kong and Japan many low skilled labor, e.g. security, cleaner who complaint that employers need them to work long time hours. (abnormal working hours) e.g. one day 12 to 15 working hour per day. Hence, if (AI) machine means invention technology success. Security or cleaning job can be worked from (AI) machine man in some hours every day in order to reduce the long time working hours cleaners or security workers, e.g. one (AI) machine man works 4 hours for cleaning or security job, one day as well as another cleaner or security labor only needs to work 8 hours one day. So total security or cleaning employers can employ 12 hours machine cleaners or security workers and human cleaners or security workers in one day. For long term benefit, Hong Kong or Japan every security or cleaning worker does not need to work 12 hours minimum working hours one day. They won't feel tried and bore and without private with whose families, so who will accept to do these cleaning or security jobs, even they can raise work efficient and performance when who feel happy and health.

So, (AI) technology of machine man invention can raise low skillful labor efficiency and it can help them to avoid abnormal working hours demand in some busy work life countries, such as Hong Kong and Japan. Before, one Japan female labor feel unhappy to work, due to who often needs to work abnormal working hours for her employer and who has less sleeping and without any private time to enjoy her life with her families every day. So this abnormal working hours factor causes her to do commit suicide behavior, then she is die unlucky. So (AI) technology of machine man invention ought avoid abnormal working hours demand for employer in any countries in the future.

The most important occurrence to any employers, some researchers had attempted to do one experiment to find that private research and development , venture capital and public research and development investment all have strong net effect or economic growth with venture capital funding further having the strongest such effect from (AI) technology. The researchers hypothesize the venture capital investment contributes to economic growth through (AI) technology innovation and by the capacity of an economy to use existing (AI) technology knowledge to increase productivity. They predict the impacts of venture capital, business-research and development and public research and development can raise multi factor productivity from (AI) technology introduction.

Can (AI) technology influence the economic development to developing countries? The developing regions of the world contain most of natural resources. If one day, (AI) technology has invent one kind of machine man which can assist any gas or oil workers to seek any new oil/gas natural resource locations easily. I believe that (AI) technology can help these natural resource exploitation countries will gain economic benefit more easily. So, (AI) driven technology can be used to change to create any new opportunities to address poor management or resources and improve human well being, such as Africa Latin America and India can use (AI) technology machine man to seek any oil/gas natural resource countries exploitation activities to attempt to gain much economic benefits.

- Why will (AI) technology grow economic development ?

Nowadays, increases in capital and labor are no longer driving the levels of economic growth, such as (AI) technology. The ability of increase in capital investment and in labor of traditional drivers of production, have no longer to be enjoyed in most developed economies ,e.g. developed country, US, UK . However, artificial intelligence has the potential to overcome the physical limitation of capital and labor to avoid missing out on this opportunity. So, policy makers and business leaders must prepare for and work toward a future with artificial intelligence. They must do with the idea that (AI) is another simply method to enhance productivity method . Rather they must see (AI) as the tool that can transform thinking about how growth is created.

Economists have always thought of new technologies are as driving growth their ability to enhancing. It can replace labor and capital factor of production. So, it brings this question: What is the factor of production (AI) technology characteristics. They key factor is to see (AI) technology as a capital-labor .

(AI) can replicate labor activities at much greater scale and speed, and to even perform some tasks began the capabilities of human. For example, by using virtual assistants , 1000 legal documents can be reviewed in a matter of days instead of taking three people six moths to complete. Some (AI) technology may be one kind of factor of production in the future. For another example, people will work in workplace digitalization environment. So, in the future, working environment and information management are automated. Such as Konica camera sale company will use workplace digitalization. So , (AI) technology can provide workplace digitalization in order to raise productivity efficiency. (AI) technology will be one kind of production which is replaced by workplace digitalization and it will grow any organization productivity efficiently. Then, (AI) technology will assist overall social economy growth , due to productivity is raised and products can be produced in short time to prepare to sell in consumption market. So, time will be shortened to increase GDP growth fast for the development of (AI) technology countries.

● How can (AI) technology impact to global
economic and social and psychological
changes?

What will be the development of (AI) technology and predictions concerning the future evolution? The computers and robots will develop conscious, intelligent and minds into humans, enhancing psychological and behavioral abilities and allowing for direct communication with (AI) minds. (AI) technology will be impacted human life by (AI) technology information communicative and environmental influence. A " world brain" and " world mind", this psychological system will be enhanced and enriched the capacities of both individual and collective cognition by (AI) technology of service industries.

(AI) technology with influence these human needs of service industries changes, such as , biological science, finance, entertainment, business, biological science, transportation, communication military etc. The personal computer evolution, the internet and the world wide web which exploded on the scene, linking business, homes, schools, social organizations which were a completely unpredicted phenomenon to influence human life. Kurzweil (1999) predicts that by 2029 year, most human communication will be with machines. According to Person, by 2100 year, there will be human machine convergence.

How can (AI) technology influence environmental protection to make benefits to farming economic growth? (AI) technology can be applied to predict how to solve environmental pollution challenge to avoid to damage any crop or vegetable or rice or fruit etc. food growth. Because environmental experts can gather global environmental pollution data from an environmental database to build a perform a systematic analysis from (AI) technology. The first step is this broad analysis can include understanding, statistical and data gathering techniques to obtain the relevant data, the correlation among the variables involved, and a list of possible models. The next step is to select a set of methods and models that cover all kinds of knowledge and functionalities needed for the decision making process. Once the models are selected, they must be fully implemented by means of machine learning , data mining, statistical or numerical technique. After that, those models must be integrated to build the whole EDSS. The EDSS must be tested to check its performance, accuracy, usefulness and reliability, both from the user's and (AI) technology/computer scientist's point of view. If these is any wrong feature in any development stage, such as model's integration, models' implementation, selection of models, database, problem analysis etc. the developers must come back in the update the required components. When the evaluation phase is all right, the EDSS is ready to be applied to the environment. The great contribution of artificial intelligence to EDSS the integration of several methods complementing the classical statistical models/simulation , statistical analysis, linear models, etc. and numerical models (control algorithms, optimization techniques etc.) .

This cooperation makes the resulting systems more reliable and powerful in coping with real world environment systems. Date interpretation has been a principal area of research in (AI) technology since the very beginning. The most demanding problem in the environmental assessment context. Knowledge representation permits the definition of the different types of data that the existing methods adapt to the process. There is also a lot of work to clean, repair and transform the huge available quantities of raw data. Apart from this, the availability of meta-information or background knowledge is required to guide the process. Data mining is multi-disciplinary: It covers expert systems, data based technology, statistics, data visualization and unsupervised machine learning. These techniques operate at the level of data and background information, where numerous and often incompatible new commensurate pieces of information from disparate sources have to be brought together (K, Fedra, 1994).

So, it seems that in the future, (AI) technology with the increasing maturity in particular those related to knowledge and engineering, new dimensions can be assisted to users in environmental decision making are available. For example, many environmental systems are characterized both by incomplete models and by limited data. Hence, in the future, (AI) technology will be applied to predict climate change to reduce crop or fruit etc. food agriculture challenge by climate change bad influence.

● Will (AI) technology influence digital economy change to manufacturing industry ?

To understand how the manufacturing business must adapt to prosper in the technology, we need to understand how (AI) technology will change us to shape our daily habits to satisfy our expectation of products to how we shop and even the immediate of the entire process. For example, taxi services are in the crosshairs as on demand transportation services like, available of the touch of a smart phone button expand. In fact, Yellow lab, US country , san Francisco city's largest taxi company is filing for bankruptcy as the industry starts to change faster than almost anyone expected. However, at this point, its more than an app that is changing, some our taxi passengers renting taxi transportation to catch consumption behavior.

(AI) technology will influence digital economy for taxi passenger's individual customer experience, offering a growing renting taxi to catch of service and feedback opportunities when any one taxi passenger who chooses to use mobile phone app online tool to prepaid to rent any taxi more easily.

Also in the long term, (AI) technology can influence vehicles drive themselves of behavior. Already, companies like Google and GM are working on projects to bring fleets of autonomous vehicles to cities at the path of a button. Moreover, this on-demand service model is beginning to appear across a much broader range of markets. For example , Amazon company is investing in its own fleet of trucks, planes and even drone at the same time as it pushes for same-day delivery of products. As some point, vehicles will be autonomous too. So, it seems that (AI) technique will influence any transportations choose to use digital autonomous driving technology in the future . For Amazon company case, it is not stopping of logistics. It is also aiming to automatically manage the supply of consumer home products with its recently launched Amazon replenishment service, Dash. Dash is a digital service that enables that connected derive to automatically order physical products from Amazon when supplies are running low. So, it seems (AI) technology will be applied to logistic function by digital technology method introduction in the future.

Hence autonomous vehicles will optimize industry supply chains and logistics operations through increased efficiency and flexibility. In fact, fully automated and lean supply chains will keep reduce load sizes and inventory by leveraging smart distribution technologies and smaller autonomous vehicles by machine man assistance. If Amazon continues to grow market share for online sales by reducing effort required by the consumer to place an order, when also contributing the almost immediate delivery of products to the doorstep. So, it will further fuel the trend toward on-demand derive. As Amazon company fuels the on-demand economy, consumers will expect immediacy in more parts of the digital economy. On top of speed, consumers increasing expect more personalization options.

So, (AI) technology will influence digital manufacturing, such as Amazon publishing to monitor every aspect of every process in real -time and communicating to self-optimized deep learning robotics, new methods of high volume and high customization will become possible. Then, as products merge into product platforms and even services, manufacturers have the opportunity to provide components and platforms used by smaller players. So, (AI) technology will influence manufacturing industry to choose automated SMI lines, robots installed, automation engineers.

Another future (AI) technology development can be applied to space science aspect, such as Automation engineering space in manufacturing process to achieve digital manufacturing benefits to any businesses in the future. Such as reducing cost, shortening manufacturing time, raising efficiency, shortening delivery products to client individual time. How can artificial intelligence give the need and advanced fast and evaluation methods benefits for space exploration? When US NASA (space exploration organization) achieves any space exploration missions, it will

answer this question:
When is it useful to have a machine use (AI) technology to achieve a decision? After all, after millions of years of space exploration and rough 10,000 years of civilization, humans are usually quite good at making decisions in complex uncertain environments. Through, Johns Hoplains University's Applied Physical Lab. Research in (AI) technology enabled systems, which has identified three general use cases for (AI) technology to explore space mission:
First, for some tasks (AI) technology is more cost effectiveness than human. Second, (AI) technology is better suited than humans at solving some, but not all problems. Third, (AI) technology allows NASA organization's space exploration mission to develop machines that ate capable of responding faster than when a human is in the decision loop (D. Scheidt, 2012, A. Castano et. al. 2008).

So, the use of (AI) technology to enable science by observing the pace of rapidly evolving phenomena was demonstrated. It is more effectively coordinating and (AI) technology utilizing to earn economic benefits to use for space exploration mission.
However, (AI) technology also have current risk for space exploration. Today (AI) technology is immature and requires further development to reach its potential. For instance, the (AI) technology algorithms that detected the dust derive could not have identified whether the Martain weather represented a threat to the cover. Also it can not yet use instrument input to determine what, where and how to autonomously make the next space science measurement. An equally important factor limiting (AI)'s deployment is that lacks the methodology and technology to effectively test (AI) technology. So, the challenge will testing (AI) enabled system is how (AI) performance can be measured. It would be NASA organization's difficulty to find (AI) technology to develop to carry on researching any space exploration missions in the future. However, (AI) technology will be a good economic benefit choice for space exploration mission in the future.

- What is artificial intelligence potential
benefits and ethical considerations?

The ability of (AI) technology systems to transform vast amounts of complex information into insight has the potential to help solve manufacturing or service challenges for human needs. However, to reap the societal benefits of (AI) systems, humans will need to trust then and make sure that which follow the same ethical principles, moral values, professional codes and social norms that we humans would follow in the same scenario, research and educational efforts as well as carefully designed regulation in order to achieve the most effort of economic benefits goals. For example, international business machines corporation (IBM) is actively engaged both competitors , in global discussions about how to make (AI) ethical and as beneficial as possible for people as social economic benefits. (AI) is usually defined as the " capability of a computer program to perform tasks or reasoning processes " that human usually associate to intelligence in a human being. Often, it has to do with the ability to make a good decision, even when there is uncertainty, too much information to handle. As an example, play chess or complex card games of entertainment activities is believed to need some form of intelligence in a human being, as well as choosing the best medical facilities in a difficult medical case, or creating something new, such as mathematical theorem or even some form of act, or even driving automatic machine man (self driving vehicle) replacing human driving in the middle of a crowded city.

(AI) needs depends on what we consider being intelligence in the behavior of a human being act a certain point in time. If human belief about human intelligence changes and we don't believe any longer that a certain task requires intelligence, then a computer program performing that task is no longer part of (AI), it becomes just another boring computer program. So, it means that (AI) technology will replace some old computer programs, if human can invent new generation of (AI) software for any functions or activities to satisfy human needs.

As IBM, it argues intelligence. This means that we aim to build systems that enhance and scale human expertise and skills rather than replacing them. We therefore focus on practical applications of (AI) capabilities that assist people in performing well-defined tasks of needs by exploiting and wide range of (AI)-based services. We also use the term " cognitive computing" it is mean a comprehensive net of capabilities based on technology. It comprises the fields of machine learning, reasoning and decision technologies, language, speech and vision recognition and processing technologies, high performance and high efficient functions for any industries or individual consumers needs. For example, robotics, which are usually very good at doing what which are supposed to in any environment, much have public shopping center, factory etc. places which need simply services from the robot (machine man), such as cleans the floor of our houses to the robot that can work together with humans in production chains, passing through the warehouse, robots can take care of the tasks of an entire warehouse and the companion robots like Nao, Pepper, Aibo and Giraff, who can entertain use, talk to use and help elderly people to stay connected to their friends, relatives and doctors.

Google company is building automatic machine (self-driving cars) and has acquired more than 10 robotics companies. Facebook had opened whole new research facility only on (AI) research. Apply computer has developed Siri. Microsoft computer company has built a similar personalized assistant. Google has Deep mind, a UK company whose long term aim is to build general (AI) and has already great potential to win game to the world champion and IBM is investing a huge amount of resources in applying its Watson cognitive computing system to the medical domains to finance and to personalized education. In Europe, IBM is establishing new centers in Munich and Milan focused in the application of cognitive computer capabilities to the internet of things and healthcare respectively.

For example, automatic machine man (self-driving cars) are all about (AI), which used to be able to see what happens in the street (signals ,lanes, other cars, pedestrians, traffic lights, which need to able predict what other cars and pedestrians will do, and who need to be able to cope with unforeseen situations. Since, most car accidents are due to human fault, it is estimated that the adoption of self-driving cars will save about half of the lives that are usually last in car accidents.

IBM Watson company has to understand spoken language, make sense of massive amount to text , respond correctly to questions in many categories, as well as assess its own confidence in responding to such questions. In the future, (AI) technology can own question/answering capabilities that would be very useful, for example, in assisting a doctor when trying to some to the correct diagnosis for a patient and to propose the best therapy .

Intelligent machines can also rely on huge amounts of data to be used to learn how to make better decisions. This data comes from all of us over the years Facebook users have uploaded more than 250 billion pictures and every day who upload about 350 million more. Every second, we submit 40,000 google search queries. So, (AI) technology will be connected through the web from appliances to traffic lights from cars to watches. Other tasks that are very easy for humans are physical and manipulation tasks, such as walking , running, picking up an object to make its shape and location, restricted environment. But (AI) machine man technology still not able to have the general physical and manipulation capabilities even of a 6 year old.

So, it brings this question: Why do (AI) scientists need to concern ethics? Because (AI) technology is complex, information into insight has the potential to reveal long held secrets and help solve some of the world's most difficult problems. (AI) systems can potentially be used to help discover insights to treat disease, predict the whether, and manage the global economy. So, ethic issues is important to and (AI) scientists . If any one new (AI) technology research investigation could success, it will be a secret to and the (AI) scientists can not permit to their loyalty to any competitors to damage the fair (AI) technology products trading market. The country (countries) (AI) technology scientists need to concern ethic issues, who need to keep secrets for their countries economic or/and social benefits. This is moral issues to any countries/country loyalty is whose countries intangible assets. They can not sell (AI) loyalty to any their countries to assist whose economic benefits immorally.

- How can (AI) technology influence to global
health care economy development?

According to (AI) lecturer analysis, when combined key clinical health (AI) application can potentially create $150 billion in annual savings for the US healthcare economy by 2026 year. (AI) technology is re-winning modern conception of healthcare delivery. It enables machines to sense, comprehend, act and learn. So which can perform administrative and clinical healthcare functions (Accenture, 2017).

It will help health care service organizations to reduce health care cost, will improve and raise service quality and access. So, (AI) health market size will be predicted growth. (AI) applications in health care include robot-assisted surgery, virtual nursing assistant, administrative workflow assistant, fraud detection, error reduction connected machines, clinical trial participant identifier, preliminary diagnosis, automated image diagnosis and cybersecurity.

What kind of benefits (AI) technology can contribute to healthcare service? (AI) technology can deliver what many health care organizations need, such as financial and operational of labor costs, digital expectations from patient consumers how to use (AI) technology to solve interoperability challenges in any healthcare organizations. Also (AI) technology can be applied to wellness an d lifestyle management, diagnostics, delivers financially but also way of organizational and workflow improvement. So, (AI) technology will be continue to become most prevalent and adoption to healthcare organizations , which must need to enhance structure to be position to take full advantages of new (AI) technological capabilities. (AI) technology can change the nature of work and employment is rapidly changing to make the best use of both humans and (AI) talent in healthcare industry in the future. For example, (AI) technology offers a way to fill in gaps and the rising labor shortage in healthcare. According to Accenture analysis, the physicians shortage is increasing. However, (AI) technology will manufacture healthcare machine men to replace physicians in future one day(2017). Hence, (AI) technology will be invented to raise health care service staffs work efficiency and performance in any hospitals or clinics in the future.

In conclusion, (AI) technology will raise efficiency for any service or manufacturing industries in the future, although, it is possible that it will also rise low skillful workers unemployment numbers. But, the most important influence to human technological innovation will be risen and it will influence human life will be changed to be better, e.g. self drive cars, health care physician machine men, machine man cleaners etc. intelligent machine men will be manufactured to serve for our daily life. Furthermore, (AI) technological products will influence countries trading, some low technological development countries manufacturing businessmen can choose to buy any (AI) products to raise whose productivity and efficiency and reducing cost to achieve economic cost saving result. Also, GDP of trading growth income will increase to the (AI) products sale countries. Hence, it will be beneficial to economic development to both developed and developing countries both in the future as well as (AI) scientists time and money spending will be valued to continue to invest (AI) technology development for human life and economy benefits for long term.

In conclusion (AI) technology will raise macro economy growth and it can create many (AI) jobs , but it also raise the low level technological worker unemployment change. In the future, (AI) technology can be applied to digital technology to attempt to invent any new undiscovered (AI) and digital technology. So, it needs any scientists to continue to research how digital and (AI) technology can be mixed to satisfy human's future undiscovered needs.

Reference

A. Castano et. al. " Automatic detection of dust devils and clouds at Mars" Machine vision and applications, Oct. 2008, vol. 19, no 5-6, pp. 467-482.

Accenture, " Why artificial intelligence is the future of growth"(2017) <http://www.accenture.com/us-en/insight-a rtificial-intelligence-future-growth>.

D. Schedidt , Unmanned Air Vehicle Command And Control, Handbook Of Unmanned Air Vehicles, Springer-Verlag, 2014. Facebook (AI) Research Available at https://research.facebook.com/ai, research at google, machine intelligence available at
http://research.google.com/pubs/machineintellige nce.html; micro soft research-machine learning and artificial intelligence available at http://research.microsoft.com/en-us/research- areas/machine-learning-ai.aspx.

International Federation Of Robotics, 2016. IFR press release world robotics report. IFR, org . 29 Sept. Accessed Feb. 01, 2017. http://www.ifr.org/news/ifr-press-release/world-robitics report -2016-8321.

K, Fedra , "GIS and environmental modelling" in environmental modelling with GIS, edited by M.F. Goodchild.B.O. Parks and L.T. Steyaert, Oxford University press, pp. 35-50, 1994.

Keynes, J.M. (1933). Economic possibilities for our
grandchildren (1930). Essays in persuasion, pp.358-73.

Mckinsey & Company (2013, May). Disruptive technologies: Advices that will transform life, business and the global economy , USA.

Ministry of economy, trade and industry, Japan, 2015, Japan's robot strategy. Ministry of economy, trade and industry.

Ray Kurzweil , The age of spiritual machines (1999) is cited numerously through this chapter: Kurzweilai.net http://www.kurzweilai.net

Rich, Elaine & Knight, Kevin, Artificial Intelligence Second Edition, 1991, New York; Mc-Graw-Hill.

CHAPTER FOUR

Competitive influences between artificial intelligence and human job to improve efficiency and performance

Although, (AI) technology will be popular to applied to different jobs, but it still needs social acceptance to replace some human jobs. Today, it is increasingly common for people to use robots in various situations at home and in retail stores, hotels and hospitals. Robots are classified into several types based on their functionality (service and utility robots or those designed to communicate with humans) and appearance (humanoid robots or mechanical robots). The types of robot to which every country attaches particular important in the advance of robotics, reflects the sense of values and preferences of its population . Thus, (AI) will be applied to replace human to do these above different kinds of job nature. For example, U.S. has the highest level of robot utilization at home and an retail stores with its people being the most enthusiastic about the future use of robots. Otherwise, Germany shows a strong tendency to consider robots for industrial purposes, and its people feel strong to the presence of robots in their households. Japanese accepts to apply" human aid robot" that can communicate with humans and they have a high level of familiarity with robots.

Hence, it implied those three countries have accept (AI) to replace human to do any these kinds of job duty and it will influence these three countries' workers lose their old occupations and who will unemployed absolutely, due to many (AI) robots replace them to do their job duties in the future. Also, US will have many retail service workers or retail warehouse workers are unemployed. Germany will have many manufacturing industry's workers are unemployed. Japanese will have many communication industry workers are unemployed, such as telephone service, shopping center services etc. different kind of service industry's service staffs . It will cause these kind of workers' competitive abilities are lost in themselves countries' jobs that require such skills include software developers, court judges, nurses, high school teachers, dentists and university lecturers, these occupations are still difficult to be replaced by (AI) robots.

Are robots taking our jobs or making them? In fact, our societies will have unemployment challenges, even (AI) technology has not created before. However, after (AI) robots invention, some of human jobs will be replaced and it can raise many low skillful and low knowledge level worker unemployment number. However, I think that high productivity driven by increasingly powerful IT -enabled machines is the causes of global labor market problems and accelerating technological change will only make those problems worse.

IT technology brings this question: Are robots killing human's jobs or benefiting human's jobs? I suppose that there is a limited amount of labor to be done. The implication is that technology can create unemployment by displacing workers, such as (AI) invention, because the more efficiently worker work (using machines or (AI) robots), the loss work there is for workers to do. Even, any new jobs will be better done by machines or (AI) robots, and unemployment will still skyrocket. How do we know that humans will always be better at some work, or more importantly, enough work, than machines or (AI) robots, e.g. human drivers drive more safe or careful to compare (AI) robot drivers. But, the challenge is that it is not ensure that (AI) robots drivers must not drive careless to cause the chance of accident occurrences more than human drivers. However, technological change can be beneficial to innovation, automation and increasing productivity for businesses.

Consequently , it may seem machines can hurt wages and job for low skillful, less educated workers. Also, high educated workers are likely as less educated workers to find themselves displaced and devalued, and more education may create as many problems as it solves. Thus, in negative influence, automation effects on particular jobs shift workers to other jobs that are equally or more desirable. Workers may be highly compensated for possessing human capital that is specialized to a labor market. If technological advance is very rapid, such as (AI) invention, causing a large and very rapid drop in demand in a large labor market, the economy may not be able to absorb the sudden surplus of labor in a short period of timer when (AI) robots are popular to replace some workers to do some occupations in global societies.

For example, self-driving vehicles threaten to send truck drivers to the unemployment office. Computer programs can now write journalistic accounts of sporting events and stock price movement. There are even computers that can grade essay revolutionize some part of teaching jobs. Hence, (AI) robots will have possible to replace human brain to do any judgement, argument, and mind job duties. It implies some occupations which need human' mind will be threaten by (AI) robots, e.g. author, accountant, nurse, engineer. Thus, (AI) robots will have possible to replace some professional and high educated workers' jobs in the future.

But, technology can create new nature of jobs in possible. For example, a 60 minutes program indicated technology is putting new categories of jobs in the sites (sic) of automation, the 60% of the workforce that makes its living gathering and analyzing information. Also, recession: technology kills middle -class jobs that overall technology is eliminating for more jobs than it is creating by (AI) technology. Hence, human's brain work may be assisted by 60% of (AI) gathering and analyzing information for some occupation , e.g. space scientists, ocean scientists, earth scientists etc.

However, I believe the (AI) invention and human job competition may influence global productivity change. Productivity is economic output per unit of input, the unit of on input can be labor hours(labor productivity), but if (AI) robots replace human job, then the unit of input may be (AI) machine hours (AI) robot productivity or all production factors including labors, machines and energy (total factor of productivity). Producing more output with less input can take several forms.

The traditional notion of productivity is a form reorganizing production and/or using better or more technology to produce more output per worker hour. But when (AI) robots invention, the form can be reorganizing production and/or using better or more (AI) robots to produce more output per (AI) robot hour. Hence, if the firm apply (AI) robots to produce its products. Then , productivity improvements in the firm may result in less workers employment, due to (AI) robots replace more worker number to achieve more productivity improvement, it has economic benefits (less factor of production) , but more production in long term.

Thus, (AI) robots can help any firm to achieve productivity improvement in long term, for example, if unproductve farmers move to the city and start working for high-tech. manufacturers. The shift effect can be more dynamic and disruptive as low-productivity industries lose out in the marketplace to high -productivity industries and the compositional mix of the economy changes. Thus, in the long term (AI) robots can also be beneficial to high productivity industries to bring the mix of economy positive changes.

Moreover, automation will also produce some new jobs in firms that sell the new robot or other labor-saving technology. This means that, in general, there will be shift in the economy in the direction of higher-skill and higher wage jobs. Even if the (AI) robot invention country, US becomes a leader in (AI) robots producing productivity-enhancing technology, it will experience a growth in jobs serving foreign (AI) robots product buyers. Hence, (AI) robots can also create (AI) salespeople, (AI) manufacturing workers , (AI) inventors, scientists, (AI) software designer etc. occupations, when if all society does is move workers from insurance firms, restaurants and car factories to robot factories, productivity will have remained the same to create job needs for insurance, restaurant and car manufacturing worker service occupations for (AI) software designer, (AI) service robots manufacturer, (AI) service robot seller etc. related (AI) service robot product occupation created in (AI) robot technology job market. Hence, (AI) invention also create new (AI) technology job chance. (AI) impacts management job market.

In future, organization management will be changed from (AI) introduction. Division of labor will change and collaboration among humans and machines will increase. Companies will have to adapt their training, performance and talent acquisition strategies to account for a new found emphasis on work that hinges on human judgement and skills, including experimentation and colloboration.

How (AI) impacts any organizational administrative management work? (AI) 's greatest impact will be on administrative coordination and control tasks, such as scheduling, resource allocation and reporting, (AI)-driven will place a higher premium on what we call " judgement work", the application of human experience and expertise to critical business decisions and practices when information available is insufficient to suggest a successful course of action. This kind of work will require new skills and mindsets; replacing people with machines is not goal in itself. When, artificial intelligence enables cost-cutting automation of routine work, it also empowers value -adding augmentation of human capabilities.

Thus, administrative and routine tasks, such as scheduling, allocation of resources, and reporting, will within intelligent machines, responsibilities that have long been reserved for humans. For instance, a typical store manager or a lead nurse at a nursing home must constantly juggle shift schedules, accounting for staff members' absense owing to illness, vaction, time or sudden departures. Many of these tasks will be automated by (AI). Imagine (AI) writing management monthly reports, it is not a distant dream. Leading news providers and Wall street banks are now using (AI) report generators to write news and analytical reports by drawing on quantitative data. The associated press, for example, expanded its quarterly earnings reporting from approximately 300 companies to nearly 3,000 with the help of (AI) powered software robots, freeing up journalists to conduct more investigative and interpretive reporting. For another example, Jobalime, a job-placement site, uses intelligent voile analysis algorithms to evaluate job applicants. The algorithm assesses paralinguistic elements of speech, such as tone and inflection, products which emotions a specific voice will elicit, and identifies the type of work at which an applicant will likely excel. In the future , (AI) machines can be applied to assist some kind of office administrative jobs duties. It's attractive to office managers to achieve more accurate judgment to do any administrative matters when who can be assisted from (AI) machines. Thus, managers need to spend time to learn how to apply (AI) machine to assist them to do more accurate judgement, and better informed choices. (AI) robots can be applied to improve the speed quality and cost of available products and services, instead of applying on productivity improvement and administrative improvement aspects. Thus, they may also displace large numbers of workers. This, possibility challenges the traditional benefits model of trying health care and retirement savings to jobs.

In an economy that employs dramatically fewer workers to deliver benefits to displaced workers. For example, the worldwide number of industrial robots has increased rapidly over the past few years. The fall prices of robots, which can operate all day without interruption, make them cost- competitive with human workers. In special consideration, in the service sector, computer algorithums can execute stock trades in a fraction of a second, much faster than any human. As those technologies become cheaper, more capable, and more widespread, they will find even more applicants in an economy.

Consequently, (AI) technology brings unemployed number increasing many businesses continued automating their operations rather than hiring additional workers. A trend among technology companies that receive massive valuations with relatively few workers. For example, in 2014 year Google was valued at $370 billion with only 55,000 employees, a tenth the size of AT & T's workforce in the 1960 year. Hence, if automation technologies like robots and artificial intelligence make jobs less secure in the future, there needs to be a way to deliver benefits outside of employment " flexi security" or flexible security is one idea for providing healthcare, education and housing assistance whether or not someone is formally employed.

In conclusion, (AI) and robots technology will raise unemployment to some occupations when (AI) replaces same industries‘ workers job duties in our societies in the future, but it also create new jobs to raise employment in any related (AI) robots and automated machine products in (AI) manufacturing. (AI) design, (AI) sale self-related industry, when (AI) replaces same industries' workers' job duties.

(AI) journalism, media publishing, digital communication technology trend

How to apply (AI) technology in digital communication journalism media, publishing industry? Some scientists indicate future (AI) and digital technology may consist such as: voice driven assistants, emerge. For example, Amazon e book publish applying digital technology and (AI) auto printing technology to sell e books to let readers to listen any e book content by (AI) voice driven speaker when they turn on computer to read e book contents; capable phones start to unlock the possibilities of 3D image of mobile story telling. New smart wearables include ear buds that handle instant translation and glasses that talk and hear. China and India will become a key focus for digital growth with innovations around payment online identity, and artificial intelligence. Thus, future (AI) technology can be applied to 3D image mobile story telling, online payment method to dealt online transaction publishing industry.

Thus, future (AI) technology can be applied to online e book publishing industry to make sound books to let readers feel more attractive . Such as Amazon publish has published sound e books to attract readers to choose to read any its books from online. Also, (AI) technology can also be applied to communication industry. For example, some online pure-play news, opinion and entertainment websites. It is a digital communication media, e.g. online journalism blog (AI) technology can be applied to visual storytellers to let online book readers to enjoy to listen to watch and send any online electronic book contents more attractive. Thus, future (AI) technology will be popular to assist any electronic book publishers to publish visual and sound talking storybook to let readers who can watch motive image and listen and read words from e books more attractive.

Thus, (AI) technology can be applied to internet ecommerce publishing or media industry to help any electronic book publishers to publish sound, image motion electronic book to attract global readers to read, even (AI) technology can be applied to digital entertainment industry, e.g. electronic 3D image virtual video games, computer games. It can be also applied to education industry, e.g. the first true digital native generation and are the native speakers of the digital language of computers to let student to learn different languages or translate words to compare to classroom learning more easily. It can be also applied to communication industry, e.g. (AI) mobile phone. Hence, it seems (AI) technology can be applied to publishing, communication, education , entertainment etc. different industries in the future. (AI) technology will be one kind of tool to satisfy human's daily life needs in the future and these industries has one characteristics is that they need to apply internet to operate to operate to do online business.

Thus, it has three trends of (AI) technology and internet technology need to be linked to achieve one kind of attractive technological business to satisfy client's needs. These three trends as below: All consumer trends involve the internet. It will be many consumer's online habits, shopping, working, socializing, watching TV, studying, travelling, listening. Thus, (AI) music, eating and exercising are just a few examples. This is happening because human usually use mobile broadband or Wi-Fi, rather than cables. Thus, (AI) technology will be applied to mobile to satisfy client's need absolutely.

The mobile phone can be more popular to be used more than computer or laptop tools. The reasons are because women dive the smartphone market by defining mass-market use. But as the speed of technology adoption increases mass market use becomes much quicker then before. Successful new technological products and services , such a (AI) mobile phone products now reach the mass market in popular use. It means that the time period when early adopters influence others is shorter than before. Also, since new products and services increasingly use the internet mass markets are not only faster , but are also more important than ever to consumer themselves. Most internet services become more valuable to individuals when many use them. Thus, it causes why (AI) mobile phone will be popular to be used.

Since, new products and services increasingly use the internet, in the future several trends focus on (AI) smart phone users. Consumers' familiarity with using smartphone apps. Essentially, the technologies will bring other related (AI) and internet service needs, e.g. sound and image emotion e book needs, (AI) mobile communication needs, e-virtual games or e-3D image virtual games etc. entertainment activities needs with such a large part of the world's population now online, it is clear that there is strength in numbers.

Thus, (AI) imagines , if future any (AI) and internet related services or products new technology is easy to use and inexpensive, when the latest products reach the mass market almost as quickly as they reach the early adopters and industry experts. I believe that any (AI) and internet related products or services must be popular to accept to consume for entertainment or useful aim. For example, with major players including Apply, Facebook and Google had invested (AI) technology to develop their businesses. (AI) technology has the potential to disrupt everything in the coming years, from the lives of connected consumers to every industry (AI) will be an alternative route for brands to reach consumers with convincing and relevant messages. Digital technology will assist of the future, then it can improve technology to bring this effect, such as sophisticated software machine learning and speech recognition effective. Hence, Google, Facebook , Yahoo web site service companies can apply (AI) technology to help other companies to advertise their businesses, such as travel, retail, and education etc. industries more attractive. (AI) technology can be applied to internet company to be aware and familiar enough to drive among mainstream consumers, it can create online experience to travel, retail , education and other entertainment needs to online consumers to seek their entertainment needs more easily. Hence, in the future (AI) technology and internet related entertainment service needs will be raised in this (AI) and online consumption market.

● (AI) healthcare service industry development

In the future, (AI) medical internet technology tool can be applied to assist individual's health at the center of their focus, e.g. smartwatch compatible mobile app. patients can let personalized reminders for taking their medication snap pictures of their prescriptions to expedite refills, and scan their insurance card. So that, store clerks are prepared with up-to-date patients' information . (AI) owned health operated technological clinics can help patients to receive treatment for minor illnesses, flu shots, cholesterol screenings and more than a dozen other medical services, all of

which can be patients who can't make it to a physical location. (AI) healthcare services organizations can provide various telemedicine services. So, patients can receive care via phone or video chat.
For example, one London-based intelligent Brewing company has developed an (AI) system to continuously collect and incorporate customer feedback, which the system itself uses to brew ne various of the company's beers. Thus, the beer clients can give feedback to talk to the algorithm (AI) machine, whenever or anywhere who're drinking the beer. It is such any healthcare services organizations can apply (AI) machine to collect patient's feedback to talk to the algorithum (AI) machine whenever or anywhere who're eating any medicines. So doctors can know every patient's health conditions any time. If the patients feel uncomfortable, the doctor can know from (AI) machine notification to decide whether the patient needs to eat another new medicine or keep to eat same medicine is better. Hence, (AI) medial internet technological body check report machine will be proper to be needed to serve any hospitals' patients in the future.

However , it brings this question. How can (AI) medical internet technological body check report machine apply to hospital more efficient? The essential new medicine co-workers for the health service digital age health service leaders need apply (AI) medical report machines and artificial intelligence to the newest recruits to the workforce bringing new skills to help health service staffs do new jobs and reinventing what's possible, building the health service workforce for today's digital health service demands for patients. Thus, technology-driven health service model innovation from the health service organization outside in and providing digital health service ecosystems for patients to use the (AI) health service equipment will be popular to be accepted to be used.

● I Robot and internet things future machine
men invention

Nowadays, there are some company, which apply internet and (AI) I Robot technology to do any similar human job nature. For fishing industry example, one company, known for creating the Roomba, I Robot is now working with marine conservationists to launch an ocean-patrolling intelligent robot to hunt and manage invasive species, protecting native populations. And evolved industries like precision agriculture are ramping of our increasing population. Area of practice that once seemed impossible to digitize are fundamentally changing because of the impacts of (AI), internet of things capabilities and big data analytics, which have many potentially positive impactions for society.

For textile industry example, automation is nothing new, it has shaped the workplace to replace human jobs to boost productivity in the textile industry. Textile machines have had a generally positive impact over gears, creating value and allowing textile workers to take up more rewarding age will likely continue to create opportunities and lead to new textile industries, companies and textile occupations. It may also compensate for a demographically driven slowdown in the growth of the textile workforce. The future impact of textile (AI) and automatic and internet link is somewhat uncertain. It seems textile industry will be trend to accept (AI) textile workers and internet of thing to replace traditional manual textile workers to produce any shirts, cloths etc. wearing products in factories popularly, during the (AI) textile machine and internet thing technology can be invented to reach the mature stage in the future.

For factory worker transportation job example, they have also expanded their influence, migrating from the factory floor to the service sector and taking the place of humans in a range of activities from financial transactions to transport route optimization. Further (AI) machines and robots are increasingly programmed to learn, meaning they improve with time and undertake cognitive activities. Hence, (AI) machines and internet technology enable automation of work activities to raise factory workers' efficient and performances, also factories can reduce manual worker numbers, due to (AI) machine workers' assistance.

Future, (AI) robotics technologies and internet technique have these different kinds of characteristics: For soft robotics example, it is non-rigid robots construct with soft and deformable materials that can manipulate items of varying size, shape and weight with a single device. For swarm robotics, it coordinated multi-robot systems often involving large numbers of mostly physical robots. For touch/factile robotic example, it robotic body pails (often biologically inspired hands) with capability to sense, touch , dexterity robots example, serpentine robots with many internal degrees of freedom to threat through tightly packed spaces for humanoid robots example, robots physical is similar to human being often bi-pedal that investigate variety capable of performing human tasks , including movement across terrains, object recognition, speech sensing etc. For autonomous cars and trucks example, it is capable of operating with a human pilot, e.g. the unarmed general atomics Predator XPUAV with roughly half the wingspan of a Boeing 737 can fly autonomously for up to 35 hours from take-off to landing, for unmanned aerial vehicles example, flying vehicles capable of operating without a human pilot, the unarmed general atomics predator -XPUAV , with roughly half the wingspan of a Boeing 737, and fly autonomously for up to 35 hours from take off to landing, for (AI) chat bots example, (AI) systems designed to simulate conversation with human users, particularly those integrated into massaging apps.

In Dec. 2015. the general service administration of the US Govt. described how it used a chat bot named Mrs. Landingham (a character from the television show the west wing) to help onboard new employees. Finally, for robotic process automation example, class of software robots that replicates the actions of a human being interacting with the user interfaces of both software systems. Enables the automation of many back-office work flows without requiring expensive IT integration . Hence, future (AI) robot machine men will have different functions to be applied to different industries to use in possible.

Statistics Denmark shows that (AI) automation potential robots will influence few jobs are completely automatable , but close to half consists of 40% automatable tasks: It showed example occupations include share of automated, such as brewing machine operators are more than 80%, logging equipment operators are more than 50%, roofers , stock tasks clerks, travel agent are more than 50%, farmers , nursing assistants are more than 30%, physicians, teachers , managers are more than 10%.

For example, humans perform a wide variety of tasks from planting corn to examine spreadsheets, meeting clients and lifting crates in a store. Each of these actions requires a combination of innate or acquired capabilities, internet technique assistance, ranging from social perceptiveness to fine motor skills and natural language understanding. To understand and map automation feasibility by existing technology. Mc Kinsey has developed a framework of 18 technical capabilities that can substitute tasks performed by humans. The capabilities are grouped in five categories: sensory, cognitive, language, social and emotional and physical. So, it seems (AI) robot machine men and internet technique will have possible combination to invent to own human's emotion , language, learning, task skill abilities.

Mckinsey global institute analysis also showed current technologies have achieved different levels of human performance across 18 capabilities include: sensory perception, autonomously infer and integrate complex input using sensors, cognitive capabilities reorganizing known patterns/categories supervised learnings, generating novel, logical reasoning/problem solving, optimization and planning, creative, information retrieval, coordination with multiple agents, output articulation/presentation, national language processing, social and emotional capabilities-natural language understanding, social and emotion sense, reasoning output, physical capabilities-fine motor skills, navigation mobility. Hence, it seems (AI) robots and internet technological will combine to invent to own human' some skills to replace human to do some kind of tasks in possible.

In conclusion, future (AI) robot and internet will be needed to link to cooperate together to raise human's work efficiency in popular.

How artificial intelligence replaces human job possibility

What is the risk of automation for jobs to replace human job? In recent years, there has been a revival of concerns that automation and digitalization night after all result in jobless future. As I argue, this might lead to an overestimation of job (AI) automate , as occupations labelled as high-risk occupations often still contain a substantial share of tasks that are hard to automate.

For example, when the share of (AI) automatable jobs is 6% in Korea, the corresponding share is 12% in Australia. Differences between countries may reflect general differences in workplace organization, differences in previous investments into (AI) automation technologies as well as differences in the education of workers across countries. I also discover that (AI) automation and digitalization are unlikely to destroy large numbers of jobs. But, however, low qualified labors are likely to raise costs as the (AI) automate of their jobs is higher compared to highly qualified workers.

In fact, (AI) technology will influence some new technology to replace some human's job, such as driverless car, the largely autonomous smart factory , service robots or 3D printing. These technologies are driven by advances in computing power, robotics and artificial intelligence and ultimately redefine what type of human capabilities machines are able to do.

Hence, question brings whether (AI) invention will influence general human jobs to be replaced by (AI) autonomous jobs? Whether will the potential foe automation with actual employment loss? In particular, the technical possibility to use (AI) machines rather tasks need not mean that the substitution of humans by machines actually takes place.

Whether (AI) technology replaces human's some job, it is beneficial to our society or not. Instead, machines are increasingly capable of performing non-routine cognitive tasks, such as driving or legal writing . In particular, advances in the field of machine learning (ML), e.g. computational statistics and visions, data mining, artificial intelligences allow for automating cognitive task, when the use of (ML) in mobile robotics (MR) also allows for automating certain manual tasks. So, it seems, (AI) technology can replace some labor job, e.g. warehouse transportation, even mind's job, e.g. legal writing, driving in possible.

For example, if (AI) automatic non-manual driving can reduce hurt or death risk, it is beneficial to our society, or (AI) automatic robots can more any heavy things (products) in warehouse safely. Then, it can reduce the warehouse labor's bodies hour risk, it is beneficial to the workers. Even, if (AI) robot can write any legal documents, no any word errors in short time. It is beneficial to the law companies , but it also bring unemployment chance, due to these jobs can be replaced by (AI) robots to do in the future. Hence, it will cause some occupation to be disappeared, due to (AI) robots can do our these kinds of jobs in the future.

Frey & Osborne (2013) reported these kinds of occupations will be replaced by (AI) robots in possible. They include computer, engineering, financial, management, legal , art and medium, community service, education, healthcare practitioners and technical service, sales and related, office and administrative support, farming, fishing and forestry, construction and extraction, installation, maintenance, and repair , production, transportation and material moving. It seems our future some professional occupations will have possible to the replaced by (AI) robots to replace, instead of labor jobs. Hence, (AI) robots technology will have much trend to replace high knowledge or low knowledge skillful labors in the future.

In conclusion, it implies that only using information on task-usage at the individual level leads to significantly lower estimates of jobs " at risk", some workers in occupations with according to high automate nevertheless often perform tasks with are hard to automate. Why can (AI) replace human to do some kinds of jobs? (AI) artificial intelligence refers to the ability of a computer or a computer enable robotic system to process information and produce outcomes in a manner similar to the thought process of humans in learning, decision making and solving problem. By extension, the goal of (AI) systems is to develop systems to capable of tasking complex problems in ways similar to human's logic and reasons who feel in our future. Hence, it means future (AI) robots has effort to replace human to do any jobs in possible.

(AI) directions for future non-manual
control road vehicles market

Future road vehicle products and technologies must meet social, economic and environmental protection and driving safety goals , and satisfying market requirements for mobility, accident reducing, performance, cost desirability. Thus, (AI) auto-non manual control vehicles need to be followed this direction to invent. To satisfy future driver's safety of needs, enhanced vehicle speed desired functional performance of road transportation system, required and desired technological response, including research needs. It is long term up to 20 years vision, for (AI) auto non-manual research. Thus, (AI) auto non manual vehicle manufacturers need often to revise their (AI) vehicles functions to raise to improve their system performance and driving industry driver's needs, e.g. private drivers need or public transportation driver's need or business client's need. Hence, future (AI) transportation will need have individual driving consumer and business driving consumer both targets.

Thus, future (AI) automation manual control vehicles need to deliver high impact technology solutions to meet social , economic and environmental and safe goals. Engine needs to be improved efficiency , performance, drivability, reliability, durability and speed-to-market together with reduced emissions and cost; hybrid, electric and alternatively fuel (AI) non manual control vehicle technology development, leading to new fuel and power systems, such as hydrogen, fuel cells and batteries, which satisfy future social, economic and environmental and safe goals. Software, sensors, electronics and telematics technology development are needed to be lead to improve vehicle performance, control and adaptability, intelligent , mobility and security, structure and materials technology development, leading to improved safety, performance and leading to flexibility with reduced cost and environmental pollution to achieve the (AI) non manual drivers to feel (AI) vehicle performance, auto control and adaptability is better to compare traditional manual driving vehicles.

In fact, in traditional manual driving market, Japan and USA had had over 80% of world car production by six major global groups. In the future, it is possible only USA can dominate (AI) non manual auto driving vehicle manufacturing market if Japan had no effort to manufacture any (AI) auto non manual control vehicles. So, it means that it is only Japan is USA potential (AI) auto non manual control vehicle manufacturing competitors. Also, it means that it is only USA has effort to export (AI) auto non manual control vehicles to global (AI) auto non manual control vehicle market.

Thus, in long term, (AI) non manual control vehicle product market development, USA (AI) vehicle manufacturers will have these requirement to win new technological competition to traditional manual control vehicle. The requirements include: low cost fuel, low carbon, fuel cell and telematics technologies, the technological roadmap function, such as detailed consideration clear provision of other important areas to the drivers. When the (AI) non manual auto drivers are sitting in the non manual control auto vehicle. Although, who does not need to drive, but who need to know how to go to anywhere by electric road map show clearly. So, the driver won't lose direction and he/she can know the (AI) non manual control vehicles is driving to anywhere in any time, even when who is sleeping.

In the future, the (AI) non manual control vehicles need to be invented to satisfy any business , transportation clients' needs, instead of individual clients needs, e.g. cans, trucks, buses, emergency and utility vehicles, trains, trams etc. Hence, technological road mapping is one important tool to help any business, transportation (AI) non manual control vehicle clients. Technological roadmap is a technique that is used in industry to support strategic planning for (AI) non manual driving vehicles in the future. Electronic road maps generally take the form of multi-layered time based charts, linking technology developments to future (AI) non manual control vehicle market requirements.

Technology road mapping is a flexible technique and the roadmap architecture and process for developing the roadmap most generally be customized to meeting the particular aims. Why technology roadmap will be popular to (AI_ non manual control vehicles. It's advantages include: It is a technology solutions and options that can enable the performance targets to be achieved engine hybrid, electric and alternatively fueled vehicles, software, sensors electronics and telematics, structures and materials design and manufacturing process. It is road transport system performance measures and targets tool, in response to the trends and get (AI) non manual control vehicle drivers to get society, economy, environment protection, low cost driving benefits, also it can help any transportation clients to know how to go to anywhere clearly. Hence, technological road map will be one good tool to assist (AI) non manual control auto vehicle to develop future road driving market.

Reference(source)

Frey & Osborne (2013), The future of employment: How susceptible are jobs to computerization? University of Oxford.

Mckinsey Global Institute Analysis

Statistics Denmark, Global automation impact model, Makinsey analysis

(AI) development second stage

(AI) -driven automation industry development

(AI) -driven automation industry will create wealth and expand economy growth to any countries, but it will be accompanied by changed in the skills that workers need to learn. One of main ways that technology increases productivity is by decreasing the number of labor hours needed to create a unit of output. It implies (AI) technology will influence low educated and low skillful labor number to be decreased (reduction employment number).

In contrast, technological change tended to work in a different direction throughout the nowadays. The advance of computer and the internet raised the relative productivity of higher skilled workers. So, routine-intensive occupations that focused on predictable tasks disappearance, such as switch board, operators, filming checkers, travel agents and assembling line workers etc. were particularly replaced by new technologies.

However, today, it may be challenging to predict exactly which jobs will be most immediately affected by (AI) driven-automation. The reason is because (AI) is not a single technology, but rather a collection of technologies that are felt unevenly through the economy to influence job changing both negatively and positively. In positively view point, (AI) driven-automation will make many workers more productive and increase demand for certain skills. Consequently, new jobs are likely to be directly create in areas , such as the development and supervision of (AI) as well as indirectly created in a range of areas throughout the economy as higher incomes lead to expanded demand. Otherwise, in negatively view point, many traditional human needed (demand) skillful jobs will be threatened by automation are highly concentrated among lower-paid, lower-skilled and less -educated workers. It means automation will cause pressure on demand for this group, pressure and employment, if (AI) can replace the low skilled and less educated workers' jobs. Thus, (AI) will have negative influence to impact on the labor market.

(AI) capabilities will enable automation of some tasks that have long required human labor. Why can (AI) replace some simple human jobs? For example, advances in robotics are expanding machines' abilities to interact with and sharp the physical world. Combined , (AI) and robotics will give rise to smarter machines that can perform more sophisticated functions than ever before and brings more advantages that humans have exercised. This will permit automation of many tasks now performed by human workers and could change the shape of the labor market and human activity.

How (AI) influences labor market

Today, it may be challenging to predict exactly which jobs will be most immediately affected by (AI)-driven automation. Because (AI) is not a single technology, but rather a collection of technologies that are applied to specific tasks.

Some specific predictions are possible based on the current (AI) technology. For example, driving jobs and house cleaning jobs, bank counter service jobs, telephone enquiry service operators. Restaurant cooking jobs, simple accounting record service jobs etc. that require relatively less education to perform. Advancements in computer vision and related technologies have made the feasibility of fully appear more likely, potentially displacing some workers in driving-dominant professions. Seemingly similar robot, for which the operational tasks is less specific of navigating to a specific destination when following a set of given rules and preserving safety.

In the future, the effects of (AI) on the labor market in the decade ahead will continue the trend toward skill-biased change that computerization and communication innovations have driven in recent decades. Thus, some human driving occupation will be disappeared or replaced by (AI) automation driven. For example, bus drivers, light truck or delivery services drivers, heavy and tractor-trailer truck drivers, school drivers, tax drivers, travel bus drivers.

However, (AI) technology could enable some workers to focus time on other job responsibilities, boosting their productivity, and actually raised wage growth among those still holding the reshaped jobs. For example, salespeople, who currently spend a considerable amount of time driving could find themselves able to do other work when a car drives them from place to place, or inspectors and appraisers could fill out paperwork, when their car drives itself. This (AI) -driven technology should make these workers more productive, with (AI) -driven technology serving as a complement, not a substitute. New jobs will also likely be created, both in existing occupations cheaper transportation costs with lower prices and increase demand for products and all the related occupations, such as service and fulfillment, and in new occupations not currently foreseeable.

What kind of jobs will be created by (AI) technology? Predicting future job growth is extremely difficult, due to it depends on technologies or substitute for existing today as well as they may complement or substitute for existing human skills and jobs. However, (AI) will also lead to substantial indirect job creation to the degree it raises productivity and wages, it may also lead to higher consumption that would support additional jobs from high-end draft production to restaurant and retail. The future(AI) " augmented intelligence", the technology's role is as assisting and expanding the productivity of individuals rather than replacing human work. Thus, based on the biased-technical change framework, demand for labor will likely increase the most in the areas where humans complement (AI) automation technologies. For example, (AI) technology , such as IBM's Watson may improve early detection of some cancers or other illnesses, but a human healthcare professional is needed to work with patients to understand and translate patients' symptoms, inform patients of treatment options, and guide patients through treatment plans. Shipping companies may also partner workers who pick up and deliver products over the last feet with (AI) enabled autonomous vehicles that move workers efficiently from site to site. In such cases, (AI) augments what a human is able to do and allows individuals to either be move effective in their specially task or to operate on a larger scale. Thus, it seems (AI) technology will also create new jobs, raise productivities and workers' efficiencies.

Redefining management in the workforce of artificial intelligence

In the future, due to artificial intelligence influences to some kind of human jobs nature. So, the kind of human jobs of management methods will also need to change to adapt the artificial intelligence technology input to their organizations. It will cause challenges for every executive and manager if who won't have effort to manage their teams how to apply artificial intelligence technology to work efficiently and easily. For example, division of labor will change among humans and machines will increase. Thus, companies will have to adapt their training performance and talent strategies how to emphasize on work that how to make human judgment and skills and experimentation. Thus, (IA)'s greatest impact will be on administrative coordination and control tasks, such as scheduling , resource allocation.

In fact, mangers will encounter this challenges: How to apply human experience and expertise to judge critical business decisions and practices when the information available is insufficient to suggest a successful course of action? Due to this kind of work will require new skills and mindsets. I shall indicate these change management methods to adapt (AI) technology. Such as: administration and routine tasks, scheduling , allocation of resources and reporting will fall within the intelligence machines, responsibilities that have long been reserved for humans. For example, a typical store manager or a lead nurse at a nursing home most constantly arrange shift schedules, accounting for staff members' absences owing to illness, vacation time or sudden departures.

Thus, the managers need to learn how to arrange new division of labor within the organizations after (AI) technology had been implemented to the organization. Artificial intelligence is currently influencing into once considered exclusive to humans: assessing and acting on human emotions and personality traits. The influences to managers need to change their strategies to adapt (AI) technology implements include such as below:

Firstly, managers need to spend the bulk of their time on coordination and control tasks from intelligent system implements. Their time spending on these major three aspects from impact of intelligent system: coordinate and control, solve problems and collaborate and people and community , strategy and innovation three aspects. Thus (AI) will influence managers need to change their judgment method to teach whose teams how to adapt the (AI) system operations in any organizations.

Secondly, (AI) will influence top, middle and low level management needs to change to adapt the (AI) technology operations to any owned (AI) technology organizations in the future. Intelligent machines must be trained in context. Just like humans , on-the-job training is a requirement for such machines because they typically arrive with only very general capabilities. To get the most from (AI), managers at all levels must participate in the instructional experience and in the learning process and provides managers' familiarity with such systems on these aspects, e.g. How the system works and generate advice, how the system has a proven track record , how the system provides convincing explanations , how the system can make simple rule- based decisions.

Thirdly, managers need to learn how to make judgment more accurate (AI) systems assistance. Although (AI) will invariably take on more routine work and even augment human decision-making, it won't judgment work, the application of human experience and expertise to critical business decisions when the information available is insufficient to suggest a successful course of action or reliable enough to suggest an obvious course of action. For a sense of the nature of judgment work, consider big data marketing and sales analytics. Such analytics often provide insights that can inform promotional campaigns, including predicting which promotions will generate desired sales brand further into the future, marketing executives need use judgment, combining analytics with their own and others' insight and experience.

The application of experience and expertise to critical business decisions and practice represents the real value of human judgment. But, when artificial intelligent machines are implemented to any organizations to assist the low, middle and top level management to make any business judgment. These forms of judgment work that managers can gather data interpretation, idea development more absolute from (AI) machine assistance. Thus, why these level management executives need to learn how to apply (AI) machines to help them to make any business judgment more accurate.

How (AI) influences organizational change

Consequently creative and social intelligence will be in even greater demand as (AI) makes in management and the workforce. This development will represent a long term trend in labor markets , one characterized by intensifying demand and reward for social skills with a growing desire for creative capabilities, managers will seek to fashion of ideas and hypotheses from inside and outside of the enterprise to shape solutions to their most pressing business problems. Thus, (AI) will influence overall organizational team members who have chance to participate any decision to make more accurate business judgment.

Many managers mistakenly view judgment work as only an individual discipline, failing to appreciate that it can also involve decide interpersonal and organizational practices. In more complex settings, judgment is typically a collective outcome of individuals' and teams' diverse perspectives, insights and experiences. And often , the resulting choices are better informed than decisions that an individual would have arrived at on his or her own.

Thus, when any organizations apply (AI) technology to assist managers to gather data and ideas to make any judgment. In these cases, organizations can create the conditions for effective collective judgment by establishing structures , such as " shadow advisory boards" that prompt managers and employees to source and synthesize multiple perspectives. Thus, a traditional organization (firm) might freshen its thinking is t put together a shadow advisory board, comprised of young, digital people who can apply (AI) machine assistance to make judgment work more accurate whether related to people development, problem-solving or strategizing and innovating for considerable degrees of creative and social intelligence.

Thus, on the one hand, (AI) technology machine augmentation and automation can give these advantages to human (organization managers) , e.g. developing people and community, solving problems and collaborating, coordinating and controlling work, shaping strategy and leading innovation. Besides, on the other hand, the next generation managers need have these individual attitude to treat intelligent machines to be as colleagues.

When, judgment is a human skill, intelligent machines can accelerate human learning that supports it, assisting in data -driven simulations, scenarios and search and discovery activities. Focuses on judgment work, some decisions require insight beyond what data can tell them. This is the sweet sport for human judgment, the application of experience and expertise to critical business decisions and practices. Thus, managers will also need to find ways to learn how to use digital (AI) technologies to tap into the knowledge and judgment of partners, customer external stakeholders and role models in other industries after the (AI) machine had been implemented to the organization.

Future works change:
Automation, employment
and productivity

Human future " micro to macro" industry trends will be affected business strategy and public policy by (AI) technology. In the future (AI) technology will influence those six themes: productivity and growth, natural resources, labor markets, the evolution of global financial markets, the economic impact of technology and innovation and urbanization. However, (AI) technology will bring economic benefits of tackling gender inequality, a new global competition, Chinese innovation and digital globalization.

Nowadays, advances in robotics artificial intelligence, and machine learning are in a new age of automation, as machines match or outperform human performance in a development to any countries. For example, automation of activities can enable businesses to improve performance by reducing errors and improving quality and speed, and in some cases achieving outcomes that go beyond human capabilities. For example, some research indicated automation could raise productivity growth globally by 0.8 to 1.4 % annually; more than 2,000 work activities across 800 occupations. When less than 5% of all occupations can be automated using demonstrated technologies about 60% of all occupations have at least 30% of constituent activities that could be automated. Many occupations will change that will be automated away: Activities most susceptible to automation involve physical activities, in highly structured and predictable environments, as well as the collection and processing of data. They are most prevalent in manufacturing , accommodation and food service and retail trade and include some middle-skill jobs. For example, such as natural language processing is a key factor. Beyond technical feasibility, the cost of technology competition with labor including skills and supply and demand dynamics, performance benefits including and beyond labor cost savings, and social and regulatory acceptance will be affected by (AI) automation technology. Thus, (AI) automation will impact to influence global employment in those aspects as below:

Firstly, assuming that people are displaced by automation will find other employment. The anticipated shift in the activities in the labor force is of a similar order as the long-term shift away from agriculture and decreases in manufacturing share of employment. Both of manufacturing and agriculture industries which would be accompanied by the creation of new types of work not foreseen at the time.

Secondly, for business, the performance benefits of automation are relatively clear. Thus, the businessmen have opportunities for their micro economies to benefits from the productivity growth potential and macro economies to benefit to encourage continued progress and innovation , investment and market incentives. At the same time, employers must innovate policies to help workers and institutions adapt to the impact on employment.

This will likely include rethinking education and training, income support and safety nets , as well as support for those dislocated, when employees need to leave themselves homes to move to other cities to learn new (AI) automation works. Thus, individuals in the workplace will need to engage move comprehensively with machines as part of their everyday activities, and acquire new skills that will be in demand in the new automation age. Consequently , the scale of shifts in the labor force over many decades that automation technologies can be a similar order to the long -term technology -enables shifts in the developed countries‘ workforces away from agriculture in the 21 th century. Those shifts did not result in long-term mass unemployment because they were accompanied by the creation of new types of work not foreseen at the time. However, human will still be needed in the workforce when the total productivity gains are caused by (AI) technology.

What occupations will be influenced by (AI) technology.

In the future, scientists predict that these occupations will be influenced by (AI) technology mostly. They include : retail salespeople, food and beverage service workers, language or translation teachers, health practitioners. Since these work activities have a more relevant occupations are made up of a range of activities with different potential for (AI) automation . For example, a retail salesperson will spend more time interacting with customers, stocking shelves , or ringing up sales. Each of these activities is distinct and requires different capabilities to perform successfully.

Thus, these job activities have similar simple control characteristics. Simple activities include greet customers, answer questions about products and services, clean and maintain work areas, demonstrate product feature process sales and transactions. All these activities can have similar simple activities in order to (AI) machines can be learn how to do these activities from (AI) technology . For example, the capability perception includes sensory perception, cognitive capabilities, such as retrieving automation, recognizing known patterns(supervised learning), logical reasoning problem solving.

Thus, (AI) machine is such human, which has feeling and emotion, such as social and emotional sensing, judgement reasoning methods, natural language understanding and physical capabilities, such as mobility , navigation, gross motor skill, fine motor skills. It seems that the future, (AI) human invents machines which will have these human characteristics to do human similar behavioral job duties more easily and efficiently. It implies these above human occupations will be replaced by (AI) human invention machines in the future. Due to (AI) creation, it is possible to cause unemployment number of these above workers will increase because (AI) machines can do their similar job behavioral activities.

Consequently, employers won't need to employ many of these skillful labor. Otherwise, they can buy less number (AI) machines to attempt to do whose job activities more easily and efficiently. So, it seems (AI) machines will have more high work performance to replace these occupation workers' work performance. Finally, these occupation worker unemployment number will only increase when the (AI) machines had been invented to achieve to do their work behavioral activities absolutely success in the future.

Whether (A) technology machine labor will replace human worker more or assist human worker more

There is no single agreed definition of a robot how outcome of a task that is completed without human intervention. When some definitions require the task to be completed by a physical machine moves and respond to its environment, other definitions use the term robot in connection with tasks completed by software , without physical embodiment.

However, to answer the question : Whether (AI) technology machine labor will replace human worker more or assist human worker more. I shall indicate some examples to let readers to judge whether (AI) technology can create new jobs or reduce old jobs.

Firstly, I shall explain what (AI) function is. (AI) is a service robot that performs useful tasks for humans or equipment excluding industrial automation application . Thus, the classification of a robot into industrial robot or service robot is done according to its intended application. It is also a personal service robot or a service robot for personal used for a non commercial task, usually by lay persons . Examples are domestic servant robot, and pet exercising robot. It is also a professional service robot or a service robot for professional used for a commercial task, usually operated by a properly trained operator. Examples, are cleaning robot for public places, delivery robot in offices or hospitals, fire-fighting robot, rehabilitation robot and surgery robot in hospitals. Thus, these functions will be future (AI) application to our daily life necessaries or business necessaries.

However, some authors agree (AI) will bring negative outcomes of automation, due to raise competiveness, reduce human job nature. Otherwise, other authors argue (AI) will bring positive outcomes of automation, due to raise productivities, job creation, assist humans work.

On the positive outcome hand, robots can increase productivity . This is particularly important for small-to medium sized businesses both are in developed and developing countries economies. It also enables large companies to increase their competitiveness through faster product development and delivery. Increased use of robot is also enabling companies in high cost countries to re shore, or bring back to their domestic base parts of the supply chain that will have previously outsourced to sources of cheaper labor. Currently , the greater threat to employment is not a automation, but an inability to remain competitive. Automation has led overall to an increase in labor demand and positive impact on wages. The reason is that the middle-income/middle-skilled jobs have reduced as a proportion of overall contribution to employment and earnings leading to fears of increasing income inequality, the skills range within the middle income bracket is large. Thus, robots are driving an increase in demand for workers at the higher -skilled and with a positive impact on wages. This issue is how to enable middle-income earners in the lower-income range to unskilled or retain. Finally, the (AI) positive impact supporter who argue the future will be robots and humans can work together.

However, on the negative outcome hand, robots can substitute labor activities, but don't replace jobs. They believe that less than 10% of jobs are fully automatable. Increasingly , robots are used to complement and augment labor activities, the net impact on jobs and the quality of work is positive. Automation can provide the opportunity for humans to focus on higher-skilled, higher-quality and higher-paid tasks. Robots can improve productivity when they are applied to tasks that which perform more efficiently and to a higher and more consistent level of quality than humans. For example, increased productivity is enabling some firms, such as Whirlpool, Caterpillar and Ford Motors company in the US restructure their supply chains, bringing back parts of the manufacturing process to the country of origin. Thus, productivity gains due to robotics and automation are important not just at the company level, but also for build industry and nation competitiveness.

I suppose that productivity can be raised. What are the impacts of robots on employment? Firstly, the main focus of development has been on personal entertainment, which does not drive worker productivity (manufacturing production). When the internet (information and communication technology (ICT)) innovation. This is borne and by findings that manufacturing productivity, which has been driven by innovations in automation rather than consumer technologies, has government strongly than productivity in the services sectors of the economy in most nature economies. It seems (AI) automation will create many jobs in internet communication entertainment game industry. For example, many young people like to use internet to play any electronic games from computer or mobile at home or outside home conveniently. Thus, (AI) automation will increase demand to be invented to any new entertainment game from internet channel. It will need to employ many (AI) entertainment game inventors to create many automation entertainment games. Thus, (AI) automation in internet entertainment game industry will need human (AI) entertainment game inventors to invent the knowledge-based capital of (AI) automation entertainment games. The (AI) entertainment game inventors will need own research and development skills, form specific skills, organizational know-how skills, databased knowledge, design and various forms of intellectual property to do these (AI) automation entertainment game invention occupations in the future.

International Federation Of Robotics(2016) indicated that China will be as a major robotics manufacturer and user of robots, benefiting from jobs created by robot manufacturing and productivity gains from robot use. Chins had sold of robots to any one single market every year since 2017 year. The Chinese government has included a focus on robotics in its 10 year strategy. In order to achieve its target of a robot density of 150 units per 10, 000 workers by 2020 year. Thus, Chinese companies will have to install around 650,000 new industrial robots between 2016 to 2020

year, 2.5 times more than installed globally in 2015 year.
Hence, China (AI) manufacturing industry will need to employ many workers . It implies (AI) manufacturing industry will create many new occupations in China. Also, ministry of economy, trade and industry (2015) also showed that Japan currently has the largest stock of industrial robots in operations, primarily in the automation industry. Driven by a rapidly aging population and low productivity rates, the Japanese government has sights on a 20-fold increase in the use of robots in the non-manufacturing sector and a three-fold growth rate of labor productivity in the service sector both by 2020 year. Thus, it also implies Japan will need many robots to be provide to service industry. Due to robots will provide to serve any businessmen's clients. Thus, it is possible that the service workers won't be dismissed as well as it is depended on the serving job nature to decide whether Japan's service workers can still serve to their employer when the service (AI) robots are applied to whose employers.

Consequently, it seems that (AI) can create employment, Ministry of economy, trade and industry (2015) showed that such as China will develop the major (AI) automation manufacturing industry. The (AI) employers will need to employ many workers to manufacture any these different kinds of (AI) robots to satisfy China or overseas individual or business buyers needs. But, (AI) can also cause unemployment to the low skillful service workers. Such as if Japan some service businesses choose to buy any (AI) service robots to replace their service staffs to serve their clients. It is possible that the service staffs will be dismissed, due to (AI) robots can do such as their same service job duties to achieve better service performance.

Thus, today, it is increasingly common for people to use robots in various situations at home and in retail stores, hotels and hospitals these service industries. Robots are classified into server types based on their functionality (service and utility robots or those designed to communicate with humans) and appearance (humanoid robots or mechanical robots). The type of robot, to which each country allocated particular importance in the advance of robotics, reflects the sense of values and preferences of its population. Thus, if the country has high population needs to use robots, then they will influence either more new jobs creation or more old job loss in the country's (AI) manufacturing or (AI) service industries both. For example, Japan respondents often associate the term " robot " with humanoid robots that can communicate with human and they have a high level of familiarity with robot. The US has the highest level of robot utilization at home and in retail stores with its people being the most enthusiastic about the future use of robots. Germany shows a strong tendency to consider robots for industrial purposes and its people feel strong effort to the presence of robots in their households.

In conclusion, to judge whether how (AI) will influence the country's employment to be better or worse. It will depend on the country home buyers (users) or business buyers (users) how to use (AI) for their daily needs. If the country , such as US retail stores need to use (AI) , it will have possible to reduce some or many retail service workers. Even, if the country , such as Japan has many home users need to use (AI) , it will not influence the employment market. Otherwise, it will raise (AI) salespeople numbers. Even, if the country, such as Germany and China will have many (AI) manufacturers, then it will create many (AI) manufacturing occupations for these (AI) manufactory workers.

Consequently, (AI) robots manufacturing and service needs will have positive or negative impact to any country's employment. It will depend on the (AI) service provision and service workers' job nature as well as the manufacturing workers of (AI) knowledge level to decide their employment chance in their country's employment market.

What does artificial intelligence(AI) development stage mean

● What (AI) function is?

Some scientists explain that artificial intelligence means which is an expert system, computer software that embodies a portion of the specialized knowledge of a human portion in a specific, narrow domain, owns decision making ability of human expert. The (AI)technology is based on the premise that what makes a person an expert is years of experience that enables who recognizes certain patterns in a problem as being similar to pattern. For example, in the future artificial intelligence system can be applied to control air traffic, design to computer configuration, medical diagnosis, instruction/training, speech/interpretation, monitoring to (nuclear plant), planning to mission, factory scheduling, prediction weather, repairing telephone, automatic driving etc. different industries.

Artificial intelligence characteristics include: creative, adaptive , common sense, fact processing, quick replication, broad focus permanent and consistent skill. Otherwise, traditional computer expert system disadvantage includes perishable, unpredictable, slow reproduction, expensive, slow reproduction, slow processing lacks inspiration, needs instruction, narrow focus only machine knowledge. So, artificial intelligence is a branch of computer science devoted to creating computer to influence software and hardware to attempt to create human intelligence or human intelligent behavior. It is learning from experience, responds flexibility in situation that are, new or not anticipated.

Thus, (AI) can be learnt programmed knowledge to solve problems, using reasoning in solving problem, understanding and inferring facts and rules, recognizing the relative importance of different elements in a situation. In summary, artificial intelligence is concerned with two basic ideas mainly: The first idea, it involves studying the thought processes of humans to understand what intelligence is; the second idea, it deals with representing thought processes using companies to create artificially intelligent entities for testing the theories of intelligence.

● Can (AI) impact human job nature?

Human need concern this question: Will artificial intelligence (AI) reduce some human jobs in order to instead of replacing machines to do? Due to artificial intelligence is the ability of machines to do thing, that people would require intelligence. For example, artificial intelligence machine man driving(self-driver), it (AI) machine man driving research is an attempt to discover and describe aspects of human intelligence that can be simulated by driving machine functions. Alternatively, (AI) mathematical research may be another viewed as an attempt to develop a mathematical theory function to describe the abilities and actions of things (natural or man-made) exhibiting intelligent behavior and server as a design of intelligent calculation machine function.

Why do humans need artificial intelligence machines to instead of traditional human service job? For example, can artificial intelligence machine man (self-driving) driver drive to replace human driver? I shall compare the differences between humans and computers : The characteristics of humans are good at recognizing various things, either seen before or not, recognizing the relationship patterns between things. Human thinking is common sense reasoning, combining all types of sensory input, acting appropriately in novel situations, learning new things and changing behavior patterns, making decisions , even when given incomplete information, working with noisy, incomplete information gathering behaviors . However, characteristics of computers are good at: The tasks humans do naturally are extremely difficult for a computer program as intelligent, which must be able to do the same kind of tack as humans do naturally.

Hence, (AI) is an combination of many different success and technologies: Linguistics - computational and socio, philosophy-logic, philosophy of mind and of language, electronical engineering -image and speech processing, pattern recognition, robotics, machine learning, neural networks, optimization scheduling, management information system and decision making. So, it is possible that (AI) can impact human job nature to instead of human working behavior in the future.

● How can human society job nature
to be changed to artificial intelligent society?

From the first intelligent perspective reason view point, artificial intelligence is making machines " intelligent" acting as humans expect people to act. Artificial intelligence has ability to distinguish computer responses from human responses, it owns knowledge to solve expert problem. From another research perspective reason view point, artificial intelligence is the study of how to make computers do things which, at the moment, people do better (Rich & Knight, 1991, p.3).

(AI) researchers are native in a variety of domains, e.g. formal tasks (mathematics, games), tasks (perception, robotics, natural language, common sense reasoning), expert tasks (financial analysis, medical diagnostics, engineering, scientific analysis and other areas).

From the second business perspective reason view point, (AI) is a set of many powerful tools, and methodologies for using those tools to solve business problems. From a programming perspective reason view point, (AI) includes the study of symbolic programming problem solving and search .

From the third human technological perspective reason view point, today's computer can do many well-defined tasks, for example, arithmetic operations, are much faster and more accurate than human beings. However, the computers' interaction with their environment is not very sophisticated yet. How can human test whether a computer has reached the general intelligence level of a human being? Can a computer convince a human interrogator that it is a human? But before thinking of such advanced kinds of machines, human will start developing our own extremely simple " intelligent" machines.

So, it is possible that human society job nature will to be changed to artificial intelligent society when (AI) technology is developed to the mature stage in the future.

● Why does human need artificial intelligence machines?

One of major division in (AI) is between humans who think (AI) is the only serious way of finding out how we (

human) work and human who want companies to do very smart things, independently of how we (human) work. This is the important distinction between cognitive scientists vs engineers. One of another major division in (AI) is between symbolic (AI), which represents information through symbols and their relationships. Specific Algorithms are used to process these symbols to solve problems or deduce new knowledge and connectionist. So (AI) , which represents information in network. Biological processes underlying learning, task performance and problem solving are imitated from human mind behaviors.

Thus, it is possible that artificial intelligence machines can do the better judgicious behavior to compare human.

● How does artificial intelligence influence future working changing in automation employment and productivity aspects?

In the automation changing influence aspect, as companies increasingly use robots on production lines or algorithms to optimize their logistics manage inventory, any carry out other core business functions. Technological advances are creating a new automation age in which ever-smarter and more flexible machines will be deployed on an ever larger scale in the marketplace. However, researching artificial intelligence with how influences human working nature. We need to answer these questions: How will automation transform the workplace? What will the implications for employment? And what is likely to be its impact both on productivity in the global economy and on employment?

Advances in robotics, artificial intelligence, and machine learning are growing in a new age of automation as machines match or outperform human performance in a range of work activities, including ones requiring cognitive capabilities. What factors are determined the changing in workplace adoption by artificial intelligence innovation? What advantages are automation? Automation of activities can be enabled businesses to improve performance by reducing errors and improving quality and speed, and achieving outcomes that go beyond human capabilities.

Some scientists indicated based on their scenario modeling. They estimated automation could raise producing growth globally by 0.8 to 1.4 percent annually. Almost, the activities people are paid almost $16 trillion in wages to do in global economy have the potential to be automated by adopting currently demonstrated technology. According to their analysis of more than 2,000 work activities across 800 occupations. When less than 5% of all occupations have of least 30% of activities that could be automated. They also indicated that technical economic and social factors will determine automation. Continued technical progress, for example, in areas such as natural language processing is a key factor beyond technical feasibility , the cost of technology, competition with labor including skills, and supply and demand dynamics, performance benefits including and beyond labor cost savings and social and regulatory acceptance will affect (alter) the scope of automation.

Other some scientists also indicate U.S. country for example, the anticipate shift in the activities in labor force of a similar order of magnitude as the long term sight away from agriculture and decreases in manufacturing. Share of employment in the United States both which were achieved. So, those factors can influence why artificial intelligence technology needs. So, it is possible that future agriculture and manufacturing both industries will apply (AI) technology manufacturer-kind of job nature to raise productivity instead of farmers, fruit picking workers, farming transportation labours as well as factory manufacturing workers and supervisors etc. human-kind of job nature.

- Is artificial intelligence possible to replace labor ?

Not just intelligence, but also debating, if machines are capable of having a conscious minds. Artificial intelligence has those characteristics as below:

On functionalism aspect, artificial intelligence inputs mental states, sensory inputs, (beliefs, desires being in pain feeling) and behavioral outputs. Since mental states are identified by a functional role, which are thoughts to be manifested in various systems. Even, perhaps computers which are physical devices with electronic substrate that inform computations on inputs to give outputs similar to brains which are artificial intelligence composed of part any intrinsic relationship to each other. Thus, artificial intelligence activities is not the whole itself, but into parts or on external influence on the parts.

On dualism aspect, artificial intelligence is a set of views about the relationship between mind are matter. On materialism aspect, it builds the only thing that exists is matter, including consciousness.

On biological naturalism aspect, it is similar a human brain than feels pains makes mental situation. So, artificial intelligence is similar biologist which might to be excited to human labor work.

Hence, it seems artificial intelligence can change (alter) or replace human labor work of nature in possible in the future.

- Can (AI) technology replace human labour nature of work?

On technological innovation reason view point, the history development of artificial intelligence studying the intelligence is one of most ancient scientific discipline. The history development of artificial intelligence what aims to achieve human use to sense, learn remember and think, logic probability, decision making and calculation develop from mathematics, instead of replacement human labor functions.

Artificial intelligence history development aim is the scientific analysis of skills in connection and practice with the appearance of computers from 1950 year beginning. The artificial intelligence (AI) can deal with the ultimate challenges. How can (either biological or electronic) mind sense, understand and manipulate a world that is much simple and more complex than itself? And what if would human like to construct something with such capabilities?

The general-purpose software of the early period of (AI) were only able to solve simple tasks effectively and failed when which should be used in a wider range or an more difficult tasks. One of the sources of difficulty was that early software had very few or mix knowledge about the problems which handled, and activities successes by simply syntactic manipulation. Moreover, the other difficulty was that many problems that were tried to solve by the (AI) were untreatable.

The early (AI) software whether trying step sequences based on the basic facts about the problem that should be solved, experimented with different combinations till which found a solution. From the end the 1960 year, developing the so-called expert systems were emphasized. These systems had (sue-based) knowledge base about the field which handled. Till to the beginning of the 1970 year, (Prolog) the logical programming language was born, which was built in the computation realization of a version of the resolution calculus. (Prolog) is a remarkably prevalent tool in developing expert systems (on medical, judiciary and other scopes), but natural language parsers were implemented in this language. Then, in 1981 s, the Japanese announced the fifth generation computer system project, a 10 years plan to build an intelligent computer system that use the (Prolog) language as a machine code. Nowadays, (AI) can be applied any industries, such as car manufacturing industry can use (AI) technological machine-men manufacture car, instead of replacing human labors in factory. Even, in the future, using (AI) machine-men drivers can drive any private cars or public transportation tools, instead of replacing human drivers, e.g. bus, train, tram, ferry etc. Also in the future, machine-men can replace housewives to serve families to do housekeeping clean job , e.g. cleaning toilets, bathrooms, kitchens, even cooking functions at home. So (AI) machine-man can reduce housewives works at home. Moreover, (AI) machine man can take care old people , when who are living at homes or elder care centers.

So, it seems artificial intelligence (AI) will be possible developed to manufacture a new generation machine-man to assist (serve) families to do any simply cleaning or cooking jobs at homes. Moreover, the overall demand of (AI) general social needs will also rise, such as security, driving transportation tools, restaurant cleaning, elder centers care service etc. So, it seems that individual or families or social needs of (AI) will be increase in the future. Thus, it will influence macro economy growth (GDP) if there are large house family consumer group and hotel or bus or taxis or ferry etc. different business consumer group demand any artificial intelligence machine numbers increasing. Then, the artificial intelligence products and material manufacturers must need to buy many artificaial intelligence materials to produce any kinds of artificial intelligence machines to prepare to satisfy consumer individual needs. Consequently, macro economy will grow to the owned artificial intelligence development countries, e.g. US, China, UK.

- Why can artificial intelligence satisfy human needs?

First, On machine-man satisfactory demand aspect view point, it makes computers that think, it is the automation of activities. We associate with human thinking: like decision making, learning. It is the act of creating machine that perform function that require intelligence when performed by people. It is the study of mental faculties through the use of computational models. It is the study of computations that make it possible to perceive, reason and act. It is a branch of computer science that is concerned with the automation of intelligent behavior. It is anything in computing service that human don't yet know how to do property.

Second, on thought aspect artificial intelligence means systems thank think like humans, systems that think rationally.

Third, on behavioral aspect, artificial intelligence systems that act like human and that systems act rationally. However, the basic objective of (AI) is to represent human's thought processes in computation . These machines are supposed to exhibit behavior that. It is performed by a human being, would be considered intelligent. However, some authors feel (AI) has disadvantages, such as it is not creative, it is excited in the use of sensory devices, it can't make use of a very wide context of experiences and it does not use common sense.

For speech recognition and understanding function needs example, (AI) can be applied in speech recognition and understanding function, which (AI) speech or voice recognition is a data input method. For example, the computer recognizes and understands one (or a few) word commands. Speech understanding on the other hand is the computer's ability to understanding a spoken language. That is , the computer understands the meaning of sentences, an paragraphs through (AI).

So, (AI) can be attempted to learn human language how to speak. It is similar to translate human language skill, instead of actual human speaking skill. Also, (AI) can assist handicap learning or language student how to listen different languages by machine-man sounds from computers more accurately.

So, it seems that it (AI) can replace human language teachers speaking function and can change teaching language nature of job in language speaking and listening education industry.

- Is artificial intelligence one good choice for human future technological benefit?

Nowadays, new technology development is popular. However, artificial intelligence is one kind of new technology choice among different technologies innovation. So it brings this question: Is artificial intelligence technology value to invest? To answer this question. I shall indicate some other new technology developments to compare (AI) technology development to judge which has urgent needs to achieve human expectation nowadays.

For example, why is green peace interested in new technologies? New technologies features prominently in our ongoing campaigns against genetic modified crops and number power. However, which are also an integral part of our solutions to environmental challenges, including renewable energy technologies, such as solar, wind and wave (water) power energy as well as waste treatment technologies, such as mechanical, biological treatment.

It seems humans need concern how to apply (AI) technology to solve environment pollution challenges in our future. So, environment protective, agriculture, natural energy technology will be popular demand to attempt to apply (AI) technology to solve their challenges or apply (AI) to assist to develop their industry.

How AI development stage to influence
economy growth

How can artificial intelligence technology influence economy?

Advances in artificial intelligence (AI) technology and related fields have opened up new markets and new opportunities progress in critical areas, such as health, education, energy, economic development, social welfare and the environment pollution.

(AI) automation will continue to create wealth and expand the global economy development in the future. However, when many will benefits that growth won't be costless and will be accompanied by changes in the skills, that workers need to increase productivity in the economy and structural changes in the economy. So, in the skills that workers need to succeed in the economy and structural changes.

I shall indicate why aggressive policy action will be needed to help Americans who are disadvantaged by these changes , due to (AI) technology is caused. For automation industry change example, artificial intelligence (AI) capabilities will enable automation of some tasks that have long required human labor. These artificial intelligence technology introduction can increase new opportunities for individuals. The economy and society, but (AI) has also the potential to disrupt be current livelihoods of many Americans. However, (AI) leads to unemployment and increase in inequality over the long run depends not only on the (AI) technology itself, but also on the institutions and policies that are changed.

Thus, it is possible that (AI) technology will raise some countries unemployment number if the employer apply (AI) technology workers to work instead of human labor in their factories, but it can also raise productivities for these employers.

- Can (AI) influence global economy growth?

Technological progress is main driver of growth of GDP per capita, allowing output to increase faster than labor and capital . However, technology can increase productivity, but also decrease the number of labor hours needed to create a unit of output. So (AI) causes unequal to labor wage decreases, even reduces the number of labor to manufacture, e.g. artificial intelligence technology of automation car manufacturing industry; clothing manufacturing industry; plane manufacturing etc. high technology of artificial intelligence manufacturing method. But (AI) should be potential environment benefit, although it raises unemployment ratio. Moreover, it can rise production , due to many skilled craft were replaced by the combination of machines and lower-skilled labor. The result of (AI) technology introduction , it causes output per hour risen when inequality declined, driving up average living standards, but the labor of some high-skill workers was no longer as valuable in the market. Otherwise, if (AI) technology is continue developed to be success. Some routine intensive occupations will be loss, which focused on predictable, e.g. easily programmable tasks, such as switchboard operators, filing clerks, travel agents, and assembly line workers would be particularly replaced by new (AI) technology. However, at the same time, (AI) technology development will bring these benefits: improvement in education (training (AI) technology scientists) , due to (AI) manufacturing technology needs are raising to businesses and institutional changes, such as the reduction in unionization and raising in the minimum wage to the (AI) manufacturing technology skilled labor in factories.

Because (AI) technology is not a single technology, but rather a collection of technologies that are applied to specific tasks, the effects of (AI) will be felt unevenly though the economy. It will bring some tasks will be most easily automated than others , and some jobs will be affected more than others, both negatively and positively. Finally, new jobs are likely to be directly created in areas , such as the development and supervision of (AI) as well as indirectly created in a range areas though out the economy as higher incomes lead to expanded demand.

However, if (AI) technology could dominate global labor markets. If labor productivity increases, do not influence into wage increases, then the large economic gains brought about by (AI) technology could be increased wealth inequality, due to employers can reduce production cost, but workers (labors) wages will not be increased, even will be decreased. Hence, it seems the (AI) technology will bring disadvantages to labor market to cause unemployment or reduce wages in possible, although it can reduce employer individual salary (wage) expenditure and it can raise productivity.

- How can artificial intelligence impact global economy growth?

Artificial intelligence (AI) technology is a branch of computer science that aims to create intelligent machines that work and react like humans. So, (AI) is a technology that appears to impact (influence) human preference by learning, understanding complex contents, enhancing humans in executing both routine and non-routine tasks. In the future, (AI) technology that can be virtual personal assistant, as well as it may exist, such as robots with human-like processing capabilities.

How can (AI) technology impact global economy growth over the next 10 years? During this time period, (AI) technology is predicted to have wide-ranging applications including: Machine learning that automates analytical model building by using algorithms that allow machines to operate without human assistance.

In global education aspect, potential applications include predicting cause-and-effect relationships from biological data, identifying new drugs, self-driving cars, and protecting against fraud, improved natural language processing that allows computers to continue to better analysis, understand and generate language to interface with humans using natural human languages. For example, transcribing notes dictated by physicians, automatically drafting articles and translating text and speech. So (AI) technology can be applied to education aspect to improve humans' knowledge level.

In visual art aspect, (AI) machine vision that allows computers to identify objects, scenes and activities in images. Current applications of (AI) machine vision include providing objective descriptions for the blind seeing(visual) needs.

We except the economic effects of (AI) technology to include both direct GDP growth from sectors that develop or manufacture. (AI) technology and indirect GDP growth through increased productivity in existing sectors that employ some form of (AI). If (AI) technology is an increasingly critical component of more products, it will become an integral part of many people's lives. Thus, (AI)'s ability to influence economic activity, rather than the economic or development status of the region. (AI) has the potential to impact income classes and to bring significant gains to both developed and developing countries. For example, (AI) has the potential to optimize good production around the world by analyzing agricultural regions and identifying what is necessary to improve crop yields.

In estimating the future economic effects by (AI) technology innovation, it is important to note that it is challenging to accurately predict which applications of (AI) will ultimately be commercially successful. In micro level economic influence, we need to apply methodologies to estimate the economic effects of investment in firms developing (AI) technology since investment levels in a technology are a telling sign of the future potential of that (AI) technology.

- How can (AI) influence GDP of high income countries in the next ten years?

How (AI)'s development may affect the global economy over the next ten years. In fact, (AI) technology has the potential to affect business across the global in a wide range of industries in ways only a number of technologies have done in the parts. For example, (AI) technology's expected to be a useful tool for enhancing human capabilities and in some instances replacing functions, such as driving a car, adoption of broadband internet, mobile telephone, industrial robotic automation have served to enhance human capabilities.

However, significant public debate has focused on projections of (AI) technology's effect on the labor force. However, large companies prefer to invest in (AI) technological industry. For example, face book's (AI) research lab., google machine intelligence lab. and micro soft machine learning and artificial intelligence research division are all making advances in (AI) technology and investing in the industry's top talent. Additionally, between 2010 year and 2015 year, nearly $5 billion in venture capital funding invested in firms across the global developing and employing (AI) technology (Facebook (AI) Research).

- How can artificial intelligence impact on workplace?

Modern information technologies and the labor economy growth of machines is powered by artificial intelligence have already strongly influenced the world of work in the 21 ST century. Computers, algorithms and software simplify every tasks and it is impossible to image how most of our life could be managed without them. How can be the information economy characterized by exponential growth replaces the most production industry based on economy of scales? What will the future world of work look like and how long will it take to get? Will the future world of work be a world where humans spend less time earning their livelihood? Alternatively, are mass unemployment, mass poverty and social distortions also possible scenario for the future, where robots, artificial intelligence systems play an increasingly central role? These questions concern how artificial intelligence further development . Can influence labor economy growth on workplace ? When the labor market has widespread impact on intelligence property, information technology, product liability, competition and labor and employment laws.

How (AI) technology impacts on labor workplace.

The future influence any organizations how labor economies use of (AI) can be analyzed, such as deep machine learning is based on a set of model high level data. Unlike human workers, the machines are connected the whole time in workplace. If one machine makes a mistake, all autonomous systems will keep this in mind and will avoid the same mistake the next time.

Over the long run intelligent machines will win against every human expert. Production robots have been replacing employees because of the (AI) technology. They work more precisely than humans and cost loss. Creative solutions like 3D printers and the self learning ability of these production robots will replace human workers, the automatic data recording and data processing, traditional back office activities are no longer in demand. Autonomous software will collect necessary information and will send it to the employee who needs it. Additionally, dematerialization leads to the phenomenon that traditional physical products are becoming software. For example, CD or DVDs are being replaced by streaming services. The replacement of traditional event ticket, e-travel ticket service products or hard cash will be the next step, due to the possibility of payment by smartphone. So, (AI) technology will impact human's daily life consumption behaviors in the future. For another example, transportation tools, such as boats and ferries and private vehicles will use sensors and navigating without human input. Taxi and truck drivers will become obsolete, the stock store applies to stock managers and postal carriers of the delivery is distributed by (AI) machine delivery method.

What is the relationship between (AI) and (CRM)?

- Can (AI) technology impact on customer relationship management (CRM) ?

Nowadays , (AI) is a technology almost as old as the computer industry itself, it is similar with the advent of personal assistants function to businesses and personal promotion channel, such as (Amazon's Alexa, Apple's Siri, Google's Assistant) image recognition (face book), personalized recommendations (Netflix , Amazon). Those innovations have been driven by a increase in processing power, lower cost hardware, and the exploding creation and availability of data. It seems, (AI) technology can impact global customer service management method.

How to forecast economic impact modeling to (AI) will affect global economy? Can human forecast business revenue growth and job creation (or destruction) based on (AI) applied to customer relationship management (CRM) activities? In addition to the economic impact on (AI) or (CRM) which can include an estimate of the economic impact attributable to sales forces customer base. What can economic benefits be brought to (CRM) from (AI) technology?

Artificial intelligence(AI) comprises a set of technologies that use natural language processing, machine learning, knowledge graphs, and other tools to answer questions, discover insights and provide recommendations. Computer systems can use (AI) hypothesize and formulate possible answers based on available evidence can be trained through the ingestion of vast amounts of content, and automatically adapt and learn from (AI) self mistakes and failures.

So, any business organizations (customer service departments) can provide efficient and effective customer relationship management of excellent customer service quality if which applied (AI) technology system. The different type of (AI) systems include: (AI) system platforms, machine learning (AI) based data preparation and enrichment tools, machine vision/image recognition, voice speech recognition, text analysis and natural language processing, bots , e.g. face book website and virtual digital assistance solutions, social media pattern analysis , sentiment analysis, advanced numerical analysis (e.g. IOT streaming , machine logs), supporting technologies, knowledge base dialog management, Q&A processing etc. different (AI) technology system customer relationship management (CRM) tools.

(AI) (CRM) of activity can include these categories, such as: corporate marketing, marketing operation, field marketing, customer support, digital commerce, customer analytics, customer influenced product or service design, product or service pricing, finance information, presentation, customer billing, inventory , logistics and fulfilment support, partner management etc. different CRM tools.

(AI) technology of CRM has been carrying on plan different stages to achieve CRM personal assistant tool for businesses. The stages are such as, in the beginning stage of (AI) projects in place, implement now, pilot phase next year in the final stage of (AI) customer relationship management tools are foreseeable future. So, this CRM technology has been improved to plan in different stages every year to prepare to achieve full capacity of CRM service quality for businesses to use in the future.

Hence, how to develop an estimate prediction of the economic impact (AI) technologies could have CRM activities, which depends on gathering macroeconomic information on business revenue and the basic marketing of business revenue and the basic markup of business expenses by major functions (customer support, marketing and sales , production etc.)

An economic impact model that can gather data together and forecast the results how (AI) artificial intelligence technology brings (CRM) customer relationship management benefits to businesses, e.g. surveys investigation includes IT spending by sample countries, GDP and population estimates and forecasts, revenue per employee and ratios of IT spend to GDP. Surveys (questionnaire questions) of forecast results are influenced by (AI) impact can include: results are projected from surveys and rely on estimates are made by respondents on the expected financial improvements in categories of (AI) –assisted customer relationship management activities. The forecast assumes that these estimates are correct; financial estimates are based on estimates of "first year" improvement from full (AI) implementation; forecasts are from planning to implement any artificial intelligence of customer relationship management (CRM) projects, the improvement forecast is of categories of activity , e.g. corporate marketing , digital commerce, and customer analytics. They are not estimates of ROI for the (AI) software. They rely on conservative estimates to which each of these entities might affect company revenue, expenses or productivity. They also rely on estimates of the penetration of software in customer relationship management activities . Net new jobs created are based on the ratio of new revenue to jobs required to support that revenue . They can assume that 50% of the net new revenue will support increases in labor and the rest will go for capital and other operating expenses that may replace jobs lost to automation.

In the future, some of the ways in micro economic benefits to any organizations. (AI) technology is expected to impact CRM activities include: Spending up sales cycles, improving lead generation and qualification solving customer support problems faster (raising service quality), helping companies improve brand campaigns and recognition, lowering costs of support calls when increasing resolution rates, lowering the cost of recruiting employees and partners, increasing revenue from optimized product marketing, optimizing price, distribution logistics and preventing loss through fraud detection. So, micro economic benefits view point, it seems that (AI) CRM technology can raise any companies economic benefits for care term.

Artificial intelligence enables machines or the in-build software to behave like human beings which allows these decisions and act. The advent of (AI) is leading , talking, making decisions and act. The advent of (AI) is leading to new technologies advances and transforming the economic and employment opportunities for humans in a positive way. (AI) related technologies can facilitate our live. For example, industrial robotics, robotic medical assistants, smart games, financial forecasting software, big data analysis, algorithms in health and bioinformatics, pilotless cargo places, drone ambulances and general purpose and workplace robots and others. (Disruptors technologies: Advances that will transform life, business and the global economy).

Artificial intelligence also known as computational intelligence is defined as " the human –like intelligence exhibited by machines or software. It is theorized that intelligence of humans can be described and intelligence machines or software can simulate it. These machines software can be reasonable , learn, perceive and process information, like human mind and thus facilitate human life. They can think and act for us. So, artificial intelligence is an interdisciplinary field of study including computer science, neuroscience, psychology, linguistics and philosophy. However, (AI) research and developments have economically impacted many industries, such as robotics, telecommunications, computer applications , health, finance, heavy manufacturing, transportation, aviation, e-service and e-commerce, military , music and movie, toys and games entertainment etc. industries.

In fact, many ideas, systems and technologies have been developing in the world of (AI) technology. However, which are net called or considered (AI) products, rather which are mentioned with their specific names, such as smart graphics, machine learning, e-commerce etc. (i.e. this is called (AI) effect).

- How can (AI) technology influence digital economy?

Nowadays, (AI) related industrial applications will replace most human power in fields, including call centers, customer services and air cargo transportation. (AI) technologies also help weather forecasting based on repeated rainfall pattern (data) recognition, through robotics (i.e. floor cleaning, moving lawns etc.) transporting people and products with unmanned vehicles, sending space unmanned smart shuttles, developing robotic arms, predicting market values in stock exchanges by internet, making homes safer, helping elderly and disabled using robotic servants etc.

Among the (AI) related technologies , there are a few that significance for the impact on society and especially on digital economy . (AI) is particularly influential in machine learning. Such as robotics, transportation, finance, health and bioinformatics, e-commerce , e-games, big online data gathering and internet-of-things. For example, machine e-learning is based in bioinformatics and robots that can learn new skills for better caregiving in healthcare. What is machine e-learning? Machines can e-learn from e-data gathering, coming up generalizations and making decisions to act in certain ways from internet.

There are important applications , such as e-machine perception, electronic online natural language learning processing, online search engines, online bioinformatics, online brain –computer interface, online game playing, online robot locomotion, online advertising, online computations finances, online health monitoring, online DNA classification and decision making, online in chemistry –cheminformatics . So, online machine learning can positively impact productivity and it can enhance information and analytical system from (AI) online channel.

What is robotics? Robotics is one of the most strongly influenced fields in (AI). For example, heavy manufacturing industries, robots and used and man power is replaced for effectiveness, precision, and accuracy, especially in respective or dangerous tasks, including welding, assembling , picking and placing .

So, robots can acquire new skills or adapt the changing dynamic environment. Also, artificial intelligence can be applied in developing transportation. For example, automated vehicles, driver assistance systems , safety systems, collision avoidance systems and public transportation. Moreover, (AI) technology has proven to produce some of the best tools to predict stock market fluctuations from internet data gathering method. It's predictions are based on ever-evolving predictions algorithms and systems learn new models and make connections between historical data and new data to measure stock market trading more accurate from internet data gathering channel.

In health field, especially in health data processing , analysis, decision making support and medical diagnosis. So, online data can show which patients will need what treatment and what alternative drugs could be used more accurate from (AI) online data gathering method. Bioinformatics is an interdisciplinary field combining statistics, (AI) online technology can help in discovering data patterns and modeling through the application of machine learning, artificial neural networks and genetic algorithms. For example, further (AI) technology development of human genome project of online data sequences.

Online shopping can be facilitated by virtual assistants developed through (AI) technology and these assistants can offer the best advice. (AI) online purchase coming after every product image recommendations and personalization bring important revenue to shopping online sites, like Amazon . Smart computer graphics and games, artificial intelligence is useful in smarter computer, graphics, scene modeling , scene rendering processes in order to create, for example, effective human –robot interactions , online machine learning, online strategic games techniques etc. online computer related (AI) software.

So, online big data analysis and big data does have a critical need in the world of online intelligence machines and software in our future. In other words, (AI) offers online technology to enable online big data analysis to provide industrial organizations with valuable information for effective decision making in short time. For example, what IBM's Watson achieved: this machine used 200 million of structured and unstructured content with a special technology of hypothesis generation, massive evidence gathering, analysis and scoring from internet channel.

Finally, (AI) online technology another related internet invention (internet of things) (IOT) is the network of machines or objects connected through internet. These connected objects can sense their internal and external environment, communicate with each other, can send critical data and finally can make decisions to act or correct their environment from (AI) online technology. For example, factories can monitor and automatically change production processes, hospitals can monitor and regulate the health conditions of their patients , schools can collect data from facilities and cars can send data to car makers from (AI) online technology.

Partner predicts that (IOT) market will create about trillion amount value by 2020 year. Although machines collect big data from their environment, whether which gain an insight or learn from these online data largely depends on the (AI) online machine learning principals and (AI) online technology. In 2013, Mckinsey estimated that disruptive technologies closely related with potential economic impact in 2025 year between $7.1 to $13.1 trillion amount (automation of knowledge work, advanced robotics, autonomous or near-autonomous vehicles).

What is the relationship between
(AI) and global digital economy development ?

● Could work activities in China be automated
making in the nation with the world's largest automation potential?

Can (AI) technology influence China economy? Could China workers be affected and jobs made up of routine work activities and predictable? Will programmable tasks be particularly impact to China employment market ?

When impact on labor market is likely to be gradual at the aggregate level, it can be sudden and dramatic at the level of specific work activities, rending some job obsolete fairly. Overall (AI) technology will raise digital skills when reducing demand for medium incomer inequality for China workers. It seems (AI) technology's effect on productivity could be crucial to China's future economic growth as the population ages are increasing.

In China, some biggest technological companies driving significant investments in research and development. Moreover, China is one of the leading global (AI) technology development county. However, China will need to focus on building its innovation capacity. For example, United States and United Kingdom are currently producing more influential (AI) technological research. However, if China planed to achieve (AI) technology success, it's traditional industries will need to develop technical know-how –to and overcoming implementation costs prepare to develop (AI) . When (AI) technology is introduced into China society, China government needs to raise concerning ethical, legal, technological security etc. business questions. Also, surrounding issues include privacy, discrimination, legal liability and regulation. It aims to encourage overseas investors to choose to invest (AI) technological industry to raise GDP growth and manufacturing industries income growth for long term in China.

If China encouraged overseas (AI) technology investment in its country. It is possible to influence China employment market to be changed. Because (AI) technology will impact to influence China people daily life. Due to (AI) technology is introduced to China society, many rich people will prefer to spend to buy any high (AI) technological products for entertainment or learning or machine man driving etc. daily necessity activities. Then it will raise GDP growth and will raise (AI) manufacturers or related-(AI) technological manufacturers profit. It is beneficial to China because it can become one high knowledgeable and (AI) technological economical society.

But it will bring bad influences to raise unemployment chance for the low skillful labor. In labor economy aspect influence , how (AI) technology can influence China low skillful labor unemployment ratio raising. The raising low skill labor unemployment reason is because China low skillful human labors are argued or are replaced by (AI) technology creating new challenges to introduce to influence China society of simply human manufacturing job nature to be changed to be high (AI) technology manufacturing job nature in any China factories. Moreover, when (AI) technology introduction to China, it will cause other related social challenges in China. The varied (AI) related challenges, including the difficulty of creating safe and reliable hardware for sensing and affecting (transportation and education), the challenges of gaining public trust, a low resource comities and public safety and security, the challenges of overcoming fears or marginalizing humans in China employment and workplace and the risk of diminishing interpersonal trust because the low skillful labors won't believe any China employers will give chance to employ them , due to (AI) technology will replace their skills and man manufacturing of productivity is much less to compare to (AI) technology manufacturing method.

- How does (AI) technology influence
the future of employment change?

Are future nature of jobs changed to computerization from (AI) technology? Where are the probability of computing occupations from (AI) technology influence? What is expected impacts of future computing on labor market from (AI) technology influence? John Maynard Keynes's frequently cited prediction of widespread technological unemployment " du to our discovery of means of economic the use of labor outrunning the pace of which we can find new used of labor" (Keynes, 1933, p.3).

In the future, (AI) technology will impact some nature of occupations to change computing. This chance will also influence some countries' economic change. For example, some factory human labors hand routine manufacturing tasks will be changed to computerization of routine manufacturing tasks by (AI) technological machine men hand manufacturing method. it will cause a structured shift in the labor market, with workers reallocating their labor supply from middle-income manufacturing to low-income service occupations.

Arguably, this is because the manual tasks of service occupations are less computerization, as who require a higher degree of flexibility and physical adaptability. So, (AI) technology will influence the human hand labor skillful occupation nature of task cheaper , such as vehicle manufacturing , ship manufacturing, computer manufacturing, steel manufacturing, television, radio etc. home electronic products of heavy machine industry change. Due to (AI) technology machine man will be proper to be used to manufacturing these electronic products when the (AI) technology innovation can develop to the mature stage. Then, any countries manufacturers will choose to use (AI) technology machine man, instead of human hand production.

Supposing the future prices of computing are fallen, seriously, problem solving skills are becoming relatively productive, explaining the substantial employment growth in manufacturing occupations, involving cognitive tasks where skilled labor has a comparative advantage, as well as the increase education needs for (AI) technology computing of machine man subject study.

Prediction of education needs for (AI) technology student numbers will increase, due to manufacturing industry needs many (AI) technology students in future employment market. Another (AI) technology influence if the future (AI) technological innovation, e.g. machine man manufacturing or machine man service industries will both increase demand, then with more sophistic software technologies will be disrupted labor markets by marketing workers redundant.

For publishing industry, what is striking about the case in paper book publishing industry will be unpopular? Due to the electronic book publishing industry will be popular, e.g. Amazon publish . (AI) technology can influence paper book manufacturing method which is replaced by machine man electronic book manufacturing method as well as it will cause the computerization is no longer confined to routine manufacturing tasks. Due to (AI) machine man manufacturing technology will be proper to be used to manufacture any products in short time efficiently and effectively , e.g. electronic book products. In the future, if it is fact to occur this case, such as (AI) technological machine man manufacturing method will be adopted (applied) to manufacture electronic books or any products in possible. (AI) technology will cause many manufacturing workers are unemployed. It is beneficial to employers, who can reduce to spend much wages expenditure to employ manufacturing workers, but it will cause many manufacturing workers loss jobs and reduce income to support whose families lives. It will cause social challenges, e.g. increasing stealing crimes if the manufacturing workers had not other skills to find other jobs to do easily. So, manufacturers need to concern over technological unemployment which will be hardly future phenomenon if who decided to dismiss all manufacturing workers, due to (AI) technology machine men replace to them.

If (AI) technology can be innovated to produce any kinds of machine man to serve any service or manufacturing industries successfully. Then, it will bring these questions: Can future that workers be influenced to be automation employment and productivity by (AI) technology influence? Does it impact to influence the (AI) technology countries‘ productivity and growth and natural resources development and labor markets and evolution of global financial markets and economic impact of technology and innovation and urbanization etc. issues? How will automation transform the workplace? What will be the implication for employment? What is likely to be its impact both on productivity in the global economy and on employment?

In fact, automatic of activities can enable businesses to improve performance by reducing errors chance and improving quality and speed, and same cases achieving outcomes that go beyond human capabilities. Some economists indicate (AI) technology would give a needed boost to economic growth and prosperity have of the working age population in many countries. Based on the scenario modeling, they estimate automation could raise productivity growth globally by 0.8 to 1.4 % annually. They also indicated that almost half the activities people are almost $1.6 trillion in wages to do in the global economy have the potential to be automated adapting current demonstrates technology, according to their analysis of more than 2,000 work activities across 800 occupations. When less than 5% of all occupations can be automated entirely using demonstrated technology, about 60% of all occupations have at least 30% of worker made activities, that would be automated. More occupation will change to be automated. They also indicated for business performance benefits of automation are relatively clear, but the issues are more complicated by policy making to attract foreign investors. Beyond technical feasibility, the cost of technology, competition labor will include skills and supply and demand dynamics, performance benefits and beyond labor cost savings and social and regulatory acceptance will affect the automation. Their predictions suggest that half of today work activities could be automated by 2055 year, but this could happen 10 to 20 years earlier or latter depending on the various factors in addition to their wider economic condition.

Some scientists suggest (AI) technology is finally starting to deliver real-life business benefits. Computer power is growing significantly , algorithms are becoming more sophisticated and perhaps most important of all, the world is generating vast quantities of the fuel that powers (AI) technology data billions of gigabytes of it every day. Also, online firms are digital natives, such as Google online search service company is investing on (AI) technology. For new though most of the news if coming from the suppliers of (AI) technologies. And many new users are only in the experimental phase. Few products are on the market or are likely to arrive these soon to drive immediate and widespread adoption. As a result, analysts believe (AI) technology's potential will give true economic benefit in the future. (AI) industry will introduce to suppliers and users to raise economic potential of (AI) technology.

In the future, (AI) technology systems can solve business problems. Some scientists categorized those into five technology systems that are key areas of (AI) technology development: robotics and autonomous vehicles, computer vision language virtual agents and machine learning , which is based on algorithms that learn from data without

replying on rules-based programming in order to draw conclusions or direct an action.
Such as computer vision and language includes natural language processing, analytics, speech recognition technology, some are about learning from information, such as about machine learning and others are related to acting on information, such as robotics, autonomous vehicles and virtual agents, which are computer programs that can converse with humans. Machine learning and a subfield called deep learning are artificial intelligence applications.

● Can artificial intelligence impact
global economy growth?

Artificial intelligence (AI) is a term first defined in 1956 year. It is a branch of computer science that aims to create intelligent machines that work and react like humans. In contrast today, 60 years later, (AI) is characterized by a number of applications, including computers playing games against humans and understanding human languages, virtual personal assistants, and robotics which involve computers seeing , hearing and reacting to sensory stimuli. In the future, technologists predict for (AI) technology ranging from (AI) being used as a tool to aid relatively simple processes for robots with human like mental capabilities, who expect (AI) technology can emulate human performance by learning, coming to mind its own conclusions, understanding complex content, engaging in dialog with people, enhancing human cognitive performance or replacing humans in executing both routine and non-routine tasks. In existing industry, (AI) technology is used , such as targeted advertising and virtual used personal assistant as well as the (AI) technology that my exist in the future, such as robots with human vehicle processing capabilities.

The range of (AI) technology's progress in the future will determine the economic impact future of (AI) technology on the global economy with more limited advances and applications (i.e. weak (AI) only) corresponding to more limited economic impacts and more substantial progress, i.e. strong (AI) technology is corresponding to more significant economic impact.

(AI) technology learning that automates analytical model, including predicting cause-and-effect relationship from biological data, identifying new drugs, self-driving cars and protecting against fraud etc. functions. Also (AI) learning can improve natural language processing that allows computers to continue to better analyze, understand and generate language to interface with human using the natural human language, virtual personal assistant, helps users by providing scheduling appointment, reminds organizing personal finance and finding providers of various services, machine vision allows (AI) machine man to identify object, scenes and activities in detect pedestrians and bicyclists.

We expect the economic effects of (AI) technology to include both direct GDP growth from sectors that develop or manufacture (AI) technology and indirect GDP growth through increased productivity in existing sectors that employ some from of (AI) technology. If (AI) producing sectors could grow, then it could lead to increase revenues and employment of (AI) technological professionals within these existing firms as well as the potential creation of entirely new economic activities to any countries' societies productivity improvement in existing sectors could be realized through faster and move efficient processes and decision making as well as increased (AI) technological knowledge and access to information available in societies easily.

In the future, if (AI) technology is an increasingly critical component of more products, it will become an integral part of necessary products of many people's lives. The extent of (AI)'s economy effort is also likely to vary from region to region, thought variation may be more dependent on the predominate economic activity of a region and the (AI) ability can influence economic activity, rather then the economic or developmental status of the regions. (AI) technology can move accessibility and can use source development to do international business between one country and another country.

So (AI) technology has the potential to give benefits to different income chooses and to bring significant gains to both developed and developing countries. For agricultural technology, (AI) has the potential to optimize food production around the world by analyzing agricultural regions and identifying what is necessary to improve crop yield. In total, (AI) technology gives greater economic impact to any countries agricultural regions if which implemented (AI) technology to grow crop , fruit etc. food production in the farms.

Investment in (AI) technology is such as capital investment to any countries' public or private enterprises. So, it will have large economic impact to the future . If the (AI) technology is reasonable invested to the different needs aspect by the public or private enterprises in the country. Then, it will have good economic impact to the country in the future. However, when (AI) technology is likely to affect both the productivity and employment components of economic growth in many sectors. Significant public debate has focused on projections of (AI)'s effect on the labor force. However, for instance, some researchers have argued that the rise of (AI) technology and automation will led to significant unemployment as capital is substituted for the low skillful labor. So, they point to the concern that the increasing sophistication of (AI) technology may balance skilled and semi-skilled workers and the reduce the size of the middle class. However, this is not a new argument, due to (AI) technology negatively affecting the labor force and leading to mass unemployment. Because the (AI) technology is the substitution of machinery for human labor. Although, employment in certain industries, has been reduced in the past due to technological advancement. For long term, the labor market has adapted to the introduction of new technology, giving rise to new jobs in new areas. (AI) technology may also be accomplished without a reduction to total employment in the long-term to some Asia countries, such as Hong Kong and Japan. Because Hong Kong and Japan many low skilled labor, e.g. security, cleaner who complaint that employers need them to work long time hours. (abnormal working hours) e.g. one day 12 to 15 working hour per day. Hence, if (AI) machine means invention technology success. Security or cleaning job can be worked from (AI) machine man in some hours every day in order to reduce the long time working hours cleaners or security workers, e.g. one (AI) machine man works 4 hours for cleaning or security job, one day as well as another cleaner or security labor only needs to work 8 hours one day. So total security or cleaning employers can employ 12 hours machine cleaners or security workers and human cleaners or security workers in one day. For long term benefit, Hong Kong or Japan every security or cleaning worker does not need to work 12 hours minimum working hours one day. They won't feel tried and bore and without private with whose families, so who will accept to do these cleaning or security jobs, even they can raise work efficient and performance when who feel happy and health.

So, (AI) technology of machine man invention can raise low skillful labor efficiency and it can help them to avoid abnormal working hours demand in some busy work life countries, such as Hong Kong and Japan. Before, one Japan female labor feel unhappy to work, due to who often needs to work abnormal working hours for her employer and who has less sleeping and without any private time to enjoy her life with her families every day. So this abnormal working hours factor causes her to do commit suicide behavior, then she is die unlucky. So (AI) technology of machine man invention ought avoid abnormal working hours demand for employer in any countries in the future.

The most important occurrence to any employers, some researchers had attempted to do one experiment to find that private research and development , venture capital and public research and development investment all have strong net effect or economic growth with venture capital funding further having the strongest such effect from (AI) technology. The researchers hypothesize the venture capital investment contributes to economic growth through (AI) technology innovation and by the capacity of an economy to use existing (AI) technology knowledge to increase productivity. They predict the impacts of venture capital, business-research and development and public research and development can raise multi factor productivity from (AI) technology introduction.

Can (AI) technology influence the economic development to developing countries? The developing regions of the world contain most of natural resources. If one day, (AI) technology has invent one kind of machine man which can assist any gas or oil workers to seek any new oil/gas natural resource locations easily. I believe that (AI) technology can help these natural resource exploitation countries will gain economic benefit more easily. So, (AI) driven technology can be used to change to create any new opportunities to address poor management or resources and improve human well being, such as Africa Latin America and India can use (AI) technology machine man to seek any oil/gas natural resource countries exploitation activities to attempt to gain much economic benefits.

● Why will (AI) technology grow economic
development ?

Nowadays, increases in capital and labor are no longer driving the levels of economic growth, such as (AI) technology. The ability of increase in capital investment and in labor of traditional drivers of production, have no longer to be enjoyed in most developed economies ,e.g. developed country, US, UK . However, artificial intelligence has the potential to overcome the physical limitation of capital and labor to avoid missing out on this opportunity. So, policy makers and business leaders must prepare for and work toward a future with artificial intelligence. They must do with the idea that (AI) is another simply method to enhance productivity method . Rather they must see (AI) as the tool that can transform thinking about how growth is created.

Economists have always thought of new technologies are as driving growth their ability to enhancing. It can replace labor and capital factor of production. So, it brings this question: What is the factor of production (AI) technology characteristics. They key factor is to see (AI) technology as a capital-labor .

(AI) can replicate labor activities at much greater scale and speed, and to even perform some tasks began the capabilities of human. For example, by using virtual assistants , 1000 legal documents can be reviewed in a matter of days instead of taking three people six moths to complete. Some (AI) technology may be one kind of factor of production in the future. For another example, people will work in workplace digitalization environment. So, in the future, working environment and information management are automated. Such as Konica camera sale company will use workplace digitalization. So , (AI) technology can provide workplace digitalization in order to raise productivity efficiency. (AI) technology will be one kind of production which is replaced by workplace digitalization and it will grow any organization productivity efficiently. Then, (AI) technology will assist overall social economy growth , due to productivity is raised and products can be produced in short time to prepare to sell in consumption market. So, time will be shortened to increase GDP growth fast for the development of (AI) technology countries.

● How can (AI) technology impact to global economic and social and psychological changes?

What will be the development of (AI) technology and predictions concerning the future evolution? The computers and robots will develop conscious, intelligent and minds into humans, enhancing psychological and behavioral abilities and allowing for direct communication with (AI) minds. (AI) technology will be impacted human life by (AI) technology information communicative and environmental influence. A " world brain" and " world mind", this psychological system will be enhanced and enriched the capacities of both individual and collective cognition by (AI) technology of service industries.

(AI) technology with influence these human needs of service industries changes, such as , biological science, finance, entertainment, business, biological science, transportation, communication military etc. The personal computer evolution, the internet and the world wide web which exploded on the scene, linking business, homes, schools, social organizations which were a completely unpredicted phenomenon to influence human life. Kurzweil (1999) predicts that by 2029 year, most human communication will be with machines. According to Person, by 2100 year, there will be human machine convergence.

How can (AI) technology influence environmental protection to make benefits to farming economic growth? (AI) technology can be applied to predict how to solve environmental pollution challenge to avoid to damage any crop or vegetable or rice or fruit etc. food growth. Because environmental experts can gather global environmental pollution data from an environmental database to build a perform a systematic analysis from (AI) technology. The first step is this broad analysis can include understanding, statistical and data gathering techniques to obtain the relevant data, the correlation among the variables involved, and a list of possible models. The next step is to select a set of methods and models that cover all kinds of knowledge and functionalities needed for the decision making process. Once the models are selected, they must be fully implemented by means of machine learning , data mining, statistical or numerical technique. After that, those models must be integrated to build the whole EDSS. The EDSS must be tested to check its performance, accuracy, usefulness and reliability, both from the user's and (AI) technology/computer scientist's point of view. If these is any wrong feature in any development stage, such as model's integration, models' implementation, selection of models, database, problem analysis etc. the developers must come back in the update th required components. When the evaluation phase is all right, the EDSS is ready to be applied to the environment. The great contribution of artificial intelligence to EDSS the integration of several methods complementing the classical statistical models/simulation , statistical analysis, linear models, etc. and numerical models (control algorithms, optimization techniques etc.) .

This cooperation makes the resulting systems more reliable and powerful in coping with real world environment systems. Date interpretation has been a principal area of research in (AI) technology since the very beginning. The most demanding problem in the environmental assessment context. Knowledge representation permits the definition of the different types of data that the existing methods adapt to the process. There is also a lot of work to clean, repair and transform the huge available quantities of raw data. Apart from this, the availability of meta-information or background knowledge is required to guide the process. Data mining is multi-disciplinary: It covers expert systems, data based technology, statistics, data visualization and unsupervised machine learning. These techniques operate at the level of data and background information, where numerous and often incompatible new commensurate pieces of information from disparate sources have to be brought together (K, Fedra, 1994).

So, it seems that in the future, (AI) technology with the increasing maturity in particular those related to knowledge and engineering, new dimensions can be assisted to users in environmental decision making are available. For example, many environmental systems are characterized both by incomplete models and by limited data. Hence, in the future, (AI) technology will be applied to predict climate change to reduce crop or fruit etc. food agriculture challenge by climate change bad influence.

- Will (AI) technology influence digital economy change to manufacturing industry ?

To understand how the manufacturing business must adapt to prosper in the technology, we need to understand how (AI) technology will change us to shape our daily habits to satisfy our expectation of products to how we shop and even the immediate of the entire process. For example, taxi services are in the crosshairs as on demand transportation services like, available of the touch of a smart phone button expand. In fact, Yellow lab, US country , san Francisco city's largest taxi company is filing for bankruptcy as the industry starts to change faster than almost anyone expected. However, at this point, its more than an app that is changing, some our taxi passengers renting taxi transportation to catch consumption behavior.

(AI) technology will influence digital economy for taxi passenger's individual customer experience, offering a growing renting taxi to catch of service and feedback opportunities when any one taxi passenger who chooses to use mobile phone app online tool to prepaid to rent any taxi more easily.

Also in the long term, (AI) technology can influence vehicles drive themselves of behavior. Already, companies like Google and GM are working on projects to bring fleets of autonomous vehicles to cities at the path of a button.
Moreover, this on-demand service model is beginning to appear across a much broader range of markets. For example , Amazon company is investing in its own fleet of trucks, planes and even drone at the same time as it pushes for same-day delivery of products. As some point, vehicles will be autonomous too. So, it seems that (AI) technique will influence any transportations choose to use digital autonomous driving technology in the future . For Amazon

company case, it is not stopping of logistics. It is also aiming to automatically manage the supply of consumer home products with its recently launched Amazon replenishment service, Dash. Dash is a digital service that enables that connected derive to automatically order physical products from Amazon when supplies are running low. So, it seems (AI) technology will be applied to logistic function by digital technology method introduction in the future.

Hence autonomous vehicles will optimize industry supply chains and logistics operations through increased efficiency and flexibility. In fact, fully automated and lean supply chains will keep reduce load sizes and inventory by leveraging smart distribution technologies and smaller autonomous vehicles by machine man assistance. If Amazon continues to grow market share for online sales by reducing effort required by the consumer to place an order, when also contributing the almost immediate delivery of products to the doorstep. So, it will further fuel the trend toward on-demand derive. As Amazon company fuels the on-demand economy, consumers will expect immediacy in more parts of the digital economy. On top of speed, consumers increasing expect more personalization options.

So, (AI) technology will influence digital manufacturing, such as Amazon publishing to monitor every aspect of every process in real -time and communicating to self-optimized deep learning robotics, new methods of high volume and high customization will become possible. Then, as products merge into product platforms and even services, manufacturers have the opportunity to provide components and platforms used by smaller players. So, (AI) technology will influence manufacturing industry to choose automated SMI lines, robots installed, automation engineers.

Another future (AI) technology development can be applied to space science aspect, such as Automation engineering space in manufacturing process to achieve digital manufacturing benefits to any businesses in the future. Such as reducing cost, shortening manufacturing time, raising efficiency, shortening delivery products to client individual time. How can artificial intelligence give the need and advanced fast and evaluation methods benefits for space exploration? When US NASA (space exploration organization) achieves any space exploration missions, it will answer this question:

When is it useful to have a machine use (AI) technology to achieve a decision? After all, after millions of years of space exploration and rough 10,000 years of civilization, humans are usually quite good at making decisions in complex uncertain environments. Through, Johns Hoplains University's Applied Physical Lab. Research in (AI) technology enabled systems, which has identified three general use cases for (AI) technology to explore space mission:

First, for some tasks (AI) technology is more cost effectiveness than human. Second, (AI) technology is better suited than humans at solving some, but not all problems. Third, (AI) technology allows NASA organization's space exploration mission to develop machines that ate capable of responding faster than when a human is in the decision loop (D. Scheidt, 2012, A. Castano et. al. 2008).

So, the use of (AI) technology to enable science by observing the pace of rapidly evolving phenomena was demonstrated. It is more effectively coordinating and (AI) technology utilizing to earn economic benefits to use for space exploration mission.

However, (AI) technology also have current risk for space exploration. Today (AI) technology is immature and requires further development to reach its potential. For instance, the (AI) technology algorithms that detected the dust derive could not have identified whether the Martain weather represented a threat to the cover. Also it can not yet use instrument input to determine what, where and how to autonomously make the next space science measurement. An equally important factor limiting (AI)'s deployment is that lacks the methodology and technology to effectively test (AI) technology. So, the challenge will testing (AI) enabled system is how (AI) performance can be measured. It would be NASA organization's difficulty to find (AI) technology to develop to carry on researching any space exploration missions in the future. However, (AI) technology will be a good economic benefit choice for space exploration mission in the future.

- What is artificial intelligence potential
benefits and ethical considerations?

The ability of (AI) technology systems to transform vast amounts of complex information into insight has the potential to help solve manufacturing or service challenges for human needs. However, to reap the societal benefits of (AI) systems, humans will need to trust then and make sure that which follow the same ethical principles, moral values, professional codes and social norms that we humans would follow in the same scenario, research and educational efforts as well as carefully designed regulation in order to achieve the most effort of economic benefits goals. For example, international business machines corporation (IBM) is actively engaged both competitors , in global discussions about how to make (AI) ethical and as beneficial as possible for people as social economic benefits. (AI) is usually defined as the " capability of a computer program to perform tasks or reasoning processes " that human usually associate to intelligence in a human being. Often, it has to do with the ability to make a good decision, even when there is uncertainty, too much information to handle. As an example, play chess or complex card games of entertainment activities is believed to need some form of intelligence in a human being, as well as choosing the best medical facilities in a difficult medical case, or creating something new, such as mathematical theorem or even some form of act, or even driving automatic machine man (self driving vehicle) replacing human driving in the middle of a crowded city.

(AI) needs depends on what we consider being intelligence in the behavior of a human being act a certain point in time. If human belief about human intelligence changes and we don't believe any longer that a certain task requires intelligence, then a computer program performing that task is no longer part of (AI), it becomes just another boring computer program. So, it means that (AI) technology will replace some old computer programs, if human can invent new generation of (AI) software for any functions or activities to satisfy human needs.

As IBM, it argues intelligence. This means that we aim to build systems that enhance and scale human expertise and skills rather than replacing them. We therefore focus on practical applications of (AI) capabilities that assist people in performing well-defined tasks of needs by exploiting and wide range of (AI)-based services. We also use the term " cognitive computing" it is mean a comprehensive net of capabilities based on technology. It comprises the fields of machine learning, reasoning and decision technologies, language, speech and vision recognition and processing technologies, high performance and high efficient functions for any industries or individual consumers needs. For example, robotics, which are usually very good at doing what which are supposed to in any environment, much have public shopping center, factory etc. places which need simply services from the robot (machine man), such as cleans the floor of our houses to the robot that can work together with humans in production chains, passing through the warehouse, robots can take care of the tasks of an entire warehouse and the companion robots like Nao, Pepper, Aibo and Giraff, who can entertain use, talk to use and help elderly people to stay connected to their friends, relatives and doctors.

Google company is building automatic machine (self-driving cars) and has acquired more than 10 robotics companies. Facebook had opened whole new research facility only on (AI) research. Apply computer has developed Siri. Microsoft computer company has built a similar personalized assistant. Google has Deep mind, a UK company whose long term aim is to build general (AI) and has already great potential to win game to the world champion and IBM is investing a huge amount of resources in applying its Watson cognitive computing system to the medical domains to finance and to personalized education. In Europe, IBM is establishing new centers in Munich and Milan focused in the application of cognitive computer capabilities to the internet of things and healthcare respectively.

For example, automatic machine man (self-driving cars) are all about (AI), which used to be able to see what happens in the street (signals ,lanes, other cars, pedestrians, traffic lights, which need to able predict what other cars and pedestrians will do, and who need to be able to cope with unforeseen situations. Since, most car accidents are due to human fault, it is estimated that the adoption of self-driving cars will save about half of the lives that are usually last in car accidents.

IBM Watson company has to understand spoken language, make sense of massive amount to text , respond correctly to questions in many categories, as well as assess its own confidence in responding to such questions. In the future, (AI) technology can own question/answering capabilities that would be very useful, for example, in assisting a doctor when trying to some to the correct diagnosis for a patient and to propose the best therapy .

Intelligent machines can also rely on huge amounts of data to be used to learn how to make better decisions. This data comes from all of us over the years Facebook users have uploaded more than 250 billion pictures and every day who upload about 350 million more. Every second, we submit 40,000 google search queries. So, (AI) technology will be connected through the web from appliances to traffic lights from cars to watches. Other tasks that are very easy for humans are physical and manipulation tasks, such as walking , running, picking up an object to make its shape and location, restricted environment. But (AI) machine man technology still not able to have the general physical and manipulation capabilities even of a 6 year old.

So, it brings this question: Why do (AI) scientists need to concern ethics? Because (AI) technology is complex, information into insight has the potential to reveal long held secrets and help solve some of the world's most difficult problems. (AI) systems can potentially be used to help discover insights to treat disease, predict the whether, and manage the global economy. So, ethic issues is important to and (AI) scientists . If any one new (AI) technology research investigation could success, it will be a secret to and the (AI) scientists can not permit to their loyalty to any competitors to damage the fair (AI) technology products trading market. The country (countries) (AI) technology scientists need to concern ethic issues, who need to keep secrets for their countries economic or/and social benefits. This is moral issues to any countries/country loyalty is whose countries intangible assets. They can not sell (AI) loyalty to any their countries to assist whose economic benefits immorally.

● How can (AI) technology influence to global health care economy development?

According to (AI) lecturer analysis, when combined key clinical health (AI) application can potentially create $150 billion in annual savings for the US healthcare economy by 2026 year. (AI) technology is re-winning modern conception of healthcare delivery. It enables machines to sense, comprehend, act and learn. So which can perform administrative and clinical healthcare functions (Accenture, 2017).

It will help health care service organizations to reduce health care cost, will improve and raise service quality and access. So, (AI) health market size will be predicted growth. (AI) applications in health care include robot-assisted surgery, virtual nursing assistant, administrative workflow assistant, fraud detection, error reduction connected machines, clinical trial participant identifier, preliminary diagnosis, automated image diagnosis and cybersecurity.

What kind of benefits (AI) technology can contribute to healthcare service? (AI) technology can deliver what many health care organizations need, such as financial and operational of labor costs, digital expectations from patient consumers how to use (AI) technology to solve interoperability challenges in any healthcare organizations. Also (AI) technology can be applied to wellness an d lifestyle management, diagnostics, delivers financially but also way of organizational and workflow improvement. So, (AI) technology will be continue to become most prevalent and adoption to healthcare organizations , which must need to enhance structure to be position to take full advantages of new (AI) technological capabilities. (AI) technology can change the nature of work and employment is rapidly changing to make the best use of both humans and (AI) talent in healthcare industry in the future. For example, (AI) technology offers a way to fill in gaps and the rising labor shortage in healthcare. According to Accenture analysis, the physicians shortage is increasing. However, (AI) technology will manufacture healthcare machine men to replace physicians in future one day(2017). Hence, (AI) technology will be invented to raise health care service staffs work efficiency and performance in any hospitals or clinics in the future.

In conclusion, (AI) technology will raise efficiency for any service or manufacturing industries in the future, although, it is possible that it will also rise low skillful workers unemployment numbers. But, the most important influence to human technological innovation will be risen and it will influence human life will be changed to be better, e.g. self drive cars, health care physician machine men, machine man cleaners etc. intelligent machine men will be manufactured to serve for our daily life. Furthermore, (AI) technological products will influence countries trading, some low technological development countries manufacturing businessmen can choose to buy any (AI) products to raise whose productivity and efficiency and reducing cost to achieve economic cost saving result. Also, GDP of trading growth income will increase to the (AI) products sale countries. Hence, it will be beneficial to economic development to both developed and developing countries both in the future as well as (AI) scientists time and money spending will be valued to continue to invest (AI) technology development for human life and economy benefits for long term.

In conclusion (AI) technology will raise macro economy growth and it can create many (AI) jobs , but it also raise the low level technological worker unemployment change. In the future, (AI) technology can be applied to digital technology to attempt to invent any new undiscovered (AI) and digital technology. So, it needs any scientists to continue to research how digital and (AI) technology can be mixed to satisfy human's future undiscovered needs.

Online technology and online book
technology influences artificial intelligence
mind development

Nowadays, online technological invention bring online book technological development. Also, artificial intelligent technological machine men had been invented to link internet to do any jobs, e.g. children can find any data from artificial intelligent machine men when the artificial intelligent machine man had been installed internet and computer function, then children can find any online books to read from the artificial intelligent machine man. Such as Japan artificial intellgent machine men had installed computer and internet function, the Japan family children can find any online books to read from the artificial intelligent machine man at Japan any families' homes conveniently. Hence, it implies that future one day, artificial intelligent machine has possible to be invented to own human's reading and/or writing abilities.

For example,online book publishing is one kind of popular internet technology. For example, Amazon publish is as a business model with many potential advantages, relative to a physical operation. It held out the potential of lower book inventing and distribution costs and reduced overhead. Consumers could find the books, they were looking for more easily and a variety book topic choices could be offered for sale. It can accept and fulfill orders from almost any domestic location with equal ease. And most purchasers made on its site would be exempt from sales tax. One Amazon strategy hand, it would have to make its returns and redress processes transparent and reliable, and offer other ways for clients to learn, as much about the book possible before buying. Future online book market development trend, such as Amazon, Barnes & Noble etc. online book shops.

Hence, online book store technology can be applied to artificial intelligent technology. Such as artificial intelligent machine men can apply computer technology to learn the abilities of reading and/or writing any books either on paper or on computer. Hence, it is possible that artificial intelligent machine men will have similar human's writing and/or reading books ability when they own human's mind ability. However, it bring this questions: Can artificial intelligent machine men own human's mind abilities? If they own human's mind abilities, is it mean that they can write and/or read any books? Can artificial intelligent machine men own human's mind abilities to create to write any books? Can artificial intelligent machine men own human's mind abilities to read and make any judgements or decisions more accurate than human's judgements or decisions? To answer these questions? I shall indicate that online book reading and writing technology can be applied to artificial intelligent machine men reading and writing technology. Because they are similiar computer mind technological development. So, I believe that future artificial intelligence machine men can be invented to own similar human's reading and writing's mind abilities in future one day.

I believe artificial intelligence and online technological reading abilities are very similiar. Nowadays, computer can be invented to attempt to read and write any books by human. Why can not artificial intelligent machine men replace computer to read and write any books? Artificial intelligent machine men can replace human to attempt to write or/and read books, due to artificial intelligent machine men had invented to own human mind to do some jobs and their mind had been invented to be similiar to human behavioral abilities to do these behaviors, e.g. cooking, driving, playing games, singing songs, speaking, listening, frighting etc. different human's abilities. So, it seems that artificial intelligent will be possible to be invented to own human's mind abilities to do any writing or reading behaviors or functions.

Prediction of artificial intelligence
reading and writing abilities

development

What is future trend of artificial intelligence reading and writing abilities development? To answer this question, we need to know what benefits of artificial intelligent machine men can attribute to human's needs when they can own any human's mind to read or/and write any books.

I shall indicate e-books reading and writing example, if artificial intelligent machine men can be invented to own human's mind to write and/or read e-books on computer. Then, it brings this question: Can artificial intelligent machine men assist human to learn to do judgement to solve any challenges?

I believe that when artificial intelligent machine men can be invented to own human mind to write or/and read any books, then they will own human's mind ability to make judgement to solve any challenges more accurately, even their decisions can be more accurate to compare to human's decisions. So, artificial intelligent machine mens' writing and reading ability is the main factor to cause their mind to do any judgement in order to make any decisions more accurately. Consequently, in future one day, artificial intelligent machine mens‘ writing and reading ability will be invented to similar human's reading and writing abilities as well as their minds can also be invented to similar human's minds as well as their judgement abilities can be invented to similar to human's judgement abilities to make any decisions more accurate.

The influences when AI is invented to own
human's mind and judgement abilities

Finally, I shall discuss what are the influences when AI is invented to own human's mind and judgement abilities in our future job market. The achievement of artificial intelligent (AI) machine men achievement requirement of owning human's mind and judgement abilities which requires extensive manual labor, and by augmenting the calling process with machine learning, the process where speed and accuracy are needed to close to human's mind and judgement abilities. Expert human race callers now have better information at artificial intelligent machine men at their fingertips faster.

Hence, if the above those requirements are achieved to satisfy artificial intelligent machine men ind and judgement abilities demand to close or exceed humans' mind and judgement abilities. Then, I believe that future human's some simple jobs must be replaced by (AI) machine men. Even, human's some professonal jobs, e.g. lawyer, accountant, administator, typing etc. professional skillful jobs, which will be either replaced or will be assisted by (AI) machine men. For example, (AI) machine men learn how to type english or other language words to do typing job ; they can learn how to apply accounting knowledge to record any firm's income and expenditure record of accounting job; they can also learn how to assist architects to design any architectural building drawing plans to do architect jobs; they can learn how to analyze any court evidences to judge any criminal or civil cases and assist lawyers to give legal advices to achieve more reasonable judgement for any legal cases; they can also learn how to assist firm's managers or administrators to manage any organization teams efficiently.

Consequently, when (AI) machine men can be invented to achieve to exceed human's mind and judgement abilities level. Then, I believe that they can do instead of human‘ simple jobs, which can do even human's more difficult and more judgement requirement of professional skillful jobs. So, (AI) machine men must need to achieve to do any jobs, they are same, even exceed to human professionals' abilities. Then, it will cause a lot of human's jobs to be disappeared or some human's jobs will be replaced by owning judgement and mind abilities of (AI) machine men to do.

Hence, future many human's jobs will be replaced by technological labors. Employers choose to buy (AI) machine men to replace human labors. The reasons include (AI) machine men have none unhappy, angry emotion to influence their low efficiencies and low productivities. Their judgement and mind abilities can exceed human's abilities or do any jobs to compare better performance to human's abilities. Consequently, different occupation labors need to prepare to learn how to co-operate with (AI) machine men to let future employers feel (AI) machine men will be human's assistant to assist human to do jobs efficiently when human and (AI) machine men work together. It aims to avoid future employers feel (AI) machine men's judgement and mind abilities can exceed any low knowledgeable and skilful occupation labors, even high knowledge and skilful occupation labors. It means that (AI) machine men are only labors‘ assistant if (AI) machine mens' judgement and mind abilities are below under to human labors‘ judgement and mind abilities.

Consequently, to avoid (AI) machine men can replace human to do any simple or complex jobs to cause any future any occupation labors' competitors. I recommend that it is right time labors ought prepare to learn different skills. So, every individual labor does not only concentrate on one kind of skill. Because supposing one kind of the occupation labor's job duties are replaced by (AI) machine men. If the employee had owned more than one kind of occupation skill. Then, I believe that who can avoid the unemployment threat more easier than the employee only owned one kind of occupation skill, when (AI) machine men had invented to own human's mind and judgement abilities in future one day. The most important, I believe that (AI) invention will be applied to be teach how to learn human's skills and mind ability. Such as education industry, teacher won't be replaced by (AI), otherwise, (AI) will be teacher's assistant to help them to do education data gather or teaching jobs. So, teachers won't be replaced by (AI), otherwise, teachers will depend on (AI) data gather or teaching job to give them opinions how to solve student's teaching challenges as well as teachers can concentrate on researching education jobs for schools' benefits if (AI) technology can be invented to on human's mind and judgement and reading and writing abilities in the future.

Online technology and online book
technology influences artificial intelligence
mind development

Nowadays, online technological invention bring online book technological development. Also, artificial intelligent technological machine men had been invented to link internet to do any jobs, e.g. children can find any data from artificial intelligent machine men when the artificial intelligent machine man had been installed internet and computer function, then children can find any online books to read from the artificial intelligent machine man. Such as Japan artificial intelligent machine men had installed computer and internet function, the Japan family children can find any online books to read from the artificial intelligent machine man at Japan any families' homes conveniently. Hence, it implies that future one day, artificial intelligent machine has possible to be invented to own human's reading and/or writing abilities.

For example,online book publishing is one kind of popular internet technology. For example, Amazon publish is as a business model with many potential advantages, relative to a physical operation. It held out the potential of lower book inventing and distribution costs and reduced overhead. Consumers could find the books, they were looking for more easily and a variety book topic choices could be offered for sale. It can accept and fulfill orders from almost any domestic location with equal ease. And most purchasers made on its site would be exempt from sales tax. One Amazon strategy hand, it would have to make its returns and redress processes transparent and reliable, and offer other ways for clients to learn, as much about the book possible before buying. Future online book market development trend, such as Amazon, Barnes & Noble etc. online book shops.

Hence, online book store technology can be applied to artificial intelligent technology. Such as artificial intelligent machine men can apply computer technology to learn the abilities of reading and/or writing any books either on paper or on computer. Hence, it is possible that artificial intelligent machine men will have similar human's writing and/or reading books ability when they own human's mind ability. However, it bring this questions: Can artificial intelligent machine men own human's mind abilities? If they own human's mind abilities, is it mean that they can write and/or read any books? Can artificial intelligent machine men own human's mind abilities to create to write any books? Can artificial intelligent machine men own human's mind abilities to read and make any judgements or decisions more accurate than human's judgements or decisions? To answer these questions? I shall indicate that online book reading and writing technology can be applied to artificial intelligent machine men reading and writing technology. Because they are similiar computer mind technological development. So, I believe that future artificial intelligence machine men can be invented to own similar human's reading and writing's mind abilities in future one day.

I believe artificial intelligence and online technological reading abilities are very similiar. Nowadays, computer can be invented to attempt to read and write any books by human. Why can not artificial intelligent machine men replace computer to read and write any books? Artificial intelligent machine men can replace human to attempt to write or/and read books, due to artificial intelligent machine men had invented to own human mind to do some jobs and their mind had been invented to be similar to human behavioral abilities to do these behaviors, e.g. cooking, driving, playing games, singing songs, speaking, listening, frighting etc. different human's abilities. So, it seems that artificial intelligent will be possible to be invented to own human's mind abilities to do any writing or reading behaviors or functions.

Prediction of artificial intelligence reading and writing abilities

development

What is future trend of artificial intelligence reading and writing abilities development? To answer this question, we need to know what benefits of artificial intelligent machine men can attribute to human's needs when they can own any human's mind to read or/and write any books.

I shall indicate e-books reading and writing example, if artificial intelligent machine men can be invented to own human's mind to write and/or read e-books on computer. Then, it brings this question: Can artificial intelligent machine men assist human to learn to do judgement to solve any challenges?

I believe that when artificial intelligent machine men can be invented to own human mind to write or/and read any books, then they will own human's mind ability to make judgement to solve any challenges more accurately, even their decisions can be more accurate to compare to human's decisions. So, artificial intelligent machine mens' writing and reading ability is the main factor to cause their mind to do any judgement in order to make any decisions more accurately. Consequently, in future one day, artificial intelligent machine mens' writing and reading ability will be invented to similar human's reading and writing abilities as well as their minds can also be invented to similar human's minds as well as their judgement abilities can be invented to similar to human's judgement abilities to make any decisions more accurate.

The influences when AI is invented to own
human's mind and judgement abilities

Finally, I shall discuss what are the influences when AI is invented to own human's mind and judgement abilities in our future job market. The achievement of artificial intelligent (AI) machine men achievement requirement of owning human's mind and judgement abilities which requires extensive manual labor, and by augmenting the calling process with machine learning, the process where speed and accuracy are needed to close to human's mind and judgement abilities. Expert human race callers now have better information at artificial intelligent machine men at their fingertips faster.

Hence, if the above those requirements are achieved to satisfy artificial intelligent machine men ind and judgement abilities demand to close or exceed humans' mind and judgement abilities. Then, I believe that future human's some simple jobs must be replaced by (AI) machine men. Even, human's some professonal jobs, e.g. lawyer, accountant, administator, typing etc. professional skillful jobs, which will be either replaced or will be assisted by (AI) machine men. For example, (AI) machine men learn how to type english or other language words to do typing job ; they can learn how to apply accounting knowledge to record any firm's income and expenditure record of accounting job; they can also learn how to assist architects to design any architectural building drawing plans to do architect jobs; they can learn how to analyze any court evidences to judge any criminal or civil cases and assist lawyers to give legal advices to achieve more reasonable judgement for any legal cases; they can also learn how to assist firm's managers or administrators to manage any organization teams efficiently.

Consequently, when (AI) machine men can be invented to achieve to exceed human's mind and judgement abilities level. Then, I believe that they can do instead of human' simple jobs, which can do even human's more difficult and more judgement requirement of professional skillful jobs. So, (AI) machine men must need to achieve to do any jobs, they are same, even exceed to human professionals' abilities. Then, it will cause a lot of human's jobs to be disappeared or some human's jobs will be replaced by owning judgement and mind abilities of (AI) machine men to do.

Hence, future many human's jobs will be replaced by technological labors. Employers choose to buy (AI) machine men to replace human labors. The reasons include (AI) machine men have none unhappy, angry emotin to influence their low efficiencies and low productivities. Their judgement and mind abilities can exceed human's abilities or do any jobs to compare better performance to human's abilities. Consequently, different occupation labors need to prepare to learn how to co-operate with (AI) machine men to let future employers feel (AI) machine men will be human's assistant to assist human to do jobs efficiently when human and (AI) machine men work together. It aims to avoid future employers feel (AI) machine men's judgement and mind abilities can exceed any low knowledgeable and skilful occupation labors, even high knowledge and skilful occupation labors. It means that (AI) machine men are only labors' assistant if (AI) machine mens‘ judgement and mind abilities are below under to human labors' judgement and mind abilities.

Consequently, to avoid (AI) machine men can replace human to do any simple or complex jobs to cause any future any occupation labors‘ competitors. I recommend that it is right time labors ought prepare to learn different skills. So, every individual labor does not only concentrate on one kind of skill. Because supposing one kind of the occupation labor's job duties are replaced by (AI) machine men. If the employee had owned more than one kind of occupation skill. Then, I believe that who can avoid the unemployment threat more easier than the employee only owned one kind of occupation skill, when (AI) machine men had invented to own human's mind and judgement abilities in future one day.

How does (AI) robots' brain invention influence our lives

(AI) research modeling the human brain has developed important technologies, and has overcome significant barriers. How will (AI) affect humanity in the near future? How will (AI) change our lives and our societies? Is the evolution of (AI) to humanity, or it represent a threat?

On white collar workers (AI) job replacement aspect, University of Tokyo, Institute of informatics, lecturers who had attempted to do experiments to take (AI) exams over a two year period. The (AI) achieved standard scores of around 50 in each subject, exceeding the norms for humans attempting the tests. The (AI)'s results in subjects emphasizing memorization, such as world history and Japanese history subjects were comparatively high, and the results of the study suggested that an appropriate selection of subjects would give at an 80% chance of passing the entrance exams of 80% of Japan's private universities.

So, if (AI) is applies to human white collar workers‘ job duties aspect, at this level, if white collar workers were replaced by (AI) in the future, around 30% of current staff would be replaced. Whatever, the outcome, large companies will be represented with two choices. One choice will be to protect their employees, but as a result lose their international competitiveness. The latter choice will enable them to reduce the cost of general duties, financial management procedures, accounting etc. general administrative job duties of cost in offices. Hence, it seems that (AI) will be possible to be invented to own human's brain ability to do some mind jobs in future on day.

However, the method called " deep learning" must be developed to cause (AI) to match human's brain ability as well as these were dramatic advances in technologies, such as image recognition and voice recognition, which form the foundation for (AI). Nowadays, this new method called" deep learning" does not reach the matured and stagnated stage. It needs to wait human to continue to invent to let (AI) to match human brain to achieve 100% owning human's mind ability. Nowadays, (AI) industry product include cleaning robots, smart TVs and future (AI) product development market. It will include self-driving vehicles, drones, and nursing robots.

On (AI) weapons applied aspect, if (AI) can be invented to own human's brain judgement and analytical abilities. Then, it is possible that it can be applied to attack enemy to cause war effect. For example, if weapons such as missiles were equipped with (AI) in the future, they would become able to decide on their own targets. Hence, human needs to apply restrictions when necessary.

On (AI) applied to analyzing information collected technological aspect, nowadays, every one will use wearable terminals to connect to the internet to obtain various types of information as well as computers will collect and analyze information on people. Our lives will probably be more reliant on these internet technologies than they are on smartphones today. When, (AI) can match human brain to own mind ability.

Then, (AI) can be applied to do any analyzing information and collection job duties aspect to raise large information restoring and remembering efficiency. For example, (AI) will be generally used and will be extremely useful in analyzing the information collected from wearable devices and stored in the cloud. (AI) will enable wearable devices to be of real assistance in our lives offering their users more intelligent support.

Rather than allowing (AI) to develop on serves, as something separate from humanity. It will be more meaningful to encourage its development via wearable devices, situating it under the control of human intelligence. The intelligence of (AI) will increase rapidly in the future. If this increase in (AI) occurs under human control, enabling humans to increase their own abilities, then surely it will be possible for us to put up a degree of resistance to the opposite scenario, the domination of (AI) over humanity. Hence, if (AI) can be invented to remember and store and make analytical judgement to collect any information from internet. Then, it will bring the effect, such as large international organizations' (AI) internet storage robots can bear in mind factors, such as competitors' privacy or business secret information, such as the loss equality between people and threats to privacy that will be stolen form the owning (AI) storing internet information remembering robots.

Consequently, what is the effect of successful invention of (AI) matching human brain's mind ability? (AI) present computers are adequately able to reproduce the emotional, conceptual and intuitive abilities of humans. Because of this, it is important that we should envision potential future problems that may manifest when we consider how to employ wearable devices. It will be essential to enhance our technologies in order to ensure that we can use (AI) under human control.

However, when a goal has been set. (AI) will implement an appropriate means for its realization. (AI) will be need as a tool by human society. If the capacities of analytical and judgement mind abilities of (AI) brain exceed those of human brain, it is difficult to imagine the type of technological, then singularity is represented by the creation of an (AI) by another (AI). It is important that we rapidly and accurately predict these developments, when image recognition and other individual technologies are functioning at a high level. There will be a considerable matter in different sectors of (AI) industry development.

Today, however, machines have become able to decide for themselves what they will learn, making it difficult to copy human's mind ability. What we must consider when machines exceed humans and (AI) surpasses human capabilities. May technologies exceed human capabilities, cars are faster than humans, planes are able to fly.

Consequently, it brings a question that human needs to consider: When does (AI) brain technology be invented to reach the most reasonable stage to be accepted or stopped by humanity?

What is artificial intelligence
human brain invention?

A machine is likely to achieve the ability of a human brain. Does it a scientific story? Some scientists has predicted that a US$1,000 personal computer will match the computing speed and capacity of the human brain by around the year 2020 year. With human reverse engineering, human should have the software insights before 2030 year. it is possible that of machine intelligence and exotic new technology for faster and more powerful computational machines from cellular automata and DNA playing cheese game competition case example, it proves that (AI) had been invented to own human's analytical and judgement ability to exceed the best cheese game human player's brain analytical and judgement ability. Then, it seems that (AI) will have possible to be built machine brains to achieve the exceed level of human brain's analytical and judgement ability in the future one day.

Supposing we scan someone's brain and restate the resulting " mind file" into suitable computing medium. Will the entity that emerges from such an operation be conscious? How have advances in electronic communications changes power relationship? For electronic book publishing case example, a book that looks at the principles companies must adopt to meet the needs and desires of this new kind of client. So, such as paper book can be changed to electronic book for human to read. Why can't human brain be changed to (AI) machine brain to do human's analytical mind and behavioral mind of activities to replace to do any human's daily analytical and behavioral mind activities?

Over the next few decades, machine achieve super intelligence, human will encounter a dramatic phase. Will it be a "WALL" a barrier as conceptually the event of a black hole in space. Such as (AI) brain invention case, an " AI singularity" ruled super-intelligence AIs, or a gentler " surge" into a post human era of agelessness and super-intelligence brain. Will future technology, such as bio-engineered pathogens, self replicating nan robots, and super smart robots run and accelerate out of control, perhaps threatening the human race?

If one day, (AI) brain is invented to achieve agelessness possibility. It means human's brain will be old to lose mind and analytical ability when human's age is increasing. Otherwise, (AI) machine brain age won't lose mind and analytical ability, due to (AI) machine is no age increasing possibility. It is a machine brain. If (AI) machine brain can be built successfully. Scientists need to consider technological ethic matter, such as the challenge of guiding nanotechnology in a constructive direction, advances in nanotechnology and related advanced technologies can not be inevitable, any broad attempt to relinquish nanotechnology would interfere with the benefits. When actually making the dangers worse.

Keeping in mind that intelligence machines are already making their way into our blood stream. There are dozens of projects underway to create blood-stream based " biological micro electronic- system" (bio MES) with a wide range of diagnostic and therapeutic applications BioMEMS devices are being designed to intelligently pathogens and deliver medications in very precise ways. For example, a researcher at the University of Illinois at Chicago has created a ting capsule with pores measuring only seven nanometers. The pores let insulin out in a controlled manner, but prevent antibodies from invading the pancreatic Islet cells inside the capsule. These nano- engineered devices have cured rated with type I diabetes, and there is no reason that the same methodology would fail to work in humans. Similar systems could precisely deliver dopamine to the brain patients, provide blood-clotting factors for patients with hemophilia and deliver cancer drugs directly to tumor sites. A new design provides up to 20 substance-containing reservoirs that can release their cargo at programmed times and locations in the body.

Another brain health technological related invention case, such as Kensall Wise, a professor of electrical engineering at the University of Michigan, who has developed a tiny neural probe that can provide precise monitoring of the electrical activity of patients with neural disease. Future designs are expected to also deliver drugs to precise locations in the brain. Also, kazushi Ishiyama at Tohoku University in Japan has developed micro machines that use microscopic-cancer tumors.

A particularly innovative micro machine developed by Sandia National labs has actual micro teach with a jaw that opens and closes to trap individual cells and then implant them with substances, such as DNA, proteins or drugs. There are already at least four major scientific conferences on bio MES and other approaches to developing micro- and nano-scale machines to go into the body and bloodstream. All these inventions are related to how to apply machines to copy human's brain knowledge in order to achieve to do any human's brain functions.

Finally, for Freitas envisions micron-sized artificial platelets invention case example, who could achieve hemostasis (bleeding control) up to 1,000 times faster than biological platelets. Freitas describes nano-robotic microbivores (white blood cell replacement) that will download software to destroy specific infections hundreds of time faster than antibiotics, and that will be effective against all bacterial, and fungal infections with no limitations of drug resistance.

Consequently, such as above machine health scientific invention cases, there were many scientists had invented any health machines to apply drugs to transfer to human's brain to attempt to reduce human's disease causing risks, such as reducing cancer cell increasing number. Why it is no possible that scientists can attempt to invent (AI) brain which can own human's mind ability to judge or analyze any matters to give opinions in order to exceed human's judgement and analytical ability.

How can artificial intelligent brain satisfy to human beneficial and natural needs?

Nowadays, new scientists' most familiar form of this vision in our times is genetic engineering. Specifically, the prospect of designing better human beings by improving their biological systems of a small, serious and accomplished group of tailors in the field of artificial intelligence and robotics. Their goal is a simply new age of post-biological life, a world of intelligence without bodies, immortal identity without the limitations of disease, death and unfulfilled desire. If human can understand why this fate is presented as both necessary and desirable, human might understand modern science can help us to enter the good life and good society stage when (AI) brain is invented by scientists successfully in our future life.

How can (AI) beneficial brain satisfy to human natural need? For relatively recent example, similarly as a long term trend beginning with the first mechanical calculators, the evaluation of computing capacity increases in speed over time and decrease in cost. From biological evolution has been invented to influence human brain, an electronic chemical machine with a great, but finite number of computer neuron connections, the product of which we call mind or consciousness. As an electro-chemical machine, the brain obeys the laws of physics, all of its functions can be understood and duplicated. And since computers already operate at far faster speeds then the brain, they soon will rival or surpass the brain in their capacity to store and process information. When happens, the computer will at the vary least, be capable of responding to stimuli in ways that are indistinguishable for human responses. At that point, we would be justified in calling the machine intelligent, we would have the same evidence to call it conscious that human now have when giving such a label to any consciousness other than our own.

At the same time, the study of human brain will allow us to duplicate its functions in machine circuitry. Advances in brain imaging will allow us to " map out" brain functions, allowing individual minds to be duplicated in some combination of hardware and software. The result, will be a world that is remade and reconstructed at the atomic level through nanotechnology, a world whose organization will be shaped by an intelligence that surpasses all human comprehension.

Whether or not today's humans are willing or able to " download" their brains into machines, there will come a time when all human beings will be intelligent machines in the future. Computer hardware will continue to get faster, cheaper and more powerful computer software will increase in sophistication. Brain research will continue to explore the " mechanics" of consciousness. Nanotechnology will continue to develop.

There are powerful incentives, commercial, military, medical and intellectual that will drive many of the advances that the extinctions desire if for very different reasons. Much of the work in artificial intelligence and robotics is open to the same defense that is made on behalf of biotechnology: If we don't do it, they will and why suffer or be unhappy when some new agent or invention is available that will solve the problem.

Finally, we already accept significant artificial argumentation and replacement of natural body parts when those parts are missing or defective. Over time indistinguishable from or " superior" to their biological counterparts as they employ increasing computer processing power. There are powerful incentives, commercial, military, medical and intellectual that will drive many of the advances that the extinctions desire, if for very different reasons. Much of the work in artificial intelligence and robotics is open to the same defense that is made on behalf of biotechnological if we don't do it, they will and why suffer or be unhappy, when some new agent or invention is available. That will cure the problem.

Finally, we already accept significant artificial augmentation and replacement of natural body parts when those parts are missing or deductive. Over time, such replacements are only likely to get more useful and perhaps eventually indistinguishable from or superior to their biological counterparts, as they employ increasing computer processing power. Nor is there an obvious distinction between using manufactured chemicals to fight disease and using " smart" nanotechnology. The extinction project is begun by offering new routes to fulfilling old promises about doing good for human beings. But, it doesn't necessary end.

In connection with machine intelligence, it does not seem very promising to try to limit the power or ability of computers. The danger (or promise) that computers might develop characteristics that lead some people to call them conscious and that this age of intelligent machines would mean our extinction seems remote when compared with their practical benefits. We already rely so heavily on computers that the incentives to make them easier to use and more powerful are very great. Computers already do a great many things better than we can, and there seems to be no natural place to enforce a stopping point to further abilities. Certainly mechanistic and reductionist assumptions about society, ethics and psychology the notion that we are atoms or animals, driven by chance or instinct, run deep in the present world.

Artificial intelligent brain future

innovation and attribution (AI) brain invention successful factors

(AI) brain will bring what attribution to influence human's positive impact. How (AI) brain will be invented to apply to any businesses' needs. By how much the (AI) brains might exceed us remain unknown, but it could potentially be by a very significant degree. Future (AI) brain invention will have noted similar growth in everything from hard-drive storage density to the price and speed of DNA sequencing. On key feature of this technological growth that has not been adequately measured in the degree to which technology is becoming more intelligent.

(AI) brain test experiment

When there is an intuitive sense that (AI) programs/brains today are more capable than those of age, and that those were considerably " smarter" than the serial instructions that passed through the first supercomputers. Scientists will carry on testing (AI) to do any experiments to improve (AI) brain development. They will assess the progress of artificial (general) intelligence, but the need for intelligence tests and tests for other cognitive abilities will be tested in the forthcoming decades for bots, robots, avators, " animats" etc. and any collective system of these and biological systems (humans and non-human animals).

When (AI) brain technology is invented successfully, the idea of a super intelligent computer means invention successfully also. Whether super intelligent computer or (AI) brain can be invented successfully. Similarly, many think that the future beyond the technological singularity is unknowable or even unimaginable. Since its conception, the idea has been examined and explored by technologists. (AI) brain invention will be mean human-equivalent (AI) vs human-level(AI). Many machine intelligence tests is the exclusive focus on identifying systems that achieve human equivalence. Considering our experience with studying non-human animal intelligence.

The need to distinguish human equivalent (AI) from human level (AI) seems critical. Perfect human intelligence , such as (AI) brain invention is likely to be extremely difficult to achieve. Potentially every aspect of the biological processes involved would need to be translated with very high fidelity.

Human level (AI), such as (AI) brain invention is another matter. Achieving capabilities that are equivalent to those of the human mind could be very feasible if it is not limited to perfectly processes involves. For instance, some pattern recognition algorithms are already superior to human abilities. This machine ability is not achieved by duplicating the processes our brains use, through some of the methods have been inspired by them. If researchers had been limited to replicating the brain's mind processes, We would still be waiting for the development of a (AI) brain machine equivalent.

Tests that would seen to have a reasonable chance of successfully testing non-human intelligence are those that use mathematics to define the value of a given challenges for (AI) brain invention. Such as Capability-test has some potential to generate meaningful data about (AI) machine intelligence, e.g. others in complexity theory test, it presents a series of abduction and prediction problems, similar to those in standard IQ tests to (AI) brain. Hence, IQ test and C-test will be potential tests to test (AI) brain ability.

(AI) brain test goals include that to identify a range of possible types of mind such as: super-fast human mind, mind with operational access to its source code, any mind capable of general intelligence and self awareness, general intelligence without self-awareness, self-awareness without general intelligence, super-logic , machine without emotion, mind capable of imaging greater mind or creating greater mind to compare human's brain abilities. Thus, any IQ or Capability test experiments aim to evaluate whether (AI) brain mind ability which can exceed to human brain mind ability. Thus, if future one day, scientists could prove (AI) brain mind ability can exceed to human brain mind ability. Then, they believe (AI) brain invention has ensured to achieve success.

(AI) brain invention successful factors

Hence, scientists want to invent (AI) brain successfully. They need to solve these challenges. Such as How robots and computers have progressively supplemented humans, initially only in relatively simple computational and manipulation tasks, but more recently in higher cognitive tasks that used to be the pre negative of the human brain, including language, mathematics, probabilistic reasoning and decision making.

An important challenge is how to enhance the productive interactions between humans and artificial intelligence? The important challenge include these major successful factors to invent (AI) brain technology, such as:

What is the state of the art in (AI) software and machine learning?

Can all aspects of brain function be manufactured by artificial system?

What is the proper form of mathematics that may capture the operation of minds and brains?

How to make (AI) brain feels consciousness?

What would it be taken for a machine to pose a sense of art in (AI) brain software and (AI) brain machine learning by artificial system?

Will machines soon surpass us in all domains of human competence?

What is the proper form of mathematic that may capture the operation of minds and brains?

What is consciousness to (AI) brain invention?

Could a machine be endowed with an artificial consciousness?
What would it take for a machine to posses a sense of self?
Will intelligence machines soon pose a danger to humanity of (AI) brain is invented to reach the mature stage successfully?
Is it possible to design and construct an intelligent robot with an artificial brain sense of ethics?
How can we enhance the humanitarian uses of artificial intelligence brain and owning mind ability of robotics, in particular in the field of education, health and emergencies?
Consequently, above all these challenges, I recommend scientists need to solve to achieve (AI) brain invention in order to reach (AI) brain invention mature stage easily.

Artificial intelligence brain invention opportunities and challenges

If scientists focus wrong direction to invent (AI) brain, it will bring wrong marketing development to attribute any benefits to human. So, they need to reduce a mismatch of timescales between the pace of commercial innovation and (AI) brain invention process, reduce an underappreciation of the fundamental unpredictability of (AI) brain autonomous systems and reduce a lack of a university agreed upon conceptual framework for any (AI) brain invention and reduce a disconnect between the (AI) brain design of any kind of (AI) autonomous robots. Thus, scientists need examine these gapes, provide a roadmap of opportunities and challenges and identify areas of any (AI) beneficial functions to be attributed to human to use for our daily life needs.

Future (AI) brain market opportunities, it is rapidly growing innovations in digital -electronic and information technology had development of new intelligence, surveillance, and reconnaissance platform and battle management capabilities, precision-strike weapons, stealth aircraft, smart weapons and sensors and tactical exploitation of space (e.g. GPS).

Any one of these aspects will be (AI) brain future marketing development opportunities. Future (AI) brain invention of so called " narrow AI", (i.e. non-sentient artificial intelligence, whose problem-solving capability is confined to one narrow task. For example, (AI) brain needs to find the best method to win any one of chess player in any both human and (AI) robot chess playing game.

In smart weapon strategy industry, (AI) brain is needed to design how to analyze or mind to protect whose country to avoid enemy attack in any war by the best weapon protection strategy. In education industry, (AI) brain is needed to design how to analyze or mind how to assist teachers to educate whose students by the best education method. In aircraft industry, how to design (AI) brain to analyze or mind to assist pilot to make the most correct flying direction judgement to fly in the most safe way. In space exploitation industry, (AI) brain is needed to design how to find undiscovered natural resources to supply to human to use in anywhere space.

Thus, future (AI) brain invention needs have these features/characteristics to be designed. They include: (AI) learned on its own, where to find the information it needs to accomplish a specific task, (AI) can predict the immediate future from studying any matter, (AI) automatically needs to be inferred the rules that govern the behavior of individual robots within a robotic swarm simply by watching, (AI) needs to be learned how to navigation the acquired memories and experiences, much like a human brain, (AI) speech recognition needs to be reach human parity in conversational speech, (AI) communication system needs to be invented its own encryption scheme, without being taught specific cryptographic algorithms (and without revealing to researchers how its method works, (AI) translation algorithm needs to be invented to remember fluent language to more effectively translate between any two languages (without being taught to do so by humans), (AI) brain system interacted with its environment (via virtual environment) to learn and solve problems in the same ways that a human child can do, (AI) based medical diagnosis system needs to be achieved 99% percent accuracy in any medical reviewing researches (at a rate minimum 30 times faster than humans), (AI) poker playing program brain development needs to be defeated some of the world's best human poker players during a minimum three-week-long tour, (AI) brain development needs to be effectively " read minds" of human test subjects looking at pictures of faces, via functional magnetic reasonable images of brain activity.

Consequently, (AI) future brain development needs to follow above directions to be invented to attribute to human's satisfactory needs.

(AI) brain legal remembering attribution

Future, (AI) robots can assist lawyers to deal any legal cases more efficient. If human understands that smart (AI) technology is not to replace human lawyers, but to make a better lawyer that forces who to use, emotional intelligence, and capital on lawyers' mind, then human lawyer have made the first step in future, proofing whose legal service business from (AI) legal robots‘ assistance. (AI) promises to be a real advantage for today and tomorrow' lawyers having to deal with the rate of legislative evolution and technological change.

In future, for the better with regard to technology in any law firms. According to the ALM 205 law tech. survey 95% of firm leaders and technologist respondents agreed with recent decisions by management regarding the firm's technology in legal service profession.

How can (AI) robots be applied to legal service industry by legal service firms? The (AI) reality in the legal world, it includes in relation to the four key elements of legal service provision, such as commodity, research, reasoning and judgement, exist and can support or replace certain aspects of every lawyer individual jobs both fee earning processes and business processes can be supported and/or replaced by expert systems, cognitive computing, robotics automated systems, (AI) and the machine learning, clients demand and expect more speedy, accurate, expert, creative, intuitive and accessible legal advice, it can assist young lawyers to innovate / tech. focused firm, reducing pressure both from within law firms (or in house teams) and from clients to respond to the demand for client-designed service from (AI) robots assistance, technology related projects that are both user and client -centric need to be implemented successfully.

Due to the deployment of (AI) robots in the legal ecosystem where lawyers, firms, general counsel and clients are beginning to believe (AI) capability technologies, (AI) robots can assist lawyers to increasingly become more productive, efficient, accurate, better quality, less labor intensive and time intensive and the role of the lawyer is gradually changing. If we break down a lawyers‘ tasks in a legal project from beginning that can handle the majority of these four tasks far more quickly and accurately than human lawyer. (AI) robots can handle legal task in the four aspects as below:

First in legal aspect, it can be used for deep research and processing, such as extracting specific pieces of information from land registry documents, (AI) technology is placed top of a document set including client guidelines and similar forms. It searches through documents and extracts key data points to provide a report of data for improved coordination with clients.

Second on managed services technology aspect, (AI) platform which could have a huge advantage for general counsel and law departments in corporations and for clients of all company sizcs.

Third on reading aspect, (AI) brain program that reads and analyzes , e.g. clauses in loan agreements. Its program helps its lawyers through transactions and points. Then toward the correct precedents of each stage of a process.

Fourth, on academics aspect, (AI) robots can assess the merits of personal injury cases. It can automatically review high volumes of contract documents to identify provisions that could potentially be impacted by contract law regulation.

Thus, all of these systems can handle large quantities of structured and unstructured data, and assist with the process management, research, and reasoning elements related to legal issues. Thus, in future, (AI) brain development can be invented to apply to knowledge research job, such as legal industry.

Artificial legal intelligence presents a thought-provoking approach to both computational models of legal reasoning and the use of evolutionary thinking about the law. The visions of computerized artificial legal intelligence , a vision of developments in both technology and legal history. A number of creative research projects have applied artificial intelligence techniques to the domain of legal reasoning.

Consequently, (AI) brain for legal industry marketing development ought concentrate on those three fields of artificial intelligence at most relevant to work in the legal areas in order to achieve the excellent attribution. Such as case-based reasoning, expert systems and neural networks. Artificial intelligence program, such as the legal reasoning programs. Thus, (AI) brain invention needs to own these human legal concept knowledge in order to achieve (AI) legal brain program development successfully.

Whether human mind can create to (AI) brain mind

How can (AI) scientists build a machine that think? In fact, there was general agreement that minds can be existence on non-biological substrates and that algorithms are of central importance to the existence of minds. However, there are much debate about the raw hardware power present in organic brains, such as (AI) brain.

I think (AI) scientists need to invent powerful hand ware, e.g. commercial digital signal processing might be, giving an appearance even to digital operations, but nothing would ever make up the intellectual runaway that is the essence of the singularity.

I also think raw hardware power is not be able to organize the parts to behave in a super-human way as well as I also think powerful software complexity is the main factor to solve (AI) brain mind invention challenge.

Hence, future super-human (AI) brain invention will ought consider how to invent superhuman software more than hardware, because software can store any memory, i.e. human mind. When, (AI) brain invention which can achieve to own human mind ability. Then, (AI) scientists need to consider ethic matter: Does the future of (AI) pose an existential threat to humanity? How do we present learning algorithms from morally objectionable biases? Should autonomous (AI) be used to kill in warfare? How should (AI) systems be in our social relations? Is it permissible to fall in love with an (AI) system? What sort of ethical rules should (AI) like a self-driving car use? Can (AI) systems suffer moral harms? All those ethic matters. I think (AI) scientists need to consider after (AI) brain owns human's mind ability because it is possible that (AI) robots will harm human if they are educated to do any wrong or illegal or immoral mind ability. Thus, (AI) scientists need to consider (AI) robot's moral mind and judgement behavior.

Brain-inspired intelligent robotics

How to solve fundamental problems in the areas of brain sciences and brain-inspired intelligence technology? One way scientists seek to accomplished the mission is to develop brain-inspired hardware including intelligent devices, chips, robotic systems and brain inspired computing systems. (AI) scientists ultimate goal, but since the robot's memory and learning after the human brain, there is still much to learn about neurobiology before that goal is attached.

In (AI) brain university research aspect, two schools of thought have emerged in robotics: bio-logically robots that include a body, sensor and actuators, and brain-inspired computing robot.

Robots have found increasing applications in industry, service and medicine, due in large part to advances achieved in robotics research over the past decades, such as the ability to accomplish complex manipulations that are essential for automated product assembly. However, robots still have these weaknesses which need to be solved if (AI) scientists expect (AI) brain invention can be success. Robots still lack truly flexible movement, have limited intellectual perception and control, and not yet able to carry out natural interactions with human. These deficits are especially critical in service robots. A critical concern of government, academia, and industry is how to advance research and development for the key technologies that can bring about the next generation of robots. Developing robots with more flexible manipulation, improved learning ability and increased intellectual perception will achieve the main goals for (AI) scientists' solutions.

Consequently future (AI) brain -inspired intelligent robotic invention needs have these competitive or attractive abilities or strengths to compare computer storage ability. Such as (AI) brain needs have perception to exceed computation ability, adaptation exceeds computing speed, flexibility exceeds, computer memory access speed, cognition exceeds computer memory lifetime, learning exceeds computer memory capacity and innovation exceeds computer memory storage abilities. Hence, (AI) brain innovation must need to exceed general computer storage, lifetime, capacity, learning abilities if (AI) scientists expect (AI) robots will be popular to be applied by any service, manufacturing, education industries.

How can web intelligence need (AI) brain informatics?

Brain informatics (BI) invention will be a new inter-disciplinary field that systematically studies the mechanisms of human information processing from both the macro and micro view points by combining experimental cognitive neuroscience with advanced information technology. (BI) studies human brain from the viewpoint of informatics (i.e. human brain is an information processing system) and uses informatics (i.e. WI centric information technology) to support brain science study. It seems that (AI) scientists need to further understand how human intelligence and brain sciences development through brain sciences fosters innovative web intelligence research and development because innovative web intelligence research will have ability to assist (AI) scientists to research how to invent (AI) brain robotic technology more easily. The synergy between (web informatics) (WI) and (brain informatics) (BI) advances our ways of analyzing and understanding of data, knowledge, intelligence, and wisdom, as well as their interrelationship, organizations and creation processes. Web intelligence is becoming a central field that information technologies and artificial intelligence to achieve human level web intelligence.

(WI) may be viewed as applying results from existing disciplines , e.g. artificial intelligence (AI) and information technology (IT) to a totally new domain the world wide web. (WI) may be considered as an entrancement or an extension of (AI) and (IT), (WI) introduces new problems and challenges to the established disciplines.

Thus, developing human-level web intelligence will be seem to develop brain level artificial intelligent technology. Because brain informatics (BI) is an interdisciplinary field to systematically investigate human information processing mechanisms from both macro and micro points of view by cooperatively using experimental, computational, cognitive neuroscience, and advanced (WI), centric information technology. It attempts to understand human intelligence in depth, towards a holistic view at a long term, global vision to understand the principles and mechanisms of human information processing system (HIPS).

So, I recommend (AI) scientists needs to research how to invent brain informatics technology and web informatics technology to achieve how to apply (AI) brain to analyze and judge or mind web informatics ability to compete internet (web site) communication tool product industry. Hence, if (AI) brain can be applied to web informatics industry will increase attraction to an internet user market in the future.

Why model of sustainable development environment industry be (AI) robot attractive market

In the world scientific literature various conceptions of sustainable development are found, however, their basis are formed by three dimensions: environmental, economic and social development. The biggest attention is concerned to the environmental dimension which forms the basis of existence of social environment and economy. The reason is emphasized that each person must preserve and manage natural resources as the basic of economic and social development. However, in the sustainable development strategy, the sustainable development is understood as among environment protection, economic and social society that form the basis to achieve the universal welfare for present and future generations, it is difficult to let human to adapt to live in pollution environment. Thus, environment pollution will be human's concerning issue. It implies how to protect environment pollution which will be human's future challenge. If (AI) brain invention can be applied to how to solve environment pollution challenge. It is very attractive attribution to human (AI) brain environment protection function can be invented to include such as : how to apply (AI) brain to do mind to give opinions or do any environment protection behaviors/activities to assist human how to effective use of natural resources, how to effective use of universal economic society's welfare, do strong social guarantees during the period of strategy's implementation (until 2020 year to achieve global (AI) robots environment protection mission from (AI) robots' behavioral assistance. Because environment pollution will influence world's climate to be worse to cause our food or vegetable can not grow up easily. Then, human will face food / vegetable shortage challenge. If (AI) brain invention can be applied to solve environment pollution aspects, then human will avoid food / vegetable shortage crisis.

How to invent (AI) brain's environment protection ability?

All similar problems significantly promotes scientists to research for new technologies of data extraction progressive software of data extraction operating on the basis of artificial neural networks allow finding the relations among various types of data in the huge data. Due to the data extraction technologies, it is possible to prove empiric observations to group, to process and model big amounts of data, distinguishing unknown schemes in data and using them in future activities (Rotman M. J., 1995).

It seems scientists believe gather every country's climate environment data to predict environment climate change will have chance to avoid environment pollution challenge. In addition, the traditional statistic methods applied to process digital data of the indicators of sustainable development environmental dimension are not able to analyze the present situation of environmental dimension in the context of sustainable development to present possible reasons of change of environmental dimension problems to generate future forecasts characterized by a high level of accuracy.

Consequently, if (AI) scientists can apply (WI) web informatics technology and (BI) brain informatics technology both to apply to (AI) brain technology to gather global climate environment daily change data. Then, I believe (AI) brain invention can assist human to solve future environment pollution challenge Then, it is possible that (AI) brain can attribute to environment protection or how to give opinions or choose to do any environment pollution activities to avoid global warming crisis.

(AI) Asia market

When (AI) brain is successful invention, I think Asia will be a new market to need (AI) brain attribution for Asia consumer needs. I believe Asia people expect (AI) can attribute to let them to use. The (AI) ability includes the ability of machines and systems to acquire and apply knowledge, and to carry out intelligent behavior.

This includes a variety of cognitive tasks (e.g. sensing, processing, oral language, reasoning, learning, making decision) and demonstrating an ability to move and manipulate objects according). So, future Asia people (AI) consumers expect (AI) robots can attribute to whose society's needs, such as: how to apply intelligent systems to use a combination of big data analytics, cloud computing, machine-to-machine communication and the internet of things (IOT) to operate and learn. Asia people expect (AI) robots can give beneficial attribution to them, e.g. talking or playing a game for Asia young entertainment market; (AI) robots need to reflected by physical substance (such as any talking or playing game robot player).

In this sense, (AI) is like a human brain. For Asia (AI) robot service industry need, Asia (AI) robot clients expect to use soft robotics (robotic process automation) can be used to meet Asia (AI) robot service consumers' expectation of automation repetitive tasks and common processor needs, such as client servicing and sales without the need to transform existing IT system maps (e.g. (AI) salespeople robots, or (AI) service robotic).

In (AI) office task aspect, Asia office consumers expect algorithmic game theory and computational social choice of (AI) robots to be attributed to replace some office human workers' tasks. Such as (AI) systems tat address the economic and social computing dimensions of (AI), such as how systems can handle potentially incentives, including self-interested human participants or firms, and the automated (AI) -based agents representing them, e.g. complex or simple office administrative tasks, e.g. typing, accounting, filing, etc. general office tasks which need human office workers who use computers to work to be replaced by (AI) robots to do. So, Asia office (AI) robots users who expect any office human administration tasks can be replaced by (AI) robots to do.

In Asia computer vision market, Asia computer vision (image analytics), users expect (AI) robots can be replaced to human computer image workers to shorten time to work, or raise image quality to be more clear in the process of pulling relevant information from an image or sets of images to advanced classification and analysis. Such as hospital or clinic x-ray image vision (AI) robot invention, photo image (AI) robot invention etc. any Asia image industry (AI) robot market need.

In Asia collaborative systems work with human (AI) robot market, Asia autonomous systems robots users who expect (AI) robots can be applied its models and collaborative systems to help them to develop autonomous systems that can work collaboratively with other systems and with humans, e.g. car manufacturing, computer manufacturing or any machine manufacturing products. So, Asia machine related manufacturing product industry manufacturers who expect to apply (AI) robots who can assist factory human manufacturing workers to manufacture any products in the short time efficiently.

In Asia language teaching (AI) robot market, language educators expect (AI) robots own natural language processing ability, algorithms that process human language input and convert it into understanding representation, such as Asia translation education market (PWC).

Thus, Asia language teaching businessmen expect (AI) brain invention which needs to be designed to own these above abilities of (AI) robot's manufacturers expectation to provide to them to use from any (AI) robots import.

Is the brain a good model for machine intelligence?

The actions of a human " computer" using paper and pencil to perform a calculation (as the world meant), into a formalized machine, manipulating symbols on an infinite paper tape. But I believe it still has challenges to influence human brain can be invented to machine intelligence successfully.

I think (AI) scientists need to solve these further challenges to influence (AI) brain invention success. The limitations include such as below:

Computation is based on functions of integers is limited. They must let computer data can change to words, then words can change images, then any images can change to storage to let (AI) brain to remember to do any analytical and judgement able tasks to make any actions in the short time finally. It is (AI) brain behavioral and analytical mind speed limitation.

Moreover, anther limitation concerns biological systems clearly difference, they must respond to varied stimuli over long period of time, those responses any changes alter their environment and subsequent stimuli. The individual behaviors of social insects, for example, are affected by the structure of the home, they build, and the change their behaviors. Nowadays, (AI) brain invention is called computational neuro science, which have assured that the brain is a computer, it means a machine that is algorithms and architectures. Second, neuro science findings may validate the energy ability of existing algorithms being integral parts of a general (AI) system. It means (AI) robots change adapting environment limitation. It means how (AI) brains adapt to choose to make any analytical mind as well as how to be influenced by external environment to do their behaviors or actions in the efficient way, e.g. how to manufacture many cars efficiently in one factory in the short time.

Consequently, to solve these limitations, we need to know how to apply correct conceptual knowledge to let (AI) robots to learn, e.g. for example, if we know how conceptual knowledge was formed from perceptual inputs, it would crucially allow for the meaning of symbols in an artificial language system to be grounded in sensory " reality". When (AI) scientists can achieve how to solve all above limitation challenges, then (AI) brain invention will achieve more easily.

Must Developed And Developing Countries
Different Artificial Intelligent Development Stages

Must developed and developing countries need artificial intelligent development? If one developed country, e.g. US, UK , Japan , Singapore it does not continue to develop artificial intelligence, robotic, then what disadvantges or weaknesses , it will encounter to compare when it chooses to continue to develop this artificial intelligent technology in society. If one developing country, e.g. China, Korea, Taiwan, it does not continue to develop artificial intelligence, robotic, then what disadantages or weaknesses, it will also encounter to compare when it chooses to continue to develop this artificial intelligent technology in in society. I shall explan the reasons why the results may cause to either the developed country, or the developing country as below:

- How AI help developing countries to communication and agriculture and learning and medical delivery development

Why can AI help developing countries ? Drones that pick inaccessible crops and mobile phones that give medical advice are two of the ways AI can transform life in the developing world. Artificial intelligence (AI) may improve the lives of the world's poor, the technology needed to revolutionise inefficient, ineffective food and healthcare systems in developing countries is well. For example, in low-income areas, agriculture and healthcare are two critical

ecosystems that we can apply AI to immediately; this is not the far future, or even in five years.
Artificial intelligence (AI) has seeped into the daily lives of people in the developed world. From virtual assistants to recommendation engines, AI is in the news, our homes and offices. There is a lot of potential in terms of AI usage, especially in humanitarian areas. The impact could have a multiplier effect in developing countries, where resources are limited.

Emergency Response to developing countries' earthquake natural damage suddence occurrence predicting

AI and machine learning are still finding importance in emerging markets, but certain applications have emerged and are now widely used. For instance, predictive models for disaster relief enable first responders to automatically analyze large-scale behavior and movement through multiple sources of data including social media platforms, web forums, news sources, etc. Based on collected data, responders can scale reconstruction efforts and distribute supplies in a timely manner.
Why and how AI can assist farmers to predict when the earthquake occurs suddenly in order to avoid or reduce the natural damage to their agriculture productive number loss. For example, In 2015, when a major earthquake hit Nepal, more than 8 million people were affected. During the aftermath, drones were used to map and assess the destruction and speed up the rescue mission. The town of Sankhu, situated about 20 kilometers northeast of Kathmandu, was among the highly affected locations. In May 2018, my company Fusemachines and GeoSpatial Systems partnered with Sankhu's city officials to use drones and artificial intelligence in an effort to automatically estimate the reconstruction need. After processing data accumulated from a drone-powered aerial mapping of the region, the team fed this data to advanced machine learning algorithms. Combining drone imagery, digital mapping and machine learning, the team configured region modeling and infrastructure development with higher accuracy. Another organization known as One Concern, a California-based startup, has created a predictive AI program called Seismic Concern to accurately predict seism and is also working on solutions for wildfires, floods and hurricanes.

Smart AI Agriculture

Another application of AI in developing countries is smart agriculture. Farmers monitor crops more effectively and make better predictions on planting, weeding and harvesting using AI tools. It can also be used to analyze one plant at a time and add pesticides only to infected plants and trees instead of spraying pesticides across large swaths of crops. One California-based tech company is an example of this use of AI. So, the developing countries farmers in rural parts of India are also using AI to increase yields through better access to information about the farming season than they would normally have. Technology-enabled process automation offers the agribusiness industry the chance for remarkable growth -- not only in developed countries but around the world. There's a unique opportunity to increase yields, cut down labor costs and improve people's health.

Medicine Delivery to developing countries' patients urgent need

Companies are also leveraging AI to improve access to health care in some of the most remote areas of the world. In Rwanda, for example, Zipline is using drones to deliver medical supplies and blood to hospitals and clinics that are difficult to access by car. This has dramatically impacted people living in remote parts of the country because they are able to get medical help when needed. The drone system in Rwanda has also helped reduce waste of blood by 95%, as noted by Zipline. One Concern has created an AI program called Seismic Concern that accurately predicts seismic events and is also working on solutions for floods, wildfires and hurricanes. The medical field may actually benefit the most from emerging technologies in developing countries.

Assistance to reduce teaching work workload or psychological pressure to teachers in developing countries' schools

Another vital area benefiting from innovative technologies like AI is education. Advanced technologies can enhance how we learn, teach and perform tasks. In most developing countries, schools lack experienced teachers and resources to enhance students' knowledge. As a result, many students still have to walk long distances to get to the nearest school, which has created education gaps, especially in rural areas. AI tools such as personalized learning assistants can simplify learning by making tutoring services and learning materials accessible to all students, wherever they are. Machines can be automated to help students learn basic concepts without a tutor, which companies like Carnegie Learning are working on. This would allow students to learn at any time from anywhere. With AI, education is made easy and accessible to more people.

The initial usage of AI in developing countries has been at a micro level -- solving small, specific problems in a defined industry. As machine learning advances and there is a higher utilization of AI, we will see more complex issues being targeted and resolved. When duly adopted, AI can positively impact future developing countries people everyday lives not just in disaster intervention, education, health care and agriculture but can also help in mitigating poverty, malnutrition and pollution. Especially, in developing nations, to leverage AI's true potential and create a snowball effect. Startups are defining a holistic and humanitarian approach to building more sophisticated, AI-ready societies. Stakeholders in the AI landscape should understand the strengths and nuances of the developing world as well as the limitations of AI and create localized solutions and applications.

Why does smart phone help developing countries communication ?

Internet Seen as Positive Influence on Education but Negative on Morality in Emerging and Developing Nations. Internet access differs substantially across the 32 emerging and developing countries polled, with the lowest rates of internet use in South Asian and sub-Saharan African nations. Within countries, computer owners, young people, the well-educated, the wealthy and those with English language ability are much more likely to access the internet than their counterparts. To access the internet, people increasingly use smartphones rather than more cumbersome fixed landline connections and computers. Around the world, both smartphones and basic-feature phones alike are used for sending messages and taking pictures.

In fact, many developing countries young people, students are popular to use smart phones for internet usage aim, instead of communication. Moreover, many developing countries working people are also popular to use smart phones for any working usage in their working time , even non working time any time. So, smart phones (AI) phones will be important communication or leisure tools to developing countries people in the future. Unless, it is one day, scientists can develop another new communication tool to replace smart phones. So, artificial intelligence will be important to influence developing countries people , how to improve or bring positive learning attitudes to students in their daily learning lives. as well as how to raise developing countries people, how to raise working people efficiency or improve performance in their daily working lives. So, AI may bring positive learning or working attitudes to developing countries working people and students both.

The Positive Impact of Mass Media in Developing Countries

Radio, newspapers, television, Internet, social media, etc., all of these are forms of mass media. Each of these outlets has the capability of bringing information to thousands of people with one device. While in some communities it is easy to take advantage of these communication outlets such as television and Internet access, not everyone has access to such outlets. Radio is one of the most common forms of mass media in developing countries because it's affordable and uses less electricity than many other forms of mass media, but only approximately 75 percent of people in developing countries have access to a radio, and roughly 77 percent of people in rural areas have access to electricity.

For developing countries that have implemented forms of mass media in their communities, there have been numerous positive outcomes are influenced to impact developing countries mass media by artificial intelligence as below:

When AI is participated to developing countries mass media, it can influence any radio, television audiences raise more attention to each other through social media platforms such as Facebook and Twitter and create, organize and initiate street protests and campaigns. Furthermore, having access to social media in developing countries, people are able to connect to those that they usually wouldn't have the chance to talk to. Moreover, AI Provides educational opportunities- In many countries, the division between local and national languages as well as issues of literacy can make communication difficult. With the use of mass media, a bridge can be built between these two gaps. In India, there is a radio station that provides information in local languages and respects local culture and traditions. One of the main ways is to create public awareness of what is going on with businesses and government officials. The media plays an important role in giving people the opportunity to act against injustice, oppression and misdeeds that they otherwise wouldn't know about. Information on available healthcare, a mass radio broadcast was sent out encouraging parents to seek treatment at local healthcare facilities for their sick children. With this mass outreach on healthcare, the encouragement of people to take their children to healthcare facilities saved thousands of lives. This easy way of encouraging others and bringing awareness about certain diseases was made possible through a simple radio broadcast. Finally, when AI is participated to media, it may bring many social issues to life that otherwise would remain unknown to many people. In developing countries and communities like Burkina Faso, when the radio broadcast was released about malaria, diarrhea and pneumonia, people were educated and moved to action and knew to take their children to healthcare facilities for preventative care. As it is seen, having access to different media outlets is vital for those in developing countries. Here are three ways that those in developing countries can implement mass media to help their people and communities.

When AI is participated to any internet radio or internet newspaper mass online listening or reading channel. It can provide online radios or newspapers in public places- By providing online radios and newspapers in public areas it gives community members to access news, information and emergency warnings. Even though radios can be on the cheaper side, there are still many people that can't afford to have a radio in their home. By providing one in a local place, not only would it better educate the community members but also it will bring the community together. So, it can make media outlets a two-way platform- Creating a two-way platform between the community and those who are behind the radio stations, newspapers or broadcasts makes the community feel involved and that their voices are being heard. An organization called Soul City in sub-Saharan Africa is showing how well two-way platforms work by engaging their listeners and having them contribute thoughts and ideas about complex issues. Because developing countries radio listening audiences or newspaper readers are popular to accept computer online radio listening channel or online newspaper reading channel to replace traditional paper newspapers or radio machines. So, AI may raise their listening news or reading news leisure feeling from online mass media channel in the future.

● Why do developed countries need to develop AI

Artificial intelligence, or AI, is driving massive shifts across the globe, and every day more questions arise. What impact will AI have on the workforce and how can we prepare for it? How can we encourage economy-boosting and job-creating technologies? How can we ensure that AI will be implemented ethically and with minimal bias? How will society benefit? For developed country, such as US example. None of the US, Israel and Russia have a formal national AI policy yet. Private sector companies such as Google, Amazon and Apple and the US department of defence are driving the bulk of AI investment in the United States. Though Israel does not have a specific policy, it is keenly focused on AI and has seen the number of AI start-ups triple since 2014.

Developed country may learn whether what weakness it is lacking when it does not continue to develop AI from one another developed country. Which countries are approaching AI most effectively, and to what degree is there opportunity for greater international collaboration? It may be too early to tell; however, when analyzing the best practices of existing national AI policies, there is much that can be learned. These are the specific areas to consider.

When one developed country continue to develop or research AI, it may bring these benefits as below:

On gathering Data aspect, from self-driving vehicles to smart cities, data is the driver behind AI. Innovation in the United States is limited without a national strategy that answers questions about protocol and ownership. France and Denmark, on the other hand, are opening government data. France is hosting troves of centrally collected public and private data that it plans to make available as part of its strategy. Conversely, by taking a restrictive position on issues of data collection (as indicated by the implementation of General Data Protection Regulation), the EU is putting manufacturers and software designers at a disadvantage while balancing the demand for privacy. On raising technologica talent aspect, the demand for AI talent far outweighs the available supply. As a result, almost every nation's strategy addresses talent development. Canada's AI strategy is distinct in that it primarily focuses on research and talent strategy. The country boasts AI degree programmes and is building a $127 million research facility in Toronto. Companies like Facebook and my own company, Uptake, are investing in Canada to access this talent pool. On AI legal technological innovation aspect, a whole host of legal questions swirl around AI. The country is developing a bill for AI liability that will be ready in March 2019. The government hopes the legal framework will attract investors by providing a simple, comprehensive guideline to enable the broad use of AI systems. So, when the developed country applied AI technology to assist any lawyers to work, then AI can help them to reduce the workload to draft any legal documents more easier. So, any developed countries lawyers' draft legal documents time must reduce if the developed countries lawyers accept to apply AI to assist their legal works. One of the great promises of AI is its potential for improving quality of life. But without the right planning and oversight, we risk exacerbating problems of inequality or marginalizing groups of people. As an example, India's AI strategy is focused on leveraging the technology not only for economic growth, but also for social inclusion.

AI may bring what benefits to developed countries

From SIRI to self-driving cars, artificial intelligence (AI) is progressing rapidly. While science fiction often portrays AI as robots with human-like characteristics, AI can encompass anything from Google's search algorithms to IBM's Watson to autonomous weapons. Artificial intelligence today is properly known as narrow AI (or weak AI), in that it is designed to perform a narrow task (e.g. only facial recognition or only internet searches or only driving a car). However, the long-term goal of many researchers is to create general AI (AGI or strong AI). While narrow AI may outperform humans at whatever its specific task is, like playing chess or solving equations, AGI would outperform humans at nearly every cognitive task.

Why research AI safety? Would AI bring war when AI is continued to develop by developed countries? In the near term, the goal of keeping AI's impact on society beneficial motivates research in many areas, from economics and law to technical topics such as verification, validity, security and control. Whereas it may be little more than a minor nuisance if your laptop crashes or gets hacked, it becomes all the more important that an AI system does what you want it to do if it controls your car, your airplane, your pacemaker, your automated trading system or your power grid. Another short-term challenge is preventing a devastating arms race in lethal autonomous weapons.

In the long term, an important question is what will happen if the quest for strong AI succeeds and an AI system becomes better than humans at all cognitive tasks. As pointed out by I.J. Good in 1965, designing smarter AI systems is itself a cognitive task. Such a system could potentially undergo recursive self-improvement, triggering an intelligence explosion leaving human intellect far behind. By inventing revolutionary new technologies, such a superintelligence might help us eradicate war, disease, and poverty, and so the creation of strong AI might be the biggest event in human history. Some experts have expressed concern, though, that it might also be the last, unless we learn to align the goals of the AI with ours before it becomes superintelligent.

There are some who question whether strong AI will ever be achieved, and others who insist that the creation of superintelligent AI is guaranteed to be beneficial. At FLI we recognize both of these possibilities, but also recognize the potential for an artificial intelligence system to intentionally or unintentionally cause great harm. We believe research today will help us better prepare for and prevent such potentially negative consequences in the future, thus enjoying the benefits of AI while avoiding pitfalls.

How can AI be dangerous when developed countries continue to develop AI to become weapon to replace soldiers?
Most researchers agree that a superintelligent AI is unlikely to exhibit human emotions like love or hate, and that there is no reason to expect AI to become intentionally benevolent or malevolent. Instead, when considering how AI might become a risk, experts think two scenarios most likely:
The AI is programmed to do something devastating: Autonomous weapons are artificial intelligence systems that are programmed to kill. In the hands of the wrong person, these weapons could easily cause mass casualties. Moreover, an AI arms race could inadvertently lead to an AI war that also results in mass casualties. To avoid being thwarted by the enemy, these weapons would be designed to be extremely difficult to simply "turn off," so humans could plausibly lose control of such a situation. This risk is one that's present even with narrow AI, but grows as levels of AI intelligence and autonomy increase.
The AI is programmed to do something beneficial, but it develops a destructive method for achieving its goal: This can happen whenever we fail to fully align the AI's goals with ours, which is strikingly difficult. If you ask an obedient intelligent car to take you to the airport as fast as possible, it might get you there chased by helicopters and covered in vomit, doing not what you wanted but literally what you asked for. If a superintelligent system is tasked with a ambitious geoengineering project, it might wreak havoc with our ecosystem as a side effect, and view human attempts to stop it as a threat to be met. So, a super-intelligent AI will be extremely good at accomplishing its goals, and if those goals aren't aligned with ours, we have a problem. You're probably not an evil ant-hater who steps on ants out of malice, but if you're in charge of a hydroelectric green energy project and there's an anthill in the region to be flooded, too bad for the ants. A key goal of AI safety research is to never place humanity in the position of those ants.
Why the recent interest in AI safety ?
Stephen Hawking, Elon Musk, Steve Wozniak, Bill Gates, and many other big names in science and technology have recently expressed concern in the media and via open letters about the risks posed by AI, joined by many leading AI researchers. The idea that the quest for strong AI would ultimately succeed was long thought of as science fiction, centuries or more away. However, thanks to recent breakthroughs, many AI milestones, which experts viewed as decades away merely five years ago, have now been reached, making many experts take seriously the possibility of superintelligence in our lifetime. While some experts still guess that human-level AI is centuries away, most AI researches at the 2015 Puerto Rico Conference guessed that it would happen before 2060. Since it may take decades to complete the required safety research, it is prudent to start it now.
Because AI has the potential to become more intelligent than any human, we have no surprise way of predicting how it will behave. We can't use past technological developments as much of a basis because we've never created anything that has the ability to, wittingly or unwittingly, outsmart us. The best example of what we could face may be our own evolution. People now control the planet, not because we're the strongest, fastest or biggest, but because we're the smartest. If we're no longer the smartest, are we assured to remain in control?

A captivating conversation is taking place about the future of artificial intelligence and what it will/should mean for humanity. There are fascinating controversies where the world's leading experts disagree, such as: AI's future impact on the job market; if/when human-level AI will be developed; whether this will lead to an intelligence explosion; and whether this is something we should welcome or fear. But there are also many examples of of boring pseudo-controversies caused by people misunderstanding and talking past each other. When one developed country continue to develop AI, can itself country's all factories workers will lose jobs, due to AI can replace them to do simple works in factories, or any public transport drivers, e.g. bus drivers, ferry , tram, train drivers, they will lose jobs, when AI (non manual driving drivers) can replace all public transport drivers. So, some occupations will lose if developed countries continue to develop or research AI to replace human to do some simple jobs, such as some cooking jobs can be done by AI. So, it is possible that future cookers won't be needed, because AI cooking skills may be better than them to cook any good taste chinese or western food in restaurants. If you drive down the road, you have a subjective experience of colors, sounds, etc. But does a self-driving car have a subjective experience? Does it feel like anything at all to be a self-driving car? Although this mystery of consciousness is interesting in its own right, it's irrelevant to AI risk. If you get struck by a driverless car, it makes no difference to you whether it subjectively feels conscious. In the same way, what will affect us humans is what superintelligent AI does, not how it subjectively feels.

In fact, AI may be make any brokers jobs in financial market. the main concern of the beneficial-AI movement isn't with robots but with intelligence itself: specifically, intelligence whose goals are misaligned with ours. To cause us trouble, such misaligned superhuman intelligence needs no robotic body, merely an internet connection – this may enable outsmarting financial markets, out-inventing human researchers, out-manipulating human leaders, and developing weapons we cannot even understand. Even if building robots were physically impossible, a super-intelligent and super-wealthy AI could easily pay or manipulate many humans to unwittingly do its bidding. So, future brokers will be replaced by AI, when AI can be made to own financial brokers' analytical mind to make more accurate whether the share price will rise up or fall down to compare human financial brokers' analytical mind. The robot misconception is related to the myth that machines can't control humans. Intelligence enables control: humans control tigers not because we are stronger, but because we are smarter. This means that if we cede our position as smartest on our planet, it's possible that we might also cede control.

Not wasting time on the above-mentioned misconceptions lets us focus on true and interesting controversies where even the experts disagree. What sort of future do you want? Should we develop lethal autonomous weapons? What would you like to happen with job automation? What career advice would you give today's kids? Do you prefer new jobs replacing the old ones, or a jobless society where everyone enjoys a life of leisure and machine-produced wealth? Further down the road, would you like us to create superintelligent life and spread it through our cosmos? Will we control intelligent machines or will they control us? Will intelligent machines replace us, coexist with us, or merge with us? What will it mean to be human in the age of artificial intelligence?

Why do developed countries people need AI ?

Why do we assume that AI will require more and more physical space and more power when human intelligence continuously manages to miniaturize and reduce power consumption of its devices. How low the power needs and how small will the machines be by the time quantum computing becomes reality? Why do we assume that AI will exist as independent machines? If so, and the AI is able to improve its Intelligence by reprogramming itself, will machines driven by slower processors feel threatened, not by mere stupid humans, but by machines with faster processors? What would drive machines to reproduce themselves when there is no biological incentive, pressure or need to do so?

Who says superior AI will need or want to have a physical existence when an immaterial AI could evolve and preserve itself better from external dangers. What will happen if AI developed by competing ideologies, liberalism vs communism, reach maturity at the same time, will they fight for hegemony by trying to destroy each other physically and/or virtually. If AI is programmed to believe in God, and competing AI emerges programmed by muslims, christians or jews, how are the different AI's going to make sense of the different religious beliefs, are we going to have AI religious wars? What if the "powers that be" greatest fear is the emergence of a super AI that police's and rationalizes the distribution of wealth and food. A friendly super AI that is programmed to help humanity by, enforcing the declaration of Human Rights (the US is the only industrialized country that to this day has not signed this declaration) ending corruption and racism and protecting the environment.Most benefits of civilization stem from intelligence, so how can we enhance these benefits with artificial intelligence without being replaced on the job market and perhaps altogether?

Key to the process of machine learning are neural networks. These are brain-inspired networks of interconnected layers of algorithms, called neurons, that feed data into each other, and which can be trained to carry out specific tasks by modifying the importance attributed to input data as it passes between the layers. During training of these neural networks, the weights attached to different inputs will continue to be varied until the output from the neural network is very close to what is desired, at which point the network will have 'learned' how to carry out a particular task. A subset of machine learning is deep learning, where neural networks are expanded into sprawling networks with a huge number of layers that are trained using massive amounts of data. It is these deep neural networks that have fuelled the current leap forward in the ability of computers to carry out task like speech recognition and computer vision.

In conclusion, when developed countries continue to develop AI, it may bring positive advantages to bring raising productivies, or efficiencies, but it may also raise unemployment ratio to any low skill or low knowledge jobs in ther societies. However, human future society will need to change to be better to raise our living standard. But AI is one kind the best choice tool to achieve this aim in our future, so I agree developed countries continue to develop or research AI to be the super -human machine.

Reference

PWC, Sizing the prize, see: https:// www. pwc. com/gx/en/issues/data-and- analytics/publications/artificial-intelligence- study.html.

Rotman M. J., Data mining-a practical approach to database marketing (1995). IBM.

CHAPTER FIVE

How robots help manufacturers to raise productivity

Some scientists explain that artificial intelligence means which is an expert system, computer software that embodies a portion of the specialized knowledge of a human portion in a specific, narrow domain, owns decision making ability of human expert. The (AI)technology is based on the premise that what makes a person an expert is years of experience that enables who recognizes certain patterns in a problem as being similar to pattern. For example, in the future artificial intelligence system can be applied to control air traffic, design to computer configuration, medical diagnosis, instruction/training, speech/interpretation, monitoring to (nuclear plant), planning to mission, factory scheduling, prediction weather, repairing telephone, automatic driving etc. different industries.

Artificial intelligence characteristics include: creative, adaptive , common sense, fact processing, quick replication, broad focus permanent and consistent skill. Otherwise, traditional computer expert system disadvantage includes perishable, unpredictable, slow reproduction, expensive, slow reproduction, slow processing lacks inspiration, needs instruction, narrow focus only machine knowledge. So, artificial intelligence is a branch of computer science devoted to creating computer to influence software and hardware to attempt to create human intelligence or human intelligent behavior. It is learning from experience, responds flexibility in situation that are, new or not anticipated.

Thus, (AI) can be learnt programmed knowledge to solve problems, using reasoning in solving problem, understanding and inferring facts and rules, recognizing the relative importance of different elements in a situation. In summary, artificial intelligence is concerned with two basic ideas mainly: The first idea, it involves studying the thought processes of humans to understand what intelligence is; the second idea, it deals with representing thought processes using companies to create artificially intelligent entities for testing the theories of intelligence.

It brings this question:

Can AI apply these strengths to assist human to raise productivity?

- Can (AI) impact human job nature?

Human need concern these two questions:

Will artificial intelligence (AI) reduce some human jobs in order to instead of replacing machines to do?
If (AI) can replace human some jobs, does it reduce productivity or raise productivity ?
Due to artificial intelligence is the ability of machines to do thing, that people would require intelligence. For example, artificial intelligence machine man driving(self-driver), it (AI) machine man driving research is an attempt to discover and describe aspects of human intelligence that can be simulated by driving machine functions. Alternatively, (AI) mathematical research may be another viewed as an attempt to develop a mathematical theory function to describe the abilities and actions of things (natural or man-made) exhibiting intelligent behavior and server as a design of intelligent calculation machine function.
Can artificial intelligence machines better than manual to replace traditional human service job? For example, can artificial intelligence machine man (self-driving) driver drive to replace human driver?
Is (AI) replaced to human any job, Can it bring positive impact to raise more productivity than human ?
I shall compare the differences between humans and computers function as below:
The characteristics of humans are good at recognizing various things, either seen before or not, recognizing the relationship patterns between things. Human thinking is common sense reasoning, combining all types of sensory input, acting appropriately in novel situations, learning new things and changing behavior patterns, making decisions , even when given incomplete information, working with noisy, incomplete information gathering behaviors . However, characteristics of computers are good at: The tasks humans do naturally are extremely difficult for a computer program as intelligent, which must be able to do the same kind of tack as humans do naturally.
Hence, it seems that (AI) can raise productivity, due to it is manufactured computer software and computer software can do complex jobs to compare human, even human can not do these complex jobs.
Hence, (AI) is an combination of many different success and technologies: Linguistics - computational and socio, philosophy-logic, philosophy of mind and of language, electronical engineering -image and speech processing, pattern recognition, robotics, machine learning, neural networks, optimization scheduling, management information system and decision making. So, it is possible that (AI) can impact human job nature to raise more productivity to do better than human working behavior in the future.

How can (AI) influence labor market?

● How can human society job nature
to be changed to artificial intelligent society?

From the first intelligent perspective reason view point, artificial intelligence is making machines " intelligent" acting as humans expect people to act. Artificial intelligence has ability to distinguish computer responses from human responses, it owns knowledge to solve expert problem. From another research perspective reason view point, artificial intelligence is the study of how to make computers do things which, at the moment, people do better (Rich & Knight, 1991, p.3).
(AI) researchers are native in a variety of domains, e.g. formal tasks (mathematics, games), tasks (perception, robotics, natural language, common sense reasoning), expert tasks (financial analysis, medical diagnostics,

engineering, scientific analysis and other areas).
From the second business perspective reason view point, (AI) is a set of many powerful tools, and methodologies for using those tools to solve business problems. From a programming perspective reason view point, (AI) includes the study of symbolic programming problem solving and search .
From the third human technological perspective reason view point, today's computer can do many well-defined tasks, for example, arithmetic operations, are much faster and more accurate than human beings. However, the computers' interaction with their environment is not very sophisticated yet. How can human test whether a computer has reached the general intelligence level of a human being? Can a computer convince a human interrogator that it is a human? But before thinking of such advanced kinds of machines, human will start developing our own extremely simple " intelligent" machines.
So, it is possible that human society job efficiency and performance and productivity will to be changed better, due to artificial intelligent robots are applied to assist human to do anything when (AI) technology is developed to the mature stage in the future.

- Why does human need artificial intelligence machines?

One of major division in (AI) is between humans who think (AI) is the only serious way of finding out how we (human) work and human who want companies to do very smart things, independently of how we (human) work. This is the important distinction between cognitive scientists vs engineers.
One of another major division in (AI) is between symbolic (AI), which represents information through symbols and their relationships. Specific Algorithms are used to process these symbols to solve problems or deduce new knowledge and connectionist. So (AI) , which represents information in network. Biological processes underlying learning, task performance and problem solving are imitated from human mind behaviors.
Thus, it is possible that artificial intelligence machines can do the better human task behavior to achieve productivity and efficiency raising.

- How does artificial intelligence influence future working changing in automation employment and productivity aspects?

(AI) influences the automation working changing aspect, as companies increasingly use robots on production lines or algorithms to optimize their logistics manage inventory, any carry out other core business functions. Technological advances are creating a new automation age in which ever-smarter and more flexible machines will be deployed on an ever larger scale in the marketplace. However, researching artificial intelligence with how influences human working nature. We need to answer these questions:
How will automation transform the workplace?
What will the implications for employment?
What is likely to be its impact both on productivity in the global economy and on employment?
Advances in robotics, artificial intelligence, and machine learning are growing in a new age of automation as machines match or outperform human performance in a range of work activities, including ones requiring cognitive

capabilities.

What factors are determined the changing in workplace adoption by artificial intelligence innovation?

What advantages are automation?

Automation of activities can be enabled businesses to improve performance by reducing errors and improving quality and speed, and achieving outcomes that go beyond human capabilities.

Some scientists indicated based on their scenario modeling. They estimated automation could raise producing growth globally by 0.8 to 1.4 percent annually. Almost, the activities people are paid almost $16 trillion in wages to do in global economy have the potential to be automated by adopting currently demonstrated technology. According to their analysis of more than 2,000 work activities across 800 occupations. When less than 5% of all occupations have of least 30% of activities that could be automated. They also indicated that technical economic and social factors will determine automation. Continued technical progress, for example, in areas such as natural language processing is a key factor beyond technical feasibility , the cost of technology, competition with labor including skills, and supply and demand dynamics, performance benefits including and beyond labor cost savings and social and regulatory acceptance will affect (alter) the scope of automation.

Other some scientists also indicate U.S. country for example, the anticipate shift in the activities in labor force of a similar order of magnitude as the long term sight away from agriculture and decreases in manufacturing. Share of employment in the United States both which were achieved. So, those factors can influence why artificial intelligence technology needs. So, it is possible that future agriculture and manufacturing both industries will apply (AI) technology manufacturer-kind of job nature to raise productivity instead of farmers, fruit picking workers, farming transportation labours as well as factory manufacturing workers and supervisors etc. human-kind of job nature.

● Is artificial intelligence possible to replace labor ?

Not just intelligence, but also debating, if machines are capable of having a conscious minds. Artificial intelligence has those characteristics as below:

On functionalism aspect, artificial intelligence inputs mental states, sensory inputs, (beliefs, desires being in pain feeling) and behavioral outputs. Since mental states are identified by a functional role, which are thoughts to be manifested in various systems. Even, perhaps computers which are physical devices with electronic substrate that inform computations on inputs to give outputs similar to brains which are artificial intelligence composed of part any intrinsic relationship to each other. Thus, artificial intelligence activities is not the whole itself, but into parts or on external influence on the parts.

On dualism aspect, artificial intelligence is a set of views about the relationship between mind are matter. On materialism aspect, it builds the only thing that exists is matter, including consciousness.

On biological naturalism aspect, it is similar a human brain than feels pains makes mental situation. So, artificial intelligence is similar biologist which might to be excited to human labor work. Hence, it seems artificial intelligence can change (alter) or replace human labor work of nature in order to raise efficiency and productivity in possible any working environment when it own human mind and effort absolutely.

- Can (AI) technology replace human labour nature of work to raise better productivity and efficiency?

On technological innovation reason view point, the history development of artificial intelligence studying the intelligence is one of most ancient scientific discipline. The history development of artificial intelligence what aims to achieve human use to sense, learn remember and think, logic probability, decision making and calculation develop from mathematics, instead of replacement human labor functions.

Artificial intelligence history development aim is the scientific analysis of skills in connection and practice with the appearance of computers from 1950 year beginning. The artificial intelligence (AI) can deal with the ultimate challenges. How can (either biological or electronic) mind sense, understand and manipulate a world that is much simple and more complex than itself? And what if would human like to construct something with such capabilities?

The general-purpose software of the early period of (AI) were only able to solve simple tasks effectively and failed when which should be used in a wider range or an more difficult tasks. One of the sources of difficulty was that early software had very few or mix knowledge about the problems which handled, and activities successes by simply syntactic manipulation. Moreover, the other difficulty was that many problems that were tried to solve by the (AI) were untreatable.

The early (AI) software whether trying step sequences based on the basic facts about the problem that should be solved, experimented with different combinations till which found a solution. From the end the 1960 year, developing the so-called expert systems were emphasized. These systems had (sue-based) knowledge base about the field which handled. Till to the beginning of the 1970 year, (Prolog) the logical programming language was born, which was built in the computation realization of a version of the resolution calculus. (Prolog) is a remarkably prevalent tool in developing expert systems (on medical, judiciary and other scopes), but natural language parsers were implemented in this language. Then, in 1981 s, the Japanese announced the fifth generation computer system project, a 10 years plan to build an intelligent computer system that use the (Prolog) language as a machine code. Nowadays, (AI) can be applied any industries, such as car manufacturing industry can use (AI) technological machine-men manufacture car, instead of replacing human labors in factory. Even, in the future, using (AI) machine-men drivers can drive any private cars or public transportation tools, instead of replacing human drivers, e.g. bus, train, tram, ferry etc. Also in the future, machine-men can replace housewives to serve families to do housekeeping clean job , e.g. cleaning toilets, bathrooms, kitchens, even cooking functions at home. So (AI) machine-man can reduce housewives works at home. Moreover, (AI) machine man can take care old people , when who are living at homes or elder care centers.

So, it seems artificial intelligence (AI) will be possible developed to manufacture a new generation machine-man to assist (serve) families to do any simply cleaning or cooking jobs at homes. Moreover, the overall demand of (AI) general social needs will also rise, such as security, driving transportation tools, restaurant cleaning, elder centers care service etc. (AI) will have chance often to practise to do its tasks every day. It is possible that it can improve worker individual simple job duty to raise productivity and efficiency to be fast.

- Why can artificial intelligence satisfy human needs?

First, On machine-man satisfactory demand aspect view point, it makes computers that think, it is the automation of activities. We associate with human thinking: like decision making, learning. It is the act of creating machine that perform function that require intelligence when performed by people. It is the study of mental faculties through the use of computational models. It is the study of computations that make it possible to perceive, reason and act. It is a branch of computer science that is concerned with the automation of intelligent behavior. It is anything in computing service that human don't yet know how to do property.

Second, on thought aspect artificial intelligence means systems thank think like humans, systems that think rationally.

Third, on behavioral aspect, artificial intelligence systems that act like human and that systems act rationally. However, the basic objective of (AI) is to represent human's thought processes in computation . These machines are supposed to exhibit behavior that. It is performed by a human being, would be considered intelligent. However, some authors feel (AI) has disadvantages, such as it is not creative, it is excited in the use of sensory devices, it can't make use of a very wide context of experiences and it does not use common sense.

For speech recognition and understanding function needs example, (AI) can be applied in speech recognition and understanding function, which (AI) speech or voice recognition is a data input method. For example, the computer recognizes and understands one (or a few) word commands. Speech understanding on the other hand is the computer's ability to understanding a spoken language. That is , the computer understands the meaning of sentences, an paragraphs through (AI).

So, (AI) can be attempted to learn human language how to speak. It is similar to translate human language skill, instead of actual human speaking skill. Also, (AI) can assist handicap learning or language student how to listen different languages by machine-man sounds from computers more accurately.

So, it seems that it (AI) can replace human language teachers speaking function and can change teaching language nature of job in language speaking and listening education industry. IT will improve its language speaking performance to teach students to raise teaching efficience and teaching performance to satisfy student learning needs in short time.

- Is artificial intelligence one good choice for human future technological benefit?

Nowadays, new technology development is popular. However, artificial intelligence is one kind of new technology choice among different technologies innovation. So it brings this question: Is artificial intelligence technology value to invest? To answer this question. I shall indicate some other new technology developments to compare (AI) technology development to judge which has urgent needs to achieve human expectation nowadays.

For example, why is green peace interested in new technologies? New technologies features prominently in our ongoing campaigns against genetic modified crops and number power. However, which are also an integral part of our solutions to environmental challenges, including renewable energy technologies, such as solar, wind and wave (water) power energy as well as waste treatment technologies, such as mechanical, biological treatment.

It seems humans need concern how to apply (AI) technology to solve environment pollution challenges in our future. So, environment protective, agriculture, natural energy technology will be popular demand to attempt to apply (AI) technology to solve their challenges or apply (AI) to assist to develop their industry. It can raise farming food and fruit number more than farmers. It seems that it can raise farming productivity.

What is relationship between
(AI) productivity and economy growth?

- How can artificial intelligence technology influence economy?

Advances in artificial intelligence (AI) technology and related fields have opened up new markets and new opportunities progress in critical areas, such as health, education, energy, economic development, social welfare and the environment pollution.
(AI) automation will continue to create wealth and expand the global economy development in the future. However, when many will benefits that growth won't be costless and will be accompanied by changes in the skills, that workers need to increase productivity in the economy and structural changes in the economy. So, in the skills that workers need to succeed in the economy and structural changes.
I shall indicate why aggressive policy action will be needed to help Americans who are disadvantaged by these changes , due to (AI) technology is caused. For automation industry change example, artificial intelligence (AI) capabilities will enable automation of some tasks that have long required human labor. These artificial intelligence technology introduction can increase new opportunities for individuals. The economy and society, but (AI) has also the potential to disrupt be current livelihoods of many Americans. However, (AI) leads to unemployment and increase in inequality over the long run depends not only on the (AI) technology itself, but also on the institutions and policies that are changed.
Thus, it is possible that (AI) technology will raise some countries unemployment number if the employer apply (AI) technology workers to replace human labor in their factories, but it can also raise productivities for these employers, when (AI) can spend less time to compare labor to do any human's same tasks in order to raise productivity.

- Can (AI) influence global economy growth?

Technological progress is main driver of growth of GDP per capita, allowing output to increase faster than labor and capital . However, technology can increase productivity, but also decrease the number of labor hours needed to create a unit of output. So (AI) causes unequal to labor wage decreases, even reduces the number of labor to manufacture, e.g. artificial intelligence technology of automation car manufacturing industry; clothing manufacturing industry; plane manufacturing etc. high technology of artificial intelligence manufacturing method. But (AI) should be potential environment benefit, although it raises unemployment ratio. Moreover, it can rise production , due to many skilled craft were replaced by the combination of machines and lower-skilled labor. The result of (AI) technology introduction , it causes output per hour risen when inequality declined, driving up average living standards, but the labor of some high-skill workers was no longer as valuable in the market. Otherwise, if (AI) technology is continue developed to be success. Some routine intensive occupations will be loss, which focused on predictable, e.g. easily programmable tasks, such as switchboard operators, filing clerks, travel agents, and assembly line workers would be particularly replaced by new (AI) technology. However, at the same time, (AI) technology development will bring these benefits: improvement in education (training (AI) technology scientists) , due to (AI) manufacturing technology needs are raising to businesses and institutional changes, such as the reduction in unionization and raising in the minimum wage to the (AI) manufacturing technology skilled labor in factories.

Because (AI) technology is not a single technology, but rather a collection of technologies that are applied to specific tasks, the effects of (AI) will be felt unevenly though the economy. It will bring some tasks will be most easily automated than others , and some jobs will be affected more than others, both negatively and positively. Finally, new jobs are likely to be directly created in areas , such as the development and supervision of (AI) as well as indirectly created in a range areas though out the economy as higher incomes lead to expanded demand.

However, if (AI) technology could dominate global labor markets. If labor productivity increases, do not influence to wage increases, then the large economic gains brought about by (AI) technology could be increased wealth inequality, due to employers can reduce production cost, but workers (labors) wages will not be increased, even will be decreased. Hence, it seems the (AI) technology will bring disadvantages to labor market to cause unemployment or reduce wages in possible, although it can reduce employer individual salary (wage) expenditure and it can raise productivity.

- How can artificial intelligence impact global economy growth?

Artificial intelligence (AI) technology is a branch of computer science that aims to create intelligent machines that work and react like humans. So, (AI) is a technology that appears to impact (influence) human preference by learning, understanding complex contents, enhancing humans in executing both routine and non-routine tasks. In the future, (AI) technology that can be virtual personal assistant, as well as it may exist, such as robots with human-like processing capabilities.

How can (AI) technology impact global economy growth over the next 10 years? During this time period, (AI) technology is predicted to have wide-ranging applications including: Machine learning that automates analytical model building by using algorithms that allow machines to operate without human assistance.

In global education aspect, potential applications include predicting cause-and-effect relationships from biological data, identifying new drugs, self-driving cars, and protecting against fraud, improved natural language processing that allows computers to continue to better analysis, understand and generate language to interface with humans using natural human languages. For example, transcribing notes dictated by physicians, automatically drafting articles and translating text and speech. So (AI) technology can be applied to education aspect to improve humans' knowledge level.

In visual art aspect, (AI) machine vision that allows computers to identify objects, scenes and activities in images. Current applications of (AI) machine vision include providing objective descriptions for the blind seeing(visual) needs.

We except the economic effects of (AI) technology to include both direct GDP growth from sectors that develop or manufacture. (AI) technology and indirect GDP growth through increased productivity in existing sectors that employ some form of (AI). If (AI) technology is an increasingly critical component of more products, it will become an integral part of many people's lives. Thus, (AI)'s ability to influence economic activity, rather than the economic or development status of the region. (AI) has the potential to impact income classes and to bring significant gains to both developed and developing countries. For example, (AI) has the potential to optimize good production around the world by analyzing agricultural regions and identifying what is necessary to improve crop yields.

In estimating the future economic effects by (AI) technology innovation, it is important to note that it is challenging to accurately predict which applications of (AI) will ultimately be commercially successful. In micro level economic influence, we need to apply methodologies to estimate the economic effects of investment in firms developing (AI) technology since investment levels in a technology are a telling sign of the future potential of that (AI) technology.

- How can (AI) influence GDP of high income countries in the next ten years?

How (AI)'s development may affect the global economy over the next ten years. In fact, (AI) technology has the potential to affect business across the global in a wide range of industries in ways only a number of technologies have done in the parts. For example, (AI) technology's expected to be a useful tool for enhancing human capabilities and in some instances replacing functions, such as driving a car, adoption of broadband internet, mobile telephone, industrial robotic automation have served to enhance human capabilities.

However, significant public debate has focused on projections of (AI) technology's effect on the labor force. However, large companies prefer to invest in (AI) technological industry. For example, face book's (AI) research lab., google machine intelligence lab. and micro soft machine learning and artificial intelligence research division are all making advances in (AI) technology and investing in the industry's top talent. Additionally, between 2010 year and 2015 year, nearly $5 billion in venture capital funding invested in firms across the global developing and employing (AI) technology (Facebook (AI) Research).

- How can artificial intelligence impact on workplace?

Modern information technologies and the labor economy growth of machines is powered by artificial intelligence have already strongly influenced the world of work in the 21 ST century. Computers, algorithms and software simplify every tasks and it is impossible to image how most of our life could be managed without them. How can be the information economy characterized by exponential growth replaces the most production industry based on economy of scales? What will the future world of work look like and how long will it take to get? Will the future world of work be a world where humans spend less time earning their livelihood? Alternatively, are mass unemployment, mass poverty and social distortions also possible scenario for the future, where robots, artificial intelligence systems play an increasingly central role? These questions concern how artificial intelligence further development . Can influence labor economy growth on workplace ? When the labor market has widespread impact on intelligence property, information technology, product liability, competition and labor and employment laws.

How (AI) technology impacts on labor workplace.

The future influence any organizations how labor economies use of (AI) can be analyzed, such as deep machine learning is based on a set of model high level data. Unlike human workers, the machines are connected the whole time in workplace. If one machine makes a mistake, all autonomous systems will keep this in mind and will avoid the same mistake the next time.

Over the long run intelligent machines will win against every human expert. Production robots have been replacing employees because of the (AI) technology. They work more precisely than humans and cost loss. Creative solutions like 3D printers and the self learning ability of these production robots will replace human workers, the automatic data recording and data processing, traditional back office activities are no longer in demand. Autonomous software will collect necessary information and will send it to the employee who needs it. Additionally, dematerialization leads to the phenomenon that traditional physical products are becoming software. For example, CD or DVDs are being replaced by streaming services. The replacement of traditional event ticket, e-travel ticket service products or hard cash will be the next step, due to the possibility of payment by smartphone. So, (AI) technology will impact human's daily life consumption behaviors in the future. For another example, transportation tools, such as boats and ferries and private vehicles will use sensors and navigating without human input. Taxi and truck drivers will become obsolete, the stock store applies to stock managers and postal carriers of the delivery is distributed by (AI) machine delivery methods.

What is the relationship between (AI) and (CRM)?

- Can (AI) technology impact on customer relationship management (CRM) to improve service performance ?

Nowadays , (AI) is a technology almost as old as the computer industry itself, it is similar with the advent of personal assistants function to businesses and personal promotion channel, such as (Amazon's Alexa, Apple's Siri, Google's Assistant) image recognition (face book), personalized recommendations (Netflix , Amazon). Those innovations have been driven by a increase in processing power, lower cost hardware, and the exploding creation and availability of data. It seems, (AI) technology can impact global customer service management method.

How to forecast economic impact modeling to (AI) will affect global economy? Can human forecast business revenue growth and job creation (or destruction) based on (AI) applied to customer relationship management (CRM) activities? In addition to the economic impact on (AI) or (CRM) which can include an estimate of the economic impact attributable to sales forces customer base. What can economic benefits be brought to (CRM) from (AI) technology?

Artificial intelligence(AI) comprises a set of technologies that use natural language processing, machine learning, knowledge graphs, and other tools to answer questions, discover insights and provide recommendations. Computer systems can use (AI) hypothesize and formulate possible answers based on available evidence can be trained through the ingestion of vast amounts of content, and automatically adapt and learn from (AI) self mistakes and failures.

So, any business organizations (customer service departments) can provide efficient and effective customer relationship management of excellent customer service quality if which applied (AI) technology system. The different type of (AI) systems include: (AI) system platforms, machine learning (AI) based data preparation and enrichment tools, machine vision/image recognition, voice speech recognition, text analysis and natural language processing, bots , e.g. face book website and virtual digital assistance solutions, social media pattern analysis , sentiment analysis, advanced numerical analysis (e.g. IOT streaming , machine logs), supporting technologies, knowledge base dialog management, Q&A processing etc. different (AI) technology system customer relationship management (CRM) tools.

(AI) (CRM) of activity can include these categories, such as: corporate marketing, marketing operation, field marketing, customer support, digital commerce, customer analytics, customer influenced product or service design, product or service pricing, finance information, presentation, customer billing, inventory , logistics and fulfilment support, partner management etc. different CRM tools.

(AI) technology of CRM has been carrying on plan different stages to achieve CRM personal assistant tool for businesses. The stages are such as, in the beginning stage of (AI) projects in place, implement now, pilot phase next year in the final stage of (AI) customer relationship management tools are foreseeable future. So, this CRM technology has been improved to plan in different stages every year to prepare to achieve full capacity of CRM service quality for businesses to use in the future.

Hence, how to develop an estimate prediction of the economic impact (AI) technologies could have CRM activities, which depends on gathering macroeconomic information on business revenue and the basic marketing of business revenue and the basic markup of business expenses by major functions (customer support, marketing and sales , production etc.)

An economic impact model that can gather data together and forecast the results how (AI) artificial intelligence technology brings (CRM) customer relationship management benefits to businesses, e.g. surveys investigation includes IT spending by sample countries, GDP and population estimates and forecasts, revenue per employee and ratios of IT spend to GDP. Surveys (questionnaire questions) of forecast results are influenced by (AI) impact can include: results are projected from surveys and rely on estimates are made by respondents on the expected financial improvements in categories of (AI) –assisted customer relationship management activities. The forecast assumes that these estimates are correct; financial estimates are based on estimates of "first year" improvement from full (AI) implementation; forecasts are from planning to implement any artificial intelligence of customer relationship management (CRM) projects, the improvement forecast is of categories of activity , e.g. corporate marketing , digital commerce, and customer analytics. They are not estimates of ROI for the (AI) software. They rely on conservative estimates to which each of these entities might affect company revenue, expenses or productivity. They also rely on estimates of the penetration of software in customer relationship management activities . Net new jobs created are based on the ratio of new revenue to jobs required to support that revenue . They can assume that 50% of the net new revenue will support increases in labor and the rest will go for capital and other operating expenses that may replace jobs lost to automation.

In the future, some of the ways in micro economic benefits to any organizations. (AI) technology is expected to impact CRM activities include: Spending up sales cycles, improving lead generation and qualification solving customer support problems faster (raising service quality), helping companies improve brand campaigns and recognition, lowering costs of support calls when increasing resolution rates, lowering the cost of recruiting employees and partners, increasing revenue from optimized product marketing, optimizing price, distribution logistics and preventing loss through fraud detection. So, micro economic benefits view point, it seems that (AI) CRM technology can raise any companies economic benefits for care term.

Artificial intelligence enables machines or the in-build software to behave like human beings which allows these decisions and act. The advent of (AI) is leading , talking, making decisions and act. The advent of (AI) is leading to new technologies advances and transforming the economic and employment opportunities for humans in a positive way. (AI) related technologies can facilitate our live. For example, industrial robotics, robotic medical assistants, smart games, financial forecasting software, big data analysis, algorithms in health and bioinformatics, pilotless cargo places, drone ambulances and general purpose and workplace robots and others. (Disruptors technologies: Advances that will transform life, business and the global economy).

Artificial intelligence also known as computational intelligence is defined as " the human –like intelligence exhibited by machines or software. It is theorized that intelligence of humans can be described and intelligence machines or software can simulate it. These machines software can be reasonable , learn, perceive and process information, like human mind and thus facilitate human life. They can think and act for us. So, artificial intelligence is an interdisciplinary field of study including computer science, neuroscience, psychology, linguistics and philosophy.

However, (AI) research and developments have economically impacted many industries, such as robotics, telecommunications, computer applications , health, finance, heavy manufacturing, transportation, aviation, e-service and e-commerce, military , music and movie, toys and games entertainment etc. industries.

In fact, many ideas, systems and technologies have been developing in the world of (AI) technology. However, which are net called or considered (AI) products, rather which are mentioned with their specific names, such as smart graphics, machine learning, e-commerce etc. (i.e. this is called (AI) effect).

What is relationship between
(AI) and digital economy?

- How can (AI) technology influence digital economy?

Nowadays, (AI) related industrial applications will replace most human power in fields, including call centers, customer services and air cargo transportation. (AI) technologies also help weather forecasting based on repeated rainfall pattern (data) recognition, through robotics (i.e. floor cleaning, moving lawns etc.) transporting people and products with unmanned vehicles, sending space unmanned smart shuttles, developing robotic arms, predicting market values in stock exchanges by internet, making homes safer, helping elderly and disabled using robotic servants etc.

Among the (AI) related technologies , there are a few that significance for the impact on society and especially on digital economy . (AI) is particularly influential in machine learning. Such as robotics, transportation, finance, health and bioinformatics, e-commerce , e-games, big online data gathering and internet-of-things. For example, machine e-learning is based in bioinformatics and robots that can learn new skills for better caregiving in healthcare. What is machine e-learning? Machines can e-learn from e-data gathering, coming up generalizations and making decisions to act in certain ways from internet.

There are important applications , such as e-machine perception, electronic online natural language learning processing, online search engines, online bioinformatics, online brain –computer interface, online game playing, online robot locomotion, online advertising, online computations finances, online health monitoring, online DNA classification and decision making, online in chemistry –cheminformatics . So, online machine learning can positively impact productivity and it can enhance information and analytical system from (AI) online channel.

What is robotics? Robotics is one of the most strongly influenced fields in (AI). For example, heavy manufacturing industries, robots and used and man power is replaced for effectiveness, precision, and accuracy, especially in respective or dangerous tasks, including welding, assembling , picking and placing .

So, robots can acquire new skills or adapt the changing dynamic environment. Also, artificial intelligence can be applied in developing transportation. For example, automated vehicles, driver assistance systems , safety systems, collision avoidance systems and public transportation. Moreover, (AI) technology has proven to produce some of the best tools to predict stock market fluctuations from internet data gathering method. It's predictions are based on ever-evolving predictions algorithms and systems learn new models and make connections between historical data and new data to measure stock market trading more accurate from internet data gathering channel.

In health field, especially in health data processing , analysis, decision making support and medical diagnosis. So, online data can show which patients will need what treatment and what alternative drugs could be used more accurate from (AI) online data gathering method. Bioinformatics is an interdisciplinary field combining statistics, (AI) online technology can help in discovering data patterns and modeling through the application of machine learning, artificial neural networks and genetic algorithms. For example, further (AI) technology development of human genome project of online data sequences.

Online shopping can be facilitated by virtual assistants developed through (AI) technology and these assistants can offer the best advice. (AI) online purchase coming after every product image recommendations and personalization bring important revenue to shopping online sites, like Amazon . Smart computer graphics and games, artificial intelligence is useful in smarter computer, graphics, scene modeling , scene rendering processes in order to create, for example, effective human –robot interactions , online machine learning, online strategic games techniques etc. online computer related (AI) software.

So, online big data analysis and big data does have a critical need in the world of online intelligence machines and software in our future. In other words, (AI) offers online technology to enable online big data analysis to provide industrial organizations with valuable information for effective decision making in short time. For example, what IBM's Watson achieved: this machine used 200 million of structured and unstructured content with a special technology of hypothesis generation, massive evidence gathering, analysis and scoring from internet channel.

Finally, (AI) online technology another related internet invention (internet of things) (IOT) is the network of machines or objects connected through internet. These connected objects can sense their internal and external environment, communicate with each other, can send critical data and finally can make decisions to act or correct their environment from (AI) online technology. For example, factories can monitor and automatically change production processes, hospitals can monitor and regulate the health conditions of their patients , schools can collect data from facilities and cars can send data to car makers from (AI) online technology.
Partner predicts that (IOT) market will create about trillion amount value by 2020 year. Although machines collect big data from their environment, whether which gain an insight or learn from these online data largely depends on the (AI) online machine learning principals and (AI) online technology. In 2013, Mckinsey estimated that disruptive technologies closely related with potential economic impact in 2025 year between $7.1 to $13.1 trillion amount (automation of knowledge work, advanced robotics, autonomous or near-autonomous vehicles).

What is the relationship between
(AI) and global digital economy development ?

● Could work activities in China be automated
making in the nation with the world's largest automation potential?

Can (AI) technology influence China economy? Could China workers be affected and jobs made up of routine work activities and predictable? Will programmable tasks be particularly impact to China employment market ? When impact on labor market is likely to be gradual at the aggregate level, it can be sudden and dramatic at the level of specific work activities, rending some job obsolete fairly. Overall (AI) technology will raise digital skills when reducing demand for medium incomer inequality for China workers. It seems (AI) technology's effect on productivity could be crucial to China's future economic growth as the population ages are increasing.
In China, some biggest technological companies driving significant investments in research and development. Moreover, China is one of the leading global (AI) technology development county. However, China will need to focus on building its innovation capacity. For example, United States and United Kingdom are currently producing more influential (AI) technological research. However, if China planed to achieve (AI) technology success, it's traditional industries will need to develop technical know-how –to and overcoming implementation costs prepare to develop (AI) . When (AI) technology is introduced into China society, China government needs to raise concerning ethical, legal, technological security etc. business questions. Also, surrounding issues include privacy, discrimination, legal liability and regulation. It aims to encourage overseas investors to choose to invest (AI) technological industry to raise GDP growth and manufacturing industries income growth for long term in China.
If China encouraged overseas (AI) technology investment in its country. It is possible to influence China employment market to be changed. Because (AI) technology will impact to influence China people daily life. Due to (AI) technology is introduced to China society, many rich people will prefer to spend to buy any high (AI) technological products for entertainment or learning or machine man driving etc. daily necessity activities. Then

it will raise GDP growth and will raise (AI) manufacturers or related-(AI) technological manufacturers profit. It is beneficial to China because it can become one high knowledgeable and (AI) technological economical society.
But it will bring bad influences to raise unemployment chance for the low skillful labor. In labor economy aspect influence , how (AI) technology can influence China low skillful labor unemployment ratio raising. The raising low skill labor unemployment reason is because China low skillful human labors are argued or are replaced by (AI) technology creating new challenges to introduce to influence China society of simply human manufacturing job nature to be changed to be high (AI) technology manufacturing job nature in any China factories. Moreover, when (AI) technology introduction to China, it will cause other related social challenges in China. The varied (AI) related challenges, including the difficulty of creating safe and reliable hardware for sensing and affecting (transportation and education), the challenges of gaining public trust, a low resource comities and public safety and security, the challenges of overcoming fears or marginalizing humans in China employment and workplace and the risk of diminishing interpersonal trust because the low skillful labors won't believe any China employers will give chance to employ them , due to (AI) technology will replace their skills and man manufacturing of productivity is much less to compare to (AI) technology manufacturing method.

How does (AI) technology influence
the future of employment change?

Are future nature of jobs changed to computerization from (AI) technology? Where are the probability of computing occupations from (AI) technology influence? What is expected impacts of future computing on labor market from (AI) technology influence? John Maynard Keynes's frequently cited prediction of widespread technological unemployment " du to our discovery of means of economic the use of labor outrunning the pace of which we can find new used of labor" (Keynes, 1933, p.3).
In the future, (AI) technology will impact some nature of occupations to change computing. This chance will also influence some countries' economic change. For example, some factory human labors hand routine manufacturing tasks will be changed to computerization of routine manufacturing tasks by (AI) technological machine men hand manufacturing method. it will cause a structured shift in the labor market, with workers reallocating their labor supply from middle-income manufacturing to low-income service occupations.
Arguably, this is because the manual tasks of service occupations are less computerization, as who require a higher degree of flexibility and physical adaptability. So, (AI) technology will influence the human hand labor skillful occupation nature of task cheaper , such as vehicle manufacturing , ship manufacturing, computer manufacturing, steel manufacturing, television, radio etc. home electronic products of heavy machine industry change. Due to (AI) technology machine man will be proper to be used to manufacturing these electronic products when the (AI) technology innovation can develop to the mature stage. Then, any countries manufacturers will choose to use (AI) technology machine man, instead of human hand production.
Supposing the future prices of computing are fallen, seriously, problem solving skills are becoming relatively productive, explaining the substantial employment growth in manufacturing occupations, involving cognitive tasks

where skilled labor has a comparative advantage, as well as the increase education needs for (AI) technology computing of machine man subject study.

Prediction of education needs for (AI) technology student numbers will increase, due to manufacturing industry needs many (AI) technology students in future employment market. Another (AI) technology influence if the future (AI) technological innovation, e.g. machine man manufacturing or machine man service industries will both increase demand, then with more sophistic software technologies will be disrupted labor markets by marketing workers redundant.

For publishing industry, what is striking about the case in paper book publishing industry will be unpopular? Due to the electronic book publishing industry will be popular, e.g. Amazon publish . (AI) technology can influence paper book manufacturing method which is replaced by machine man electronic book manufacturing method as well as it will cause the computerization is no longer confined to routine manufacturing tasks. Due to (AI) machine man manufacturing technology will be proper to be used to manufacture any products in short time efficiently and effectively , e.g. electronic book products. In the future, if it is fact to occur this case, such as (AI) technological machine man manufacturing method will be adopted (applied) to manufacture electronic books or any products in possible. (AI) technology will cause many manufacturing workers are unemployed. It is beneficial to employers, who can reduce to spend much wages expenditure to employ manufacturing workers, but it will cause many manufacturing workers loss jobs and reduce income to support whose families lives. It will cause social challenges, e.g. increasing stealing crimes if the manufacturing workers had not other skills to find other jobs to do easily. So, manufacturers need to concern over technological unemployment which will be hardly future phenomenon if who decided to dismiss all manufacturing workers, due to (AI) technology machine men replace to them.

If (AI) technology can be innovated to produce any kinds of machine man to serve any service or manufacturing industries successfully. Then, it will bring these questions: Can future that workers be influenced to be automation employment and productivity by (AI) technology influence? Does it impact to influence the (AI) technology countries' productivity and growth and natural resources development and labor markets and evolution of global financial markets and economic impact of technology and innovation and urbanization etc. issues? How will automation transform the workplace? What will be the implication for employment? What is likely to be its impact both on productivity in the global economy and on employment?

In fact, automatic of activities can enable businesses to improve performance by reducing errors chance and improving quality and speed, and same cases achieving outcomes that go beyond human capabilities. Some economists indicate (AI) technology would give a needed boost to economic growth and prosperity have of the working age population in many countries. Based on the scenario modeling, they estimate automation could raise productivity growth globally by 0.8 to 1.4 % annually. They also indicated that almost half the activities people are almost $1.6 trillion in wages to do in the global economy have the potential to be automated adapting current demonstrates technology, according to their analysis of more than 2,000 work activities across 800 occupations. When less than 5% of all occupations can be automated entirely using demonstrated technology, about 60% of all occupations have at least 30% of worker made activities, that would be automated. More occupation will change to be automated. They also indicated for business performance benefits of automation are relatively clear, but the issues are more complicated by policy making to attract foreign investors. Beyond technical feasibility, the cost of technology, competition labor will include skills and supply and demand dynamics, performance benefits and beyond labor cost savings and social and regulatory acceptance will affect the automation. Their predictions suggest that half of today work activities could be automated by 2055 year, but this could happen 10 to 20 years earlier or latter depending on the various factors in addition to their wider economic condition.

Some scientists suggest (AI) technology is finally starting to deliver real-life business benefits. Computer power is growing significantly , algorithms are becoming more sophisticated and perhaps most important of all, the world is generating vast quantities of the fuel that powers (AI) technology data billions of gigabytes of it every day. Also, online firms are digital natives, such as Google online search service company is investing on (AI) technology. For new though most of the news if coming from the suppliers of (AI) technologies. And many new users are only in the experimental phase. Few products are on the market or are likely to arrive these soon to drive immediate and widespread adoption. As a result, analysts believe (AI) technology's potential will give true economic benefit in the future. (AI) industry will introduce to suppliers and users to raise economic potential of (AI) technology.

In the future, (AI) technology systems can solve business problems. Some scientists categorized those into five technology systems that are key areas of (AI) technology development: robotics and autonomous vehicles, computer vision language virtual agents and machine learning , which is based on algorithms that learn from data without replying on rules-based programming in order to draw conclusions or direct an action.

Such as computer vision and language includes natural language processing, analytics, speech recognition technology, some are about learning from information, such as about machine learning and others are related to acting on information, such as robotics, autonomous vehicles and virtual agents, which are computer programs that can converse with humans. Machine learning and a subfield called deep learning are artificial intelligence applications.

- Can artificial intelligence impact
global economy growth?

Artificial intelligence (AI) is a term first defined in 1956 year. It is a branch of computer science that aims to create intelligent machines that work and react like humans. In contrast today, 60 years later, (AI) is characterized by a number of applications, including computers playing games against humans and understanding human languages, virtual personal assistants, and robotics which involve computers seeing , hearing and reacting to sensory stimuli. In the future, technologists predict for (AI) technology ranging from (AI) being used as a tool to aid relatively simple processes for robots with human like mental capabilities, who expect (AI) technology can emulate human performance by learning, coming to mind its own conclusions, understanding complex content, engaging in dialog with people, enhancing human cognitive performance or replacing humans in executing both routine and non-routine tasks. In existing industry, (AI) technology is used , such as targeted advertising and virtual used personal assistant as well as the (AI) technology that my exist in the future, such as robots with human vehicle processing capabilities.

The range of (AI) technology's progress in the future will determine the economic impact future of (AI) technology on the global economy with more limited advances and applications (i.e. weak (AI) only) corresponding to more limited economic impacts and more substantial progress, i.e. strong (AI) technology is corresponding to more significant economic impact.

(AI) technology learning that automates analytical model, including predicting cause-and-effect relationship from biological data, identifying new drugs, self-driving cars and protecting against fraud etc. functions. Also (AI) learning can improve natural language processing that allows computers to continue to better analyze, understand and generate language to interface with human using the natural human language, virtual personal assistant, helps users by providing scheduling appointment, reminds organizing personal finance and finding providers of various services, machine vision allows (AI) machine man to identify object, scenes and activities in detect pedestrians and bicyclists.

We expect the economic effects of (AI) technology to include both direct GDP growth from sectors that develop or manufacture (AI) technology and indirect GDP growth through increased productivity in existing sectors that employ some from of (AI) technology. If (AI) producing sectors could grow, then it could lead to increase revenues and employment of (AI) technological professionals within these existing firms as well as the potential creation of entirely new economic activities to any countries' societies productivity improvement in existing sectors could be realized through faster and move efficient processes and decision making as well as increased (AI) technological knowledge and access to information available in societies easily.

In the future, if (AI) technology is an increasingly critical component of more products, it will become an integral part of necessary products of many people's lives. The extent of (AI)'s economy effort is also likely to vary from region to region, thought variation may be more dependent on the predominate economic activity of a region and the (AI) ability can influence economic activity, rather then the economic or developmental status of the regions. (AI) technology can move accessibility and can use source development to do international business between one country and another country.

So (AI) technology has the potential to give benefits to different income chooses and to bring significant gains to both developed and developing countries. For agricultural technology, (AI) has the potential to optimize food production around the world by analyzing agricultural regions and identifying what is necessary to improve crop yield. In total, (AI) technology gives greater economic impact to any countries agricultural regions if which implemented (AI) technology to grow crop , fruit etc. food production in the farms.

Investment in (AI) technology is such as capital investment to any countries' public or private enterprises. So, it will have large economic impact to the future . If the (AI) technology is reasonable invested to the different needs aspect by the public or private enterprises in the country. Then, it will have good economic impact to the country in the future. However, when (AI) technology is likely to affect both the productivity and employment components of economic growth in many sectors. Significant public debate has focused on projections of (AI)'s effect on the labor force. However, for instance, some researchers have argued that the rise of (AI) technology and automation will led to significant unemployment as capital is substituted for the low skillful labor. So, they point to the concern that the increasing sophistication of (AI) technology may balance skilled and semi-skilled workers and the reduce the size of the middle class. However, this is not a new argument, due to (AI) technology negatively affecting the labor force and leading to mass unemployment. Because the (AI) technology is the substitution of machinery for human labor. Although, employment in certain industries, has been reduced in the past due to technological advancement. For long term, the labor market has adapted to the introduction of new technology, giving rise to new jobs in new areas. (AI) technology may also be accomplished without a reduction to total employment in the long-term to some Asia countries, such as Hong Kong and Japan. Because Hong Kong and Japan many low skilled labor, e.g. security, cleaner who complaint that employers need them to work long time hours. (abnormal working hours) e.g. one day 12 to 15 working hour per day. Hence, if (AI) machine means invention technology success. Security or cleaning job can be worked from (AI) machine man in some hours every day in order to reduce the long time working hours cleaners or security workers, e.g. one (AI) machine man works 4 hours for cleaning or security job, one day as well as another cleaner or security labor only needs to work 8 hours one day. So total security or cleaning employers can employ 12 hours machine cleaners or security workers and human cleaners or security workers in one day. For long term benefit, Hong Kong or Japan every security or cleaning worker does not need to work 12 hours minimum working hours one day. They won't feel tried and bore and without private with whose families, so who will accept to do these cleaning or security jobs, even they can raise work efficient and performance when who feel happy and health.

So, (AI) technology of machine man invention can raise low skillful labor efficiency and it can help them to avoid abnormal working hours demand in some busy work life countries, such as Hong Kong and Japan. Before, one Japan female labor feel unhappy to work, due to who often needs to work abnormal working hours for her employer and who has less sleeping and without any private time to enjoy her life with her families every day. So this abnormal working hours factor causes her to do commit suicide behavior, then she is die unlucky. So (AI) technology of machine man invention ought avoid abnormal working hours demand for employer in any countries in the future.

The most important occurrence to any employers, some researchers had attempted to do one experiment to find that private research and development , venture capital and public research and development investment all have strong net effect or economic growth with venture capital funding further having the strongest such effect from (AI) technology. The researchers hypothesize the venture capital investment contributes to economic growth through (AI) technology innovation and by the capacity of an economy to use existing (AI) technology knowledge to increase productivity. They predict the impacts of venture capital, business-research and development and public research and development can raise multi factor productivity from (AI) technology introduction.

Can (AI) technology influence the economic development to developing countries? The developing regions of the world contain most of natural resources. If one day, (AI) technology has invent one kind of machine man which can assist any gas or oil workers to seek any new oil/gas natural resource locations easily. I believe that (AI) technology can help these natural resource exploitation countries will gain economic benefit more easily. So, (AI) driven technology can be used to change to create any new opportunities to address poor management or resources and improve human well being, such as Africa Latin America and India can use (AI) technology machine man to seek any oil/gas natural resource countries exploitation activities to attempt to gain much economic benefits.

- Why will (AI) technology grow economic development ?

Nowadays, increases in capital and labor are no longer driving the levels of economic growth, such as (AI) technology. The ability of increase in capital investment and in labor of traditional drivers of production, have no longer to be enjoyed in most developed economies ,e.g. developed country, US, UK . However, artificial intelligence has the potential to overcome the physical limitation of capital and labor to avoid missing out on this opportunity. So, policy makers and business leaders must prepare for and work toward a future with artificial intelligence. They must do with the idea that (AI) is another simply method to enhance productivity method . Rather they must see (AI) as the tool that can transform thinking about how growth is created.

Economists have always thought of new technologies are as driving growth their ability to enhancing. It can replace labor and capital factor of production. So, it brings this question: What is the factor of production (AI) technology characteristics. They key factor is to see (AI) technology as a capital-labor .

(AI) can replicate labor activities at much greater scale and speed, and to even perform some tasks began the capabilities of human. For example, by using virtual assistants , 1000 legal documents can be reviewed in a matter of days instead of taking three people six moths to complete. Some (AI) technology may be one kind of factor of production in the future. For another example, people will work in workplace digitalization environment. So, in the future, working environment and information management are automated. Such as Konica camera sale company will use workplace digitalization. So , (AI) technology can provide workplace digitalization in order to raise productivity efficiency. (AI) technology will be one kind of production which is replaced by workplace digitalization and it will grow any organization productivity efficiently. Then, (AI) technology will assist overall social economy growth , due to productivity is raised and products can be produced in short time to prepare to sell in consumption market. So, time will be shortened to increase GDP growth fast for the development of (AI) technology countries.

- How can (AI) technology impact to global economic and social and psychological changes?

What will be the development of (AI) technology and predictions concerning the future evolution? The computers and robots will develop conscious, intelligent and minds into humans, enhancing psychological and behavioral abilities and allowing for direct communication with (AI) minds. (AI) technology will be impacted human life by (AI) technology information communicative and environmental influence. A " world brain" and " world mind", this psychological system will be enhanced and enriched the capacities of both individual and collective cognition by (AI) technology of service industries.

(AI) technology with influence these human needs of service industries changes, such as , biological science, finance, entertainment, business, biological science, transportation, communication military etc. The personal computer evolution, the internet and the world wide web which exploded on the scene, linking business, homes, schools, social organizations which were a completely unpredicted phenomenon to influence human life. Kurzweil (1999) predicts that by 2029 year, most human communication will be with machines. According to Person, by 2100 year, there will be human machine convergence.

How can (AI) technology influence environmental protection to make benefits to farming economic growth? (AI) technology can be applied to predict how to solve environmental pollution challenge to avoid to damage any crop or vegetable or rice or fruit etc. food growth. Because environmental experts can gather global environmental pollution data from an environmental database to build a perform a systematic analysis from (AI) technology. The first step is this broad analysis can include understanding, statistical and data gathering techniques to obtain the relevant data, the correlation among the variables involved, and a list of possible models. The next step is to select a set of methods and models that cover all kinds of knowledge and functionalities needed for the decision making process. Once the models are selected, they must be fully implemented by means of machine learning , data mining, statistical or numerical technique. After that, those models must be integrated to build the whole EDSS. The EDSS must be tested to check its performance, accuracy, usefulness and reliability, both from the user's and (AI) technology/computer scientist's point of view. If these is any wrong feature in any development stage, such as model's integration, models' implementation, selection of models, database, problem analysis etc. the developers must come back in the update th required components. When the evaluation phase is all right, the EDSS is ready to be applied to the environment. The great contribution of artificial intelligence to EDSS the integration of several methods complementing the classical statistical models/simulation , statistical analysis, linear models, etc. and numerical models (control algorithms, optimization techniques etc.) .

This cooperation makes the resulting systems more reliable and powerful in coping with real world environment systems. Date interpretation has been a principal area of research in (AI) technology since the very beginning. The most demanding problem in the environmental assessment context. Knowledge representation permits the definition of the different types of data that the existing methods adapt to the process. There is also a lot of work to clean, repair and transform the huge available quantities of raw data. Apart from this, the availability of meta-information or background knowledge is required to guide the process. Data mining is multi-disciplinary: It covers expert systems, data based technology, statistics, data visualization and unsupervised machine learning. These techniques operate at the level of data and background information, where numerous and often incompatible new commensurate pieces of information from disparate sources have to be brought together (K, Fedra, 1994).

So, it seems that in the future, (AI) technology with the increasing maturity in particular those related to knowledge and engineering, new dimensions can be assisted to users in environmental decision making are available. For example, many environmental systems are characterized both by incomplete models and by limited data. Hence, in the future, (AI) technology will be applied to predict climate change to reduce crop or fruit etc. food agriculture challenge by climate change bad influence.

- Will (AI) technology influence digital economy change to manufacturing industry ?

To understand how the manufacturing business must adapt to prosper in the technology, we need to understand how (AI) technology will change us to shape our daily habits to satisfy our expectation of products to how we shop and even the immediate of the entire process. For example, taxi services are in the crosshairs as on demand transportation services like, available of the touch of a smart phone button expand. In fact, Yellow lab, US country , san Francisco city's largest taxi company is filing for bankruptcy as the industry starts to change faster than almost anyone expected. However, at this point, its more than an app that is changing, some our taxi passengers renting taxi transportation to catch consumption behavior.

(AI) technology will influence digital economy for taxi passenger's individual customer experience, offering a growing renting taxi to catch of service and feedback opportunities when any one taxi passenger who chooses to use mobile phone app online tool to prepaid to rent any taxi more easily.

Also in the long term, (AI) technology can influence vehicles drive themselves of behavior. Already, companies like Google and GM are working on projects to bring fleets of autonomous vehicles to cities at the path of a button. Moreover, this on-demand service model is beginning to appear across a much broader range of markets. For example , Amazon company is investing in its own fleet of trucks, planes and even drone at the same time as it pushes for same-day delivery of products. As some point, vehicles will be autonomous too. So, it seems that (AI) technique will influence any transportations choose to use digital autonomous driving technology in the future . For Amazon company case, it is not stopping of logistics. It is also aiming to automatically manage the supply of consumer home products with its recently launched Amazon replenishment service, Dash. Dash is a digital service that enables that connected derive to automatically order physical products from Amazon when supplies are running low. So, it seems (AI) technology will be applied to logistic function by digital technology method introduction in the future.

Hence autonomous vehicles will optimize industry supply chains and logistics operations through increased efficiency and flexibility. In fact, fully automated and lean supply chains will keep reduce load sizes and inventory by leveraging smart distribution technologies and smaller autonomous vehicles by machine man assistance. If Amazon continues to grow market share for online sales by reducing effort required by the consumer to place an order, when also contributing the almost immediate delivery of products to the doorstep. So, it will further fuel the trend toward on-demand derive. As Amazon company fuels the on-demand economy, consumers will expect immediacy in more parts of the digital economy. On top of speed, consumers increasing expect more personalization options.

So, (AI) technology will influence digital manufacturing, such as Amazon publishing to monitor every aspect of every process in real -time and communicating to self-optimized deep learning robotics, new methods of high volume and high customization will become possible. Then, as products merge into product platforms and even services, manufacturers have the opportunity to provide components and platforms used by smaller players. So, (AI) technology will influence manufacturing industry to choose automated SMI lines, robots installed, automation engineers.

Another future (AI) technology development can be applied to space science aspect, such as Automation engineering space in manufacturing process to achieve digital manufacturing benefits to any businesses in the future. Such as reducing cost, shortening manufacturing time, raising efficiency, shortening delivery products to client individual time. How can artificial intelligence give the need and advanced fast and evaluation methods benefits for space exploration? When US NASA (space exploration organization) achieves any space exploration missions, it will answer this question:

When is it useful to have a machine use (AI) technology to achieve a decision? After all, after millions of years of space exploration and rough 10,000 years of civilization, humans are usually quite good at making decisions in complex uncertain environments. Through, Johns Hoplains University's Applied Physical Lab. Research in (AI) technology enabled systems, which has identified three general use cases for (AI) technology to explore space mission:

First, for some tasks (AI) technology is more cost effectiveness than human. Second, (AI) technology is better suited than humans at solving some, but not all problems. Third, (AI) technology allows NASA organization's space exploration mission to develop machines that ate capable of responding faster than when a human is in the decision loop (D. Scheidt, 2012, A. Castano et. al. 2008).

So, the use of (AI) technology to enable science by observing the pace of rapidly evolving phenomena was demonstrated. It is more effectively coordinating and (AI) technology utilizing to earn economic benefits to use for space exploration mission.

However, (AI) technology also have current risk for space exploration. Today (AI) technology is immature and requires further development to reach its potential. For instance, the (AI) technology algorithms that detected the dust derive could not have identified whether the Martain weather represented a threat to the cover. Also it can not yet use instrument input to determine what, where and how to autonomously make the next space science measurement. An equally important factor limiting (AI)'s deployment is that lacks the methodology and technology to effectively test (AI) technology. So, the challenge will testing (AI) enabled system is how (AI) performance can be measured. It would be NASA organization's difficulty to find (AI) technology to develop to carry on researching any space exploration missions in the future. However, (AI) technology will be a good economic benefit choice for space exploration mission in the future.

- What is artificial intelligence potential
benefits and ethical considerations?

The ability of (AI) technology systems to transform vast amounts of complex information into insight has the potential to help solve manufacturing or service challenges for human needs. However, to reap the societal benefits of (AI) systems, humans will need to trust then and make sure that which follow the same ethical principles, moral values, professional codes and social norms that we humans would follow in the same scenario, research and educational efforts as well as carefully designed regulation in order to achieve the most effort of economic benefits goals. For example, international business machines corporation (IBM) is actively engaged both competitors , in global discussions about how to make (AI) ethical and as beneficial as possible for people as social economic benefits. (AI) is usually defined as the " capability of a computer program to perform tasks or reasoning processes " that human usually associate to intelligence in a human being. Often, it has to do with the ability to make a good decision, even when there is uncertainty, too much information to handle. As an example, play chess or complex card games of entertainment activities is believed to need some form of intelligence in a human being, as well as choosing the best medical facilities in a difficult medical case, or creating something new, such as mathematical theorem or even some form of act, or even driving automatic machine man (self driving vehicle) replacing human driving in the middle of a crowded city.

(AI) needs depends on what we consider being intelligence in the behavior of a human being act a certain point in time. If human belief about human intelligence changes and we don't believe any longer that a certain task requires intelligence, then a computer program performing that task is no longer part of (AI), it becomes just another boring computer program. So, it means that (AI) technology will replace some old computer programs, if human can invent new generation of (AI) software for any functions or activities to satisfy human needs.

As IBM, it argues intelligence. This means that we aim to build systems that enhance and scale human expertise and skills rather than replacing them. We therefore focus on practical applications of (AI) capabilities that assist people in performing well-defined tasks of needs by exploiting and wide range of (AI)-based services. We also use the term " cognitive computing" it is mean a comprehensive net of capabilities based on technology. It comprises the fields of machine learning, reasoning and decision technologies, language, speech and vision recognition and processing technologies, high performance and high efficient functions for any industries or individual consumers needs. For example, robotics, which are usually very good at doing what which are supposed to in any environment, much have public shopping center, factory etc. places which need simply services from the robot (machine man), such as cleans the floor of our houses to the robot that can work together with humans in production chains, passing through the warehouse, robots can take care of the tasks of an entire warehouse and the companion robots like Nao, Pepper, Aibo and Giraff, who can entertain use, talk to use and help elderly people to stay connected to their friends, relatives and doctors.

Google company is building automatic machine (self-driving cars) and has acquired more than 10 robotics companies. Facebook had opened whole new research facility only on (AI) research. Apply computer has developed Siri. Microsoft computer company has built a similar personalized assistant. Google has Deep mind, a UK company whose long term aim is to build general (AI) and has already great potential to win game to the world champion and IBM is investing a huge amount of resources in applying its Watson cognitive computing system to the medical domains to finance and to personalized education. In Europe, IBM is establishing new centers in Munich and Milan focused in the application of cognitive computer capabilities to the internet of things and healthcare respectively.

For example, automatic machine man (self-driving cars) are all about (AI), which used to be able to see what happens in the street (signals ,lanes, other cars, pedestrians, traffic lights, which need to able predict what other cars and pedestrians will do, and who need to be able to cope with unforeseen situations. Since, most car accidents are due to human fault, it is estimated that the adoption of self-driving cars will save about half of the lives that are usually last in car accidents.

IBM Watson company has to understand spoken language, make sense of massive amount to text , respond correctly to questions in many categories, as well as assess its own confidence in responding to such questions. In the future, (AI) technology can own question/answering capabilities that would be very useful, for example, in assisting a doctor when trying to some to the correct diagnosis for a patient and to propose the best therapy .

Intelligent machines can also rely on huge amounts of data to be used to learn how to make better decisions. This data comes from all of us over the years Facebook users have uploaded more than 250 billion pictures and every day who upload about 350 million more. Every second, we submit 40,000 google search queries. So, (AI) technology will be connected through the web from appliances to traffic lights from cars to watches. Other tasks that are very easy for humans are physical and manipulation tasks, such as walking , running, picking up an object to make its shape and location, restricted environment. But (AI) machine man technology still not able to have the general physical and manipulation capabilities even of a 6 year old.

So, it brings this question: Why do (AI) scientists need to concern ethics? Because (AI) technology is complex, information into insight has the potential to reveal long held secrets and help solve some of the world's most difficult problems. (AI) systems can potentially be used to help discover insights to treat disease, predict the whether, and manage the global economy. So, ethic issues is important to and (AI) scientists . If any one new (AI) technology research investigation could success, it will be a secret to and the (AI) scientists can not permit to their loyalty to any competitors to damage the fair (AI) technology products trading market. The country (countries) (AI) technology scientists need to concern ethic issues, who need to keep secrets for their countries economic or/and social benefits. This is moral issues to any countries/country loyalty is whose countries intangible assets. They can not sell (AI) loyalty to any their countries to assist whose economic benefits immorally.

- How can (AI) technology influence to global
health care economy development?

According to (AI) lecturer analysis, when combined key clinical health (AI) application can potentially create $150 billion in annual savings for the US healthcare economy by 2026 year. (AI) technology is re-winning modern conception of healthcare delivery. It enables machines to sense, comprehend, act and learn. So which can perform administrative and clinical healthcare functions (Accenture, 2017).

It will help health care service organizations to reduce health care cost, will improve and raise service quality and access. So, (AI) health market size will be predicted growth. (AI) applications in health care include robot-assisted surgery, virtual nursing assistant, administrative workflow assistant, fraud detection, error reduction connected machines, clinical trial participant identifier, preliminary diagnosis, automated image diagnosis and cybersecurity.

What kind of benefits (AI) technology can contribute to healthcare service? (AI) technology can deliver what many health care organizations need, such as financial and operational of labor costs, digital expectations from patient consumers how to use (AI) technology to solve interoperability challenges in any healthcare organizations. Also (AI) technology can be applied to wellness an d lifestyle management, diagnostics, delivers financially but also way of organizational and workflow improvement. So, (AI) technology will be continue to become most prevalent and adoption to healthcare organizations , which must need to enhance structure to be position to take full advantages of new (AI) technological capabilities. (AI) technology can change the nature of work and employment is rapidly changing to make the best use of both humans and (AI) talent in healthcare industry in the future. For example, (AI) technology offers a way to fill in gaps and the rising labor shortage in healthcare. According to Accenture analysis, the physicians shortage is increasing. However, (AI) technology will manufacture healthcare machine men to replace physicians in future one day(2017). Hence, (AI) technology will be invented to raise health care service staffs work efficiency and performance in any hospitals or clinics in the future.

In conclusion, (AI) technology will raise efficiency for any service or manufacturing industries in the future, although, it is possible that it will also rise low skillful workers unemployment numbers. But, the most important influence to human technological innovation will be risen and it will influence human life will be changed to be better, e.g. self drive cars, health care physician machine men, machine man cleaners etc. intelligent machine men will be manufactured to serve for our daily life. Furthermore, (AI) technological products will influence countries trading, some low technological development countries manufacturing businessmen can choose to buy any (AI) products to raise whose productivity and efficiency and reducing cost to achieve economic cost saving result. Also, GDP of trading growth income will increase to the (AI) products sale countries. Hence, it will be beneficial to economic development to both developed and developing countries both in the future as well as (AI) scientists time and money spending will be valued to continue to invest (AI) technology development for human life and economy benefits for long term.

In consequent, when (AI) can spend more time to attempt to do human's any tasks every day, then it have possible to raise productivity and efficiency better than human.

Reference

A. Castano et. al. " Automatic detection of dust devils and clouds at Mars" Machine vision and applications, Oct. 2008, vol. 19, no 5-6, pp. 467-482.

Accenture, " Why artificial intelligence is the future of growth"(2017) <http://www.accenture.com/us-en/insight-a rtificial-intelligence-future-growth>.

D. Schedidt , Unmanned Air Vehicle Command And Control, Handbook Of Unmanned Air Vehicles, Springer-Verlag, 2014. Facebook (AI) Research Available at https://research.facebook.com/ai, research at google, machine intelligence available at
http://research.google.com/pubs/machineintellige nce.html; micro soft research-machine learning and artificial intelligence available at http://research.microsoft.com/en-us/research- areas/machine-learning-ai.aspx.

K, Fedra , "GIS and environmental modelling" in environmental modelling with GIS, edited by M.F. Goodchild.B.O. Parks and L.T. Steyaert, Oxford University press, pp. 35-50, 1994.

Keynes, J.M. (1933). Economic possibilities for our grandchildren (1930). Essays in persuasion, pp.358-73.

Mckinsey & Company (2013, May). Disruptive technologies: Advices that will transform life, business and the global economy , USA.

Ray Kurzweil , The age of spiritual machines (1999) is cited numerously through this chapter: Kurzweilai.net http://www.kurzweilai.net

Rich, Elaine & Knight, Kevin, Artificial Intelligence Second Edition, 1991, New York; Mc-Graw-Hill.

How Robots bring positive impact to raise organizational productive change

What is (AI) consumer behavioral prediction tool? How any why will (AI) tool assist manufactures to attempt to predict consumer behavior before and after consumption occurrence? First, I shall indicate how to apply (AI) tool to predict vehicle product consumer behavior case example.

Nowadays, many vehicle manufacturers hope their vehicles can attract to vehicle buyers to choose to buy their vehicles. However, there are many different brands of vehicles to provide to them to choose, so the vehicle market competition is very serious.

How to judge their different kinds of vehicle price which is reasonable acceptance to attract vehicle buyers to choose to buy the brand of vehicle manufacturers' any kinds of vehicles, e.g. fast speed sport style vehicles, comfortable and slow speed common cars, for four passengers common small size or more than four passengers common large car size? How to evaluate the vehicle prices issue is important factor to influence vehicle buyers‘ choices. Either if the brand of vehicle price is too high to compare brands, it will influence many vehicle buyers choose to buy other brands' vehicles or if the brand of vehicle price is too low, it will influence vehicle buyers feel this brand's vehicle's quality is worse to compare to other vehicle brands‘ similar vehicle products.

Thus, if the brand of vehicle manufacturers can predict how to design vehicles which can attract many vehicle buyers to choose to buy whose any vehicle products. What are future vehicle buyers' favorable vehicle styles? Then, the vehicle manufacturer can concentrate on manufacturing the kind style of vehicle products to sell already. It will reduce its vehicle manufacturing investment risk.

How to apply (AI) tools to predict vehicle buyers‘ behavioral consumption model? Whether artificial intelligent tools can predict automotive buyers' behavioral consumption model and predict future trend. In fact, automotive brands and dealerships are facing an increasingly competition when attempting to manually gathering the vast quantities of data required to create customer focused programs that increase retention, ultimately new sales and service automotive business. Building a based on that client's intrinsic needs and interests to any kinds of automotive vehicles at any given time. This is especially true in the automotive industry where the time span between purchases is measured in years. Because vehicle buyers would not like often to change their old vehicle to another new one. So, their decisions to buying another new vehicle, the time is usually after one year, even longer time. Hence, it seems any vehicles won't be frequent consumption products to the owned at least one vehicle family consumers (vehicle buyers).

Hence, how to predict vehicle consumers' taste or preferable which styles of vehicle choices issues is very important. If the vehicle manufacturers can not manufacture any attractive vehicles to sell easily in this year. Then, it will lose time, money in this year because it won't know when the owned least one vehicle users or non-owned any vehicle users who will decide to buy one new vehicle or change another new vehicle ensure. The different brand vehicle dealers will possible wait more than one year to attract them to buy their vehicles if their styles are not attractive to compare other brands of vehicle competitors.

However, artificial intelligence and machine learning can help any vehicle manufacturers to find solution to solve patterns in highly to solve patterns in highly complex data-sets that are beyond the capability of a human brain, and then building and automatically acting on the customer insights it generates.

Given the automotive customer need for individualized communications, this technology is positioned to become a critical component of any successful vehicle retailer's domestic or/and overseas vehicle markets. How can vehicle manufacturers and retailers use (AI) to enhance their vehicle marketing campaigns? How will (AI) affect their vehicle sale marketing strategy? What criteria would they use when selecting on (AI) solution?

Vehicle consumers today are able to quickly access different brands of vehicle information, research vehicle products and reviews, negotiate prices and compare one vehicle brand or retailer to another resulting of the brands of vehicle customers. At the same time, the rise of " big -data mining", wearable devices that track user's every move and preference and greater contextualization in advertising and social media has resulted in consumer expectations of individualized. Thus, it seems that (AI) tools can be used to gather " big-data" and then they can make human's mind to analyze how to design kinds of vehicles to satisfy vehicle buyers' needs.

As automotive vehicle marketers can apply (AI) tools to achieve messaging strategies to meet the needs of this new generation of informed vehicle consumers, using data from a variety of sources to move from a variety of sources to move from mass- messaging to more personalized messages aimed at particular vehicle buyer segments, e.g. fast speed sport vehicle buyer segment, slow speed comfortable small size or large size of buyer segment. However, when 90% of vehicle marketers believe having a single vehicle buyer view is important, only 6% have achieved it.

However, one of the main issues vehicle marketers facing is the lack of capacity to efficiently sift through and analyze the massive vehicle buyer amounts of data required to create vehicle buyer individualized vehicle customer experiences easily. This is especially difficult for automotive dealers, the long periods between purchase cycles, and the highly considered nature of the vehicle purchase means that each vehicle dealer needs to not only track a large number of potential vehicle customers for an extremely long period of time, but each of those vehicle customers will generate a huge amount of different kinds of vehicle behavioral consumption data as they research their next vehicle purchase. However, by choosing the right (AI) technological tools and programs , vehicle dealers can solve this big data gathering challenge into a major advantage.

For Forrester vehicle brand example, vehicle consumers have more power over the Forrester vehicle brand's reputation than ever before. Mayne, L. (2014) indicated that Forrester calls this new (AI) tools is the " age of the vehicle customer", a 20 year business cycle in which the most successful vehicle enterprises will reinvent themselves to systematically understand and serve increasingly powerful vehicle consumers. To win in this new age, Forrester declares companies must become vehicle customer obsessed and the only sustainable competitive advantage is knowledge and engagement with customers, such as (AI) gathering data knowledge.

Thus, the biggest challenge vehicle businesses currently face is not the collection of a large quantity of vehicle consumer data, but what to do with that data once they have it. Even at a large vehicle data research firm, the data sets are often too big for a single analyze, or even a team of analysts to sort through and draw conclusion from. However, enter artificial intelligence and machine learning , an efficient technology solution that can continuously find patterns in highly complex data sets that are way beyond the capacity of a human brain and then automatic drive action based on the customer insights is generated.

What is (AI) machine learning tool? Machine learning is a type of (AI) that learns from data and is not explicitly program. Think Amazon, face book. Machine learning serves up relevant content based on an individual vehicle purchase behavior and experiences. More simply, machine learning is a computer program that can learn relationships between data, subject those learnings to errors functions, and then learn from its errors. The program in effect, trains itself.

Lee, T. (2016) explained that "Thus, (AI) tools can learn deep a more advanced branch of machine learning inspired by how our brain's nervous function, has also been found to be especial effective in identifying patterns from data."

When this way sound is complicated from a vehicle dealer perspective, the implementation of a marketing program driven by artificial intelligence can take care of these tasks in an automatic vehicle fashion with little to no manual intervention required from the staff at time vehicle stores.

In practice at a vehicle dealership, the program will continue track vehicle customer behavior online, merging that data with any offline source (like CRM or DMS data) and then analyze this aggregated vehicle buyer data set to predict what vehicle customer may be shopping for and what information they might like to
relevance from different kinds style of vehicle design photos.

1.1 Why can (AI) be applied to predict consumer behaviors?

Artificial intelligence refers to complex in vehicle market, machine learning that posses the same characteristics of human intelligence and that have all our sense, all our reason and think just like human do. Besides, machine learning is the practice of using algorithms to collect and examine data, learn from it, and then make a determination or prediction about something in the world.

The machine is " trained" using large amounts of data and algorithms that give it the ability to learn how to automatically perform a task with increasing accuracy. Otherwise, deep learning is primarily based on artificial neural networks inspired by our understanding of the biology of human's brains.

Deep learning breaks down tasks in ways that enables machines to assist us with increasingly complex tasks, driverless cars, better preventive healthcare and more accurate product recommendation (including vehicle recommendations). So, such as why (AI) technology can be applied to predict how vehicle consumer behavior changes to bring to judge whether vehicle consumer will like what kinds of vehicle styles next year. Then, vehicle manufacturers can gather overall vehicle consumer data to analyze and conclude the more accurate vehicle design direction for next year any new design vehicle manufacturing products.

Thus, (AI) machine learning can help vehicle manufacturers to solve how to design any new vehicle products challenge. A vehicle is both one of the most important and carefully considered purchases the majority of people will ever make in their lifetime. It is also a purchase that tends to be fundamentally tied to a person's identify and view of themselves. As the same time, vehicle consumers changing lifestyles result in changing vehicle needs, e.g. the young sport car enthusiast matures into the family driver.

Automotive dealers need to remember that vehicle customers and prospects are individual human beings with risk, complex and ever-changing lives factors, these factors will influence every vehicle consumer why who feels has vehicle purchase need, and how who choose to buy the first vehicle if who decided to buy the first vehicle.

The (AI) technological customer behavioral prediction tool seems to be the best vehicle salespeople in the world are those that know every one of their vehicle customers. Their likes and dislikes which style of vehicle design, preferences and changing tastes to vehicle choices. The capacity of the human brain, however, limits us from achieving this type of vehicle sales and frequent turnover at vehicle dealerships often results in the further loss of vehicle salespeople along with their vehicle customer relationships and knowledge. In this competitive vehicle environment, machine learning enables platforms to assist the vehicle sales team by tracking the vehicle consumer behaviors of each vehicle customer, learning and memorizing their preferences and predicting their future vehicle purchase needs.

Finally, I recommend that for a vehicle dealerships marketing platform to make their customer engagement efficient and fully-functional, I should be able to: applying (AI) tools to track every vehicle customer behavior across the web, connecting to a society of data sources, CRM, DMS, third-party, web vehicle brands, social email, click etc., aggregating and accurately cross-reference data from a variety of sources, leveraging this data to drive insights on a mass scale, as well as on an individualized basis, driving actions and automatically direct customer engagement via multiple channels based on where each customer is in their individual lifecycle.

How can (AI) provide businesses with better-informed decisions

I shall explain how (AI) technology can provide businesses with better-informed decisions to drive top-line growth, deliver meaningful experience for customers and smooth their path along the consumer journey. The widely understood definition of (AI) involves the ability of machines or computers to learn human thinking, reasoning and decision-making abilities.

A Narrative science study in 2015 year identified that (AI) was being used primarily in voice recognition, machine learning virtual assistants and decision support. This study also highlighted the many branches of (AI) and that techniques and their definition are used interchangeably. It is possible that (AI) can be used to gather big data , then to analyze to help businesses to predict consumer behaviors. For example, one of the most common techniques is machine learning, where algorithms are used to perform tasks

● How (AI) influences organizational change

Consequently creative and social intelligence will be in even greater demand as (AI) makes in management and the workforce. This development will represent a long term trend in labor markets , one characterized by intensifying demand and reward for social skills with a growing desire for creative capabilities, managers will seek to fashion of ideas and hypotheses from inside and outside of the enterprise to shape solutions to their most pressing business problems. Thus, (AI) will influence overall organizational team members who have chance to participate any decision to make more accurate business judgment.

Many managers mistakenly view judgment work as only an individual discipline, failing to appreciate that it can also involve decide interpersonal and organizational practices. In more complex settings, judgment is typically a collective outcome of individuals‘ and teams' diverse perspectives, insights and experiences. And often , the resulting choices are better informed than decisions that an individual would have arrived at on his or her own.

Thus, when any organizations apply (AI) technology to assist managers to gather data and ideas to make any judgment. In these cases, organizations can create the conditions for effective collective judgment by establishing structures , such as " shadow advisory boards" that prompt managers and employees to source and synthesize multiple perspectives. Thus, a traditional organization (firm) might freshen its thinking is t put together a shadow advisory board, comprised of young, digital people who can apply (AI) machine assistance to make judgment work more accurate whether related to people development, problem-solving or strategizing and innovating for considerable degrees of creative and social intelligence.

Thus, on the one hand, (AI) technology machine augmentation and automation can give these advantages to human (organization managers) , e.g. developing people and community, solving problems and collaborating, coordinating and controlling work, shaping strategy and leading innovation. Besides, on the other hand, the next generation managers need have these individual attitude to treat intelligent machines to be as colleagues.

When, judgment is a human skill, intelligent machines can accelerate human learning that supports it, assisting in data -driven simulations, scenarios and search and discovery activities. Focuses on judgment work, some decisions require insight beyond what data can tell them. This is the sweet sport for human judgment, the application of experience and expertise to critical business decisions and practices. Thus, managers will also need to find ways to learn how to use digital (AI) technologies to tap into the knowledge and judgment of partners, customer external stakeholders and role models in other industries after the (AI) machine had been implemented to the organization.

Future works change:Automation, employment
and productivity

2.1 How (AI) influences employment

Human future " micro to macro" industry trends will be affected business strategy and public policy by (AI) technology. In the future (AI) technology will influence those six themes: productivity and growth, natural resources, labor markets, the evolution of global financial markets, the economic impact of technology and innovation and urbanization. However, (AI) technology will bring economic benefits of tackling gender inequality, a new global competition, Chinese innovation and digital globalization.

Nowadays, advances in robotics artificial intelligence, and machine learning are in a new age of automation, as machines match or outperform human performance in a development to any countries. For example, automation of activities can enable businesses to improve performance by reducing errors and improving quality and speed, and in some cases achieving outcomes that go beyond human capabilities. For example, some research indicated automation could raise productivity growth globally by 0.8 to 1.4 % annually; more than 2,000 work activities across 800 occupations. When less than 5% of all occupations can be automated using demonstrated technologies about 60% of all occupations have at least 30% of constituent activities that could be automated. Many occupations will change that will be automated away: Activities most susceptible to automation involve physical activities, in highly structured and predictable environments, as well as the collection and processing of data. They are most prevalent in manufacturing , accommodation and food service and retail trade and include some middle-skill jobs. For example, such as natural language processing is a key factor. Beyond technical feasibility, the cost of technology competition with labor including skills and supply and demand dynamics, performance benefits including and beyond labor cost savings, and social and regulatory acceptance will be affected by (AI) automation technology. Thus, (AI) automation will impact to influence global employment in those aspects as below:

Firstly, assuming that people are displaced by automation will find other employment. The anticipated shift in the activities in the labor force is of a similar order as the long-term shift away from agriculture and decreases in manufacturing share of employment. Both of manufacturing and agriculture industries which would be accompanied by the creation of new types of work not foreseen at the time.

Secondly, for business, the performance benefits of automation are relatively clear. Thus, the businessmen have opportunities for their micro economies to benefits from the productivity growth potential and macro economies to benefit to encourage continued progress and innovation , investment and market incentives. At the same time, employers must innovate policies to help workers and institutions adapt to the impact on employment.

This will likely include rethinking education and training, income support and safety nets , as well as support for those dislocated, when employees need to leave themselves homes to move to other cities to learn new (AI) automation works. Thus, individuals in the workplace will need to engage move comprehensively with machines as part of their everyday activities, and acquire new skills that will be in demand in the new automation age. Consequently , the scale of shifts in the labor force over many decades that automation technologies can be a similar order to the long -term technology -enables shifts in the developed countries‘ workforces away from agriculture in the 21 th century. Those shifts did not result in long-term mass unemployment because they were accompanied by the creation of new types of work not foreseen at the time. However, human will still be needed in the workforce when the total productivity gains are caused by (AI) technology.

2.2 What occupations will be influenced by (AI) technology.

In the future, scientists predict that these occupations will be influenced by (AI) technology mostly. They include : retail salespeople, food and beverage service workers, language or translation teachers, health practitioners. Since these work activities have a more relevant occupations are made up of a range of activities with different potential for (AI) automation . For example, a retail salesperson will spend more time interacting with customers, stocking shelves , or ringing up sales. Each of these activities is distinct and requires different capabilities to perform successfully.

Thus, these job activities have similar simple control characteristics. Simple activities include greet customers, answer questions about products and services, clean and maintain work areas, demonstrate product feature process sales and transactions. All these activities can have similar simple activities in order to (AI) machines can be learn how to do these activities from (AI) technology . For example, the capability perception includes sensory perception, cognitive capabilities, such as retrieving automation, recognizing known patterns(supervised learning), logical reasoning problem solving.

Thus, (AI) machine is such human, which has feeling and emotion, such as social and emotional sensing, judgement reasoning methods, natural language understanding and physical capabilities, such as mobility , navigation, gross motor skill, fine motor skills. It seems that the future, (AI) human invents machines which will have these human characteristics to do human similar behavioral job duties more easily and efficiently. It implies these above human occupations will be replaced by (AI) human invention machines in the future. Due to (AI) creation, it is possible to cause unemployment number of these above workers will increase because (AI) machines can do their similar job behavioral activities.

Consequently, employers won't need to employ many of these skillful labor. Otherwise, they can buy less number (AI) machines to attempt to do whose job activities more easily and efficiently. So, it seems (AI) machines will have more high work performance to replace these occupation workers' work performance. Finally, these occupation worker unemployment number will only increase when the (AI) machines had been invented to achieve to do their work behavioral activities absolutely success in the future.

2.3 Whether (A) technology machine labor
will replace human worker more or assist
human worker more

There is no single agreed definition of a robot how outcome of a task that is completed without human intervention. When some definitions require the task to be completed by a physical machine moves and respond to its environment, other definitions use the term robot in connection with tasks completed by software , without physical embodiment.

However, to answer the question : Whether (AI) technology machine labor will replace human worker more or assist human worker more. I shall indicate some examples to let readers to judge whether (AI) technology can create new jobs or reduce old jobs.

Firstly, I shall explain what (AI) function is. (AI) is a service robot that performs useful tasks for humans or equipment excluding industrial automation application . Thus, the classification of a robot into industrial robot or service robot is done according to its intended application. It is also a personal service robot or a service robot for personal used for a non commercial task, usually by lay persons . Examples are domestic servant robot, and pet exercising robot. It is also a professional service robot or a service robot for professional used for a commercial task, usually operated by a properly trained operator. Examples, are cleaning robot for public places, delivery robot in offices or hospitals, fire-fighting robot, rehabilitation robot and surgery robot in hospitals. Thus, these functions will be future (AI) application to our daily life necessaries or business necessaries.

However, some authors agree (AI) will bring negative outcomes of automation, due to raise competiveness, reduce human job nature. Otherwise, other authors argue (AI) will bring positive outcomes of automation, due to raise productivities, job creation, assist humans work.

On the positive outcome hand, robots can increase productivity . This is particularly important for small-to medium sized businesses both are in developed and developing countries economies. It also enables large companies to increase their competitiveness through faster product development and delivery. Increased use of robot is also enabling companies in high cost countries to re shore, or bring back to their domestic base parts of the supply chain that will have previously outsourced to sources of cheaper labor. Currently , the greater threat to employment is not a automation, but an inability to remain competitive. Automation has led overall to an increase in labor demand and positive impact on wages. The reason is that the middle-income/middle-skilled jobs have reduced as a proportion of overall contribution to employment and earnings leading to fears of increasing income inequality, the skills range within the middle income bracket is large. Thus, robots are driving an increase in demand for workers at the higher -skilled and with a positive impact on wages. This issue is how to enable middle-income earners in the lower-income range to unskilled or retain. Finally, the (AI) positive impact supporter who argue the future will be robots and humans can work together.

However, on the negative outcome hand, robots can substitute labor activities, but don't replace jobs. They believe that less than 10% of jobs are fully automatable. Increasingly , robots are used to complement and augment labor activities, the net impact on jobs and the quality of work is positive. Automation can provide the opportunity for humans to focus on higher-skilled, higher-quality and higher-paid tasks. Robots can improve productivity when they are applied to tasks that which perform more efficiently and to a higher and more consistent level of quality than humans. For example, increased productivity is enabling some firms, such as Whirlpool, Caterpillar and Ford Motors company in the US restructure their supply chains, bringing back parts of the manufacturing process to the country of origin. Thus, productivity gains due to robotics and automation are important not just at the company level, but also for build industry and nation competitiveness.

I suppose that productivity can be raised. What are the impacts of robots on employment? Firstly, the main focus of development has been on personal entertainment, which does not drive worker productivity (manufacturing production). When the internet (information and communication technology (ICT)) innovation. This is borne and by findings that manufacturing productivity, which has been driven by innovations in automation rather than consumer technologies, has government strongly than productivity in the services sectors of the economy in most nature economies. It seems (AI) automation will create many jobs in internet communication entertainment game industry. For example, many young people like to use internet to play any electronic games from computer or mobile at home or outside home conveniently. Thus, (AI) automation will increase demand to be invented to any new entertainment game from internet channel. It will need to employ many (AI) entertainment game inventors to create many automation entertainment games. Thus, (AI) automation in internet entertainment game industry will need human (AI) entertainment game inventors to invent the knowledge-based capital of (AI) automation entertainment games. The (AI) entertainment game inventors will need own research and development skills, form specific skills, organizational know-how skills, databased knowledge, design and various forms of intellectual property to do these (AI) automation entertainment game invention occupations in the future.

International Federation Of Robotics(2016) indicated that China will be as a major robotics manufacturer and user of robots, benefiting from jobs created by robot manufacturing and productivity gains from robot use. Chins had sold of robots to any one single market every year since 2017 year. The Chinese government has included a focus on robotics in its 10 year strategy. In order to achieve its target of a robot density of 150 units per 10, 000 workers by 2020 year. Thus, Chinese companies will have to install around 650,000 new industrial robots between 2016 to 2020 year, 2.5 times more than installed globally in 2015 year.

Hence, China (AI) manufacturing industry will need to employ many workers . It implies (AI) manufacturing industry will create many new occupations in China. Also, ministry of economy, trade and industry (2015) also showed that Japan currently has the largest stock of industrial robots in operations, primarily in the automation industry. Driven by a rapidly aging population and low productivity rates, the Japanese government has sights on a 20-fold increase in the use of robots in the non-manufacturing sector and a three-fold growth rate of labor productivity in the service sector both by 2020 year. Thus, it also implies Japan will need many robots to be provide to service industry. Due to robots will provide to serve any businessmen's clients. Thus, it is possible that the service workers won't be dismissed as well as it is depended on the serving job nature to decide whether Japan's service workers can still serve to their employer when the service (AI) robots are applied to whose employers.

Consequently, it seems that (AI) can create employment, Ministry of economy, trade and industry (2015) showed that such as China will develop the major (AI) automation manufacturing industry. The (AI) employers will need to employ many workers to manufacture any these different kinds of (AI) robots to satisfy China or overseas individual or business buyers needs. But, (AI) can also cause unemployment to the low skillful service workers. Such as if Japan some service businesses choose to buy any (AI) service robots to replace their service staffs to serve their clients. It is possible that the service staffs will be dismissed, due to (AI) robots can do such as their same service job duties to achieve better service performance.

Thus, today, it is increasingly common for people to use robots in various situations at home and in retail stores, hotels and hospitals these service industries. Robots are classified into server types based on their functionality (service and utility robots or those designed to communicate with humans) and appearance (humanoid robots or mechanical robots). The type of robot, to which each country allocated particular importance in the advance of robotics, reflects the sense of values and preferences of its population. Thus, if the country has high population needs to use robots, then they will influence either more new jobs creation or more old job loss in the country's (AI) manufacturing or (AI) service industries both. For example, Japan respondents often associate the term " robot " with humanoid robots that can communicate with human and they have a high level of familiarity with robot. The US has the highest level of robot utilization at home and in retail stores with its people being the most enthusiastic about the future use of robots. Germany shows a strong tendency to consider robots for industrial purposes and its people feel strong effort to the presence of robots in their households.

In conclusion, to judge whether how (AI) will influence the country's employment to be better or worse. It will depend on the country home buyers (users) or business buyers (users) how to use (AI) for their daily needs. If the country , such as US retail stores need to use (AI) , it will have possible to reduce some or many retail service workers. Even, if the country , such as Japan has many home users need to use (AI) , it will not influence the employment market. Otherwise, it will raise (AI) salespeople numbers. Even, if the country, such as Germany and China will have many (AI) manufacturers, then it will create many (AI) manufacturing occupations for these (AI) manufactory workers.

Consequently, (AI) robots manufacturing and service needs will have positive or negative impact to any country's employment. It will depend on the (AI) service provision and service workers' job nature as well as the manufacturing workers of (AI) knowledge level to decide their employment chance in their country's employment market.

AI brings what jobs change

What does artificial intelligence(AI) mean?

- What (AI) function is?

Some scientists explain that artificial intelligence means which is an expert system, computer software that embodies a portion of the specialized knowledge of a human portion in a specific, narrow domain, owns decision making ability of human expert. The (AI)technology is based on the premise that what makes a person an expert is years of experience that enables who recognizes certain patterns in a problem as being similar to pattern. For example, in the future artificial intelligence system can be applied to control air traffic, design to computer configuration, medical diagnosis, instruction/training, speech/interpretation, monitoring to (nuclear plant), planning to mission, factory scheduling, prediction weather, repairing telephone, automatic driving etc. different industries.

Artificial intelligence characteristics include: creative, adaptive , common sense, fact processing, quick replication, broad focus permanent and consistent skill. Otherwise, traditional computer expert system disadvantage includes perishable, unpredictable, slow reproduction, expensive, slow reproduction, slow processing lacks inspiration, needs instruction, narrow focus only machine knowledge. So, artificial intelligence is a branch of computer science devoted to creating computer to influence software and hardware to attempt to create human intelligence or human intelligent behavior. It is learning from experience, responds flexibility in situation that are, new or not anticipated. Thus, (AI) can be learnt programmed knowledge to solve problems, using reasoning in solving problem, understanding and inferring facts and rules, recognizing the relative importance of different elements in a situation. In summary, artificial intelligence is concerned with two basic ideas mainly: The first idea, it involves studying the thought processes of humans to understand what intelligence is; the second idea, it deals with representing thought processes using companies to create artificially intelligent entities for testing the theories of intelligence.

- Can (AI) impact human job nature?

Human need concern this question: Will artificial intelligence (AI) reduce some human jobs in order to instead of replacing machines to do? Due to artificial intelligence is the ability of machines to do thing, that people would require intelligence. For example, artificial intelligence machine man driving(self-driver), it (AI) machine man driving research is an attempt to discover and describe aspects of human intelligence that can be simulated by driving machine functions. Alternatively, (AI) mathematical research may be another viewed as an attempt to develop a mathematical theory function to describe the abilities and actions of things (natural or man-made) exhibiting intelligent behavior and server as a design of intelligent calculation machine function.

Why do humans need artificial intelligence machines to instead of traditional human service job? For example, can artificial intelligence machine man (self-driving) driver drive to replace human driver? I shall compare the differences between humans and computers : The characteristics of humans are good at recognizing various things, either seen before or not, recognizing the relationship patterns between things. Human thinking is common sense reasoning, combining all types of sensory input, acting appropriately in novel situations, learning new things and changing behavior patterns, making decisions , even when given incomplete information, working with noisy, incomplete information gathering behaviors . However, characteristics of computers are good at: The tasks humans do naturally are extremely difficult for a computer program as intelligent, which must be able to do the same kind of tack as humans do naturally.

Hence, (AI) is an combination of many different success and technologies: Linguistics - computational and socio, philosophy-logic, philosophy of mind and of language, electronical engineering -image and speech processing, pattern recognition, robotics, machine learning, neural networks, optimization scheduling, management information system and decision making. So, it is possible that (AI) can impact human job nature to instead of human working behavior in the future.

How can (AI) influence labor market?

- How can human society job nature

to be changed to artificial intelligent society?

From the first intelligent perspective reason view point, artificial intelligence is making machines " intelligent" acting as humans expect people to act. Artificial intelligence has ability to distinguish computer responses from human responses, it owns knowledge to solve expert problem. From another research perspective reason view point, artificial intelligence is the study of how to make computers do things which, at the moment, people do better (Rich & Knight, 1991, p.3).

(AI) researchers are native in a variety of domains, e.g. formal tasks (mathematics, games), tasks (perception, robotics, natural language, common sense reasoning), expert tasks (financial analysis, medical diagnostics, engineering, scientific analysis and other areas).

From the second business perspective reason view point, (AI) is a set of many powerful tools, and methodologies for using those tools to solve business problems. From a programming perspective reason view point, (AI) includes the study of symbolic programming problem solving and search .

From the third human technological perspective reason view point, today's computer can do many well-defined tasks, for example, arithmetic operations, are much faster and more accurate than human beings. However, the computers' interaction with their environment is not very sophisticated yet. How can human test whether a computer has reached the general intelligence level of a human being? Can a computer convince a human interrogator that it is a human? But before thinking of such advanced kinds of machines, human will start developing our own extremely simple " intelligent" machines.

So, it is possible that human society job nature will to be changed to artificial intelligent society when (AI) technology is developed to the mature stage in the future.

● Why does human need artificial intelligence machines?

One of major division in (AI) is between humans who think (AI) is the only serious way of finding out how we (human) work and human who want companies to do very smart things, independently of how we (human) work. This is the important distinction between cognitive scientists vs engineers. One of another major division in (AI) is between symbolic (AI), which represents information through symbols and their relationships. Specific Algorithms are used to process these symbols to solve problems or deduce new knowledge and connectionist. So (AI) , which represents information in network. Biological processes underlying learning, task performance and problem solving are imitated from human mind behaviors. Thus, it is possible that artificial intelligence machines can do the better judgicious behavior to compare human.

● How does artificial intelligence influence future working changing in automation employment and productivity aspects?

In the automation changing influence aspect, as companies increasingly use robots on production lines or algorithms to optimize their logistics manage inventory, any carry out other core business functions. Technological advances are creating a new automation age in which ever-smarter and more flexible machines will be deployed on an ever larger scale in the marketplace. However, researching artificial intelligence with how influences human working nature.

We need to answer these questions: How will automation transform the workplace? What will the implications for employment? And what is likely to be its impact both on productivity in the global economy and on employment? Advances in robotics, artificial intelligence, and machine learning are growing in a new age of automation as machines match or outperform human performance in a range of work activities, including ones requiring cognitive capabilities. What factors are determined the changing in workplace adoption by artificial intelligence innovation? What advantages are automation? Automation of activities can be enabled businesses to improve performance by reducing errors and improving quality and speed, and achieving outcomes that go beyond human capabilities.

Some scientists indicated based on their scenario modeling. They estimated automation could raise producing growth globally by 0.8 to 1.4 percent annually. Almost, the activities people are paid almost $16 trillion in wages to do in global economy have the potential to be automated by adopting currently demonstrated technology. According to their analysis of more than 2,000 work activities across 800 occupations. When less than 5% of all occupations have of least 30% of activities that could be automated. They also indicated that technical economic and social factors will determine automation. Continued technical progress, for example, in areas such as natural language processing is a key factor beyond technical feasibility , the cost of technology, competition with labor including skills, and supply and demand dynamics, performance benefits including and beyond labor cost savings and social and regulatory acceptance will affect (alter) the scope of automation.

Other some scientists also indicate U.S. country for example, the anticipate shift in the activities in labor force of a similar order of magnitude as the long term sight away from agriculture and decreases in manufacturing. Share of employment in the United States both which were achieved. So, those factors can influence why artificial intelligence technology needs. So, it is possible that future agriculture and manufacturing both industries will apply (AI) technology manufacturer-kind of job nature to raise productivity instead of farmers, fruit picking workers, farming transportation labours as well as factory manufacturing workers and supervisors etc. human-kind of job nature.

- Is artificial intelligence possible to replace labor ?

Not just intelligence, but also debating, if machines are capable of having a conscious minds. Artificial intelligence has those characteristics as below:

On functionalism aspect, artificial intelligence inputs mental states, sensory inputs, (beliefs, desires being in pain feeling) and behavioral outputs. Since mental states are identified by a functional role, which are thoughts to be manifested in various systems. Even, perhaps computers which are physical devices with electronic substrate that inform computations on inputs to give outputs similar to brains which are artificial intelligence composed of part any intrinsic relationship to each other. Thus, artificial intelligence activities is not the whole itself, but into parts or on external influence on the parts.

On dualism aspect, artificial intelligence is a set of views about the relationship between mind are matter. On materialism aspect, it builds the only thing that exists is matter, including consciousness.

On biological naturalism aspect, it is similar a human brain than feels pains makes mental situation. So, artificial intelligence is similar biologist which might to be excited to human labor work. Hence, it seems artificial intelligence can change (alter) or replace human labor work of nature in possible in the future.

- Can (AI) technology replace human labour nature of work?

On technological innovation reason view point, the history development of artificial intelligence studying the intelligence is one of most ancient scientific discipline. The history development of artificial intelligence what aims to achieve human use to sense, learn remember and think, logic probability, decision making and calculation develop from mathematics, instead of replacement human labor functions.

Artificial intelligence history development aim is the scientific analysis of skills in connection and practice with the appearance of computers from 1950 year beginning. The artificial intelligence (AI) can deal with the ultimate challenges. How can (either biological or electronic) mind sense, understand and manipulate a world that is much simple and more complex than itself? And what if would human like to construct something with such capabilities?

The general-purpose software of the early period of (AI) were only able to solve simple tasks effectively and failed when which should be used in a wider range or an more difficult tasks. One of the sources of difficulty was that early software had very few or mix knowledge about the problems which handled, and activities successes by simply syntactic manipulation. Moreover, the other difficulty was that many problems that were tried to solve by the (AI) were untreatable.

The early (AI) software whether trying step sequences based on the basic facts about the problem that should be solved, experimented with different combinations till which found a solution. From the end the 1960 year, developing the so-called expert systems were emphasized. These systems had (sue-based) knowledge base about the field which handled. Till to the beginning of the 1970 year, (Prolog) the logical programming language was born, which was built in the computation realization of a version of the resolution calculus. (Prolog) is a remarkably prevalent tool in developing expert systems (on medical, judiciary and other scopes), but natural language parsers were implemented in this language. Then, in 1981 s, the Japanese announced the fifth generation computer system project, a 10 years plan to build an intelligent computer system that use the (Prolog) language as a machine code. Nowadays, (AI) can be applied any industries, such as car manufacturing industry can use (AI) technological machine-men manufacture car, instead of replacing human labors in factory. Even, in the future, using (AI) machine-men drivers can drive any private cars or public transportation tools, instead of replacing human drivers, e.g. bus, train, tram, ferry etc. Also in the future, machine-men can replace housewives to serve families to do housekeeping clean job , e.g. cleaning toilets, bathrooms, kitchens, even cooking functions at home. So (AI) machine-man can reduce housewives works at home. Moreover, (AI) machine man can take care old people , when who are living at homes or elder care centers.

So, it seems artificial intelligence (AI) will be possible developed to manufacture a new generation machine-man to assist (serve) families to do any simply cleaning or cooking jobs at homes. Moreover, the overall demand of (AI) general social needs will also rise, such as security, driving transportation tools, restaurant cleaning, elder centers care service etc. So, it seems that individual or families or social needs of (AI) will be increase in the future. Thus, it will influence macro economy growth (GDP) if there are large house family consumer group and hotel or bus or taxis or ferry etc. different business consumer group demand any artificial intelligence machine numbers increasing. Then, the artificial intelligence products and material manufacturers must need to buy many artificaial intelligence materials to produce any kinds of artificial intelligence machines to prepare to satisfy consumer individual needs. Consequently, macro economy will grow to the owned artificial intelligence development countries, e.g. US, China, UK.

● Why can artificial intelligence satisfy human needs?

First, On machine-man satisfactory demand aspect view point, it makes computers that think, it is the automation of activities. We associate with human thinking: like decision making, learning. It is the act of creating machine that perform function that require intelligence when performed by people. It is the study of mental faculties through the use of computational models. It is the study of computations that make it possible to perceive, reason and act. It is a branch of computer science that is concerned with the automation of intelligent behavior. It is anything in computing service that human don't yet know how to do property.

Second, on thought aspect artificial intelligence means systems thank think like humans, systems that think rationally.

Third, on behavioral aspect, artificial intelligence systems that act like human and that systems act rationally. However, the basic objective of (AI) is to represent human's thought processes in computation . These machines are supposed to exhibit behavior that. It is performed by a human being, would be considered intelligent. However, some authors feel (AI) has disadvantages, such as it is not creative, it is excited in the use of sensory devices, it can't make use of a very wide context of experiences and it does not use common sense.

For speech recognition and understanding function needs example, (AI) can be applied in speech recognition and understanding function, which (AI) speech or voice recognition is a data input method. For example, the computer recognizes and understands one (or a few) word commands. Speech understanding on the other hand is the computer's ability to understanding a spoken language. That is , the computer understands the meaning of sentences, an paragraphs through (AI).

So, (AI) can be attempted to learn human language how to speak. It is similar to translate human language skill, instead of actual human speaking skill. Also, (AI) can assist handicap learning or language student how to listen different languages by machine-man sounds from computers more accurately. So, it seems that it (AI) can replace human language teachers speaking function and can change teaching language nature of job in language speaking and listening education industry.

● Is artificial intelligence one good choice for human future technological benefit?

Nowadays, new technology development is popular. However, artificial intelligence is one kind of new technology choice among different technologies innovation. So it brings this question: Is artificial intelligence technology value to invest? To answer this question. I shall indicate some other new technology developments to compare (AI) technology development to judge which has urgent needs to achieve human expectation nowadays.

For example, why is green peace interested in new technologies? New technologies features prominently in our ongoing campaigns against genetic modified crops and number power. However, which are also an integral part of our solutions to environmental challenges, including renewable energy technologies, such as solar, wind and wave (water) power energy as well as waste treatment technologies, such as mechanical, biological treatment.
It seems humans need concern how to apply (AI) technology to solve environment pollution challenges in our future. So, environment protective, agriculture, natural energy technology will be popular demand to attempt to apply (AI) technology to solve their challenges or apply (AI) to assist to develop their industry.

What is influence to (AI) job nature change to change developed and countries economy

- How can artificial intelligence technology influence economy?

Advances in artificial intelligence (AI) technology and related fields have opened up new markets and new opportunities progress in critical areas, such as health, education, energy, economic development, social welfare and the environment pollution.
(AI) automation will continue to create wealth and expand the global economy development in the future. However, when many will benefits that growth won't be costless and will be accompanied by changes in the skills, that workers need to increase productivity in the economy and structural changes in the economy. So, in the skills that workers need to succeed in the economy and structural changes.
I shall indicate why aggressive policy action will be needed to help Americans who are disadvantaged by these changes , due to (AI) technology is caused. For automation industry change example, artificial intelligence (AI) capabilities will enable automation of some tasks that have long required human labor. These artificial intelligence technology introduction can increase new opportunities for individuals. The economy and society, but (AI) has also the potential to disrupt be current livelihoods of many Americans. However, (AI) leads to unemployment and increase in inequality over the long run depends not only on the (AI) technology itself, but also on the institutions and policies that are changed.
Thus, it is possible that (AI) technology will raise some countries unemployment number if the employer apply (AI) technology workers to work instead of human labor in their factories, but it can also raise productivities for these employers.

- Can (AI) influence global economy growth?

Technological progress is main driver of growth of GDP per capita, allowing output to increase faster than labor and capital . However, technology can increase productivity, but also decrease the number of labor hours needed to create a unit of output. So (AI) causes unequal to labor wage decreases, even reduces the number of labor to manufacture, e.g. artificial intelligence technology of automation car manufacturing industry; clothing manufacturing industry; plane manufacturing etc. high technology of artificial intelligence manufacturing method. But (AI) should be potential environment benefit, although it raises unemployment ratio. Moreover, it can rise production , due to many skilled craft were replaced by the combination of machines and lower-skilled labor. The result of (AI) technology introduction , it causes output per hour risen when inequality declined, driving up average living standards, but the labor of some high-skill workers was no longer as valuable in the market. Otherwise, if (AI) technology is continue developed to be success. Some routine intensive occupations will be loss, which focused on predictable, e.g. easily programmable tasks, such as switchboard operators, filing clerks, travel agents, and assembly line workers would be particularly replaced by new (AI) technology. However, at the same time, (AI) technology development will bring these benefits: improvement in education (training (AI) technology scientists) , due to (AI) manufacturing technology needs are raising to businesses and institutional changes, such as the reduction in unionization and raising in the minimum wage to the (AI) manufacturing technology skilled labor in factories.

Because (AI) technology is not a single technology, but rather a collection of technologies that are applied to specific tasks, the effects of (AI) will be felt unevenly though the economy. It will bring some tasks will be most easily automated than others , and some jobs will be affected more than others, both negatively and positively. Finally, new jobs are likely to be directly created in areas , such as the development and supervision of (AI) as well as indirectly created in a range areas though out the economy as higher incomes lead to expanded demand.

However, if (AI) technology could dominate global labor markets. If labor productivity increases, do not influence into wage increases, then the large economic gains brought about by (AI) technology could be increased wealth inequality, due to employers can reduce production cost, but workers (labors) wages will not be increased, even will be decreased. Hence, it seems the (AI) technology will bring disadvantages to labor market to cause unemployment or reduce wages in possible, although it can reduce employer individual salary (wage) expenditure and it can raise productivity.

- How can artificial intelligence impact global economy growth?

Artificial intelligence (AI) technology is a branch of computer science that aims to create intelligent machines that work and react like humans. So, (AI) is a technology that appears to impact (influence) human preference by learning, understanding complex contents, enhancing humans in executing both routine and non-routine tasks. In the future, (AI) technology that can be virtual personal assistant, as well as it may exist, such as robots with human-like processing capabilities.

How can (AI) technology impact global economy growth over the next 10 years? During this time period, (AI) technology is predicted to have wide-ranging applications including: Machine learning that automates analytical model building by using algorithms that allow machines to operate without human assistance.

In global education aspect, potential applications include predicting cause-and-effect relationships from biological data, identifying new drugs, self-driving cars, and protecting against fraud, improved natural language processing that allows computers to continue to better analysis, understand and generate language to interface with humans using natural human languages. For example, transcribing notes dictated by physicians, automatically drafting articles and translating text and speech. So (AI) technology can be applied to education aspect to improve humans‘ knowledge level.

In visual art aspect, (AI) machine vision that allows computers to identify objects, scenes and activities in images. Current applications of (AI) machine vision include providing objective descriptions for the blind seeing(visual) needs.

We except the economic effects of (AI) technology to include both direct GDP growth from sectors that develop or manufacture. (AI) technology and indirect GDP growth through increased productivity in existing sectors that employ some form of (AI). If (AI) technology is an increasingly critical component of more products, it will become an integral part of many people's lives. Thus, (AI)'s ability to influence economic activity, rather than the economic or development status of the region. (AI) has the potential to impact income classes and to bring significant gains to both developed and developing countries. For example, (AI) has the potential to optimize good production around the world by analyzing agricultural regions and identifying what is necessary to improve crop yields.

In estimating the future economic effects by (AI) technology innovation, it is important to note that it is challenging to accurately predict which applications of (AI) will ultimately be commercially successful. In micro level economic influence, we need to apply methodologies to estimate the economic effects of investment in firms developing (AI) technology since investment levels in a technology are a telling sign of the future potential of that (AI) technology.

● How can (AI) influence GDP of high income countries in the next ten years?

How (AI)'s development may affect the global economy over the next ten years. In fact, (AI) technology has the potential to affect business across the global in a wide range of industries in ways only a number of technologies have done in the parts. For example, (AI) technology's expected to be a useful tool for enhancing human capabilities and in some instances replacing functions, such as driving a car, adoption of broadband internet, mobile telephone, industrial robotic automation have served to enhance human capabilities.

However, significant public debate has focused on projections of (AI) technology's effect on the labor force. However, large companies prefer to invest in (AI) technological industry. For example, face book's (AI) research lab., google machine intelligence lab. and micro soft machine learning and artificial intelligence research division are all making advances in (AI) technology and investing in the industry's top talent. Additionally, between 2010 year and 2015 year, nearly $5 billion in venture capital funding invested in firms across the global developing and employing (AI) technology (Facebook (AI) Research).

● How can artificial intelligence impact on workplace?

Modern information technologies and the labor economy growth of machines is powered by artificial intelligence have already strongly influenced the world of work in the 21 ST century. Computers, algorithms and software simplify every tasks and it is impossible to image how most of our life could be managed without them. How can be the information economy characterized by exponential growth replaces the most production industry based on economy of scales? What will the future world of work look like and how long will it take to get? Will the future world of work be a world where humans spend less time earning their livelihood? Alternatively, are mass unemployment, mass poverty and social distortions also possible scenario for the future, where robots, artificial intelligence systems play an increasingly central role? These questions concern how artificial intelligence further development . Can influence labor economy growth on workplace ? When the labor market has widespread impact on intelligence property, information technology, product liability, competition and labor and employment laws.

How (AI) technology impacts on labor workplace.
The future influence any organizations how labor economies use of (AI) can be analyzed, such as deep machine learning is based on a set of model high level data. Unlike human workers, the machines are connected the whole time in workplace. If one machine makes a mistake, all autonomous systems will keep this in mind and will avoid the same mistake the next time.
Over the long run intelligent machines will win against every human expert. Production robots have been replacing employees because of the (AI) technology. They work more precisely than humans and cost loss. Creative solutions like 3D printers and the self learning ability of these production robots will replace human workers, the automatic data recording and data processing, traditional back office activities are no longer in demand. Autonomous software will collect necessary information and will send it to the employee who needs it. Additionally, dematerialization leads to the phenomenon that traditional physical products are becoming software. For example, CD or DVDs are being replaced by streaming services. The replacement of traditional event ticket, e-travel ticket service products or hard cash will be the next step, due to the possibility of payment by smartphone. So, (AI) technology will impact human's daily life consumption behaviors in the future. For another example, transportation tools, such as boats and ferries and private vehicles will use sensors and navigating without human input. Taxi and truck drivers will become obsolete, the stock store applies to stock managers and postal carriers of the delivery is distributed by (AI) machine delivery method.

What is the relationship between (AI) and (CRM)?
● Can (AI) technology impact on customer relationship management (CRM) ?
Nowadays , (AI) is a technology almost as old as the computer industry itself, it is similar with the advent of personal assistants function to businesses and personal promotion channel, such as (Amazon's Alexa, Apple's Siri, Google's Assistant) image recognition (face book), personalized recommendations (Netflix , Amazon). Those innovations have been driven by a increase in processing power, lower cost hardware, and the exploding creation and availability of data. It seems, (AI) technology can impact global customer service management method.
How to forecast economic impact modeling to (AI) will affect global economy? Can human forecast business revenue growth and job creation (or destruction) based on (AI) applied to customer relationship management (CRM) activities? In addition to the economic impact on (AI) or (CRM) which can include an estimate of the economic impact attributable to sales forces customer base. What can economic benefits be brought to (CRM) from (AI) technology?
Artificial intelligence(AI) comprises a set of technologies that use natural language processing, machine learning, knowledge graphs, and other tools to answer questions, discover insights and provide recommendations. Computer systems can use (AI) hypothesize and formulate possible answers based on available evidence can be trained through the ingestion of vast amounts of content, and automatically adapt and learn from (AI) self mistakes and failures.
So, any business organizations (customer service departments) can provide efficient and effective customer

relationship management of excellent customer service quality if which applied (AI) technology system. The different type of (AI) systems include: (AI) system platforms, machine learning (AI) based data preparation and enrichment tools, machine vision/image recognition, voice speech recognition, text analysis and natural language processing, bots , e.g. face book website and virtual digital assistance solutions, social media pattern analysis , sentiment analysis, advanced numerical analysis (e.g. IOT streaming , machine logs), supporting technologies, knowledge base dialog management, Q&A processing etc. different (AI) technology system customer relationship management (CRM) tools.

(AI) (CRM) of activity can include these categories, such as: corporate marketing, marketing operation, field marketing, customer support, digital commerce, customer analytics, customer influenced product or service design, product or service pricing, finance information, presentation, customer billing, inventory , logistics and fulfilment support, partner management etc. different CRM tools.

(AI) technology of CRM has been carrying on plan different stages to achieve CRM personal assistant tool for businesses. The stages are such as, in the beginning stage of (AI) projects in place, implement now, pilot phase next year in the final stage of (AI) customer relationship management tools are foreseeable future. So, this CRM technology has been improved to plan in different stages every year to prepare to achieve full capacity of CRM service quality for businesses to use in the future.

Hence, how to develop an estimate prediction of the economic impact (AI) technologies could have CRM activities, which depends on gathering macroeconomic information on business revenue and the basic marketing of business revenue and the basic markup of business expenses by major functions (customer support, marketing and sales , production etc.)

An economic impact model that can gather data together and forecast the results how (AI) artificial intelligence technology brings (CRM) customer relationship management benefits to businesses, e.g. surveys investigation includes IT spending by sample countries, GDP and population estimates and forecasts, revenue per employee and ratios of IT spend to GDP. Surveys (questionnaire questions) of forecast results are influenced by (AI) impact can include: results are projected from surveys and rely on estimates are made by respondents on the expected financial improvements in categories of (AI) –assisted customer relationship management activities. The forecast assumes that these estimates are correct; financial estimates are based on estimates of "first year" improvement from full (AI) implementation; forecasts are from planning to implement any artificial intelligence of customer relationship management (CRM) projects, the improvement forecast is of categories of activity , e.g. corporate marketing , digital commerce, and customer analytics. They are not estimates of ROI for the (AI) software. They rely on conservative estimates to which each of these entities might affect company revenue, expenses or productivity. They also rely on estimates of the penetration of software in customer relationship management activities . Net new jobs created are based on the ratio of new revenue to jobs required to support that revenue . They can assume that 50% of the net new revenue will support increases in labor and the rest will go for capital and other operating expenses that may replace jobs lost to automation.

In the future, some of the ways in micro economic benefits to any organizations. (AI) technology is expected to impact CRM activities include: Spending up sales cycles, improving lead generation and qualification solving customer support problems faster (raising service quality), helping companies improve brand campaigns and recognition, lowering costs of support calls when increasing resolution rates, lowering the cost of recruiting employees and partners, increasing revenue from optimized product marketing, optimizing price, distribution logistics and preventing loss through fraud detection. So, micro economic benefits view point, it seems that (AI) CRM technology can raise any companies economic benefits for care term.

Artificial intelligence enables machines or the in-build software to behave like human beings which allows these decisions and act. The advent of (AI) is leading , talking, making decisions and act. The advent of (AI) is leading to new technologies advances and transforming the economic and employment opportunities for humans in a positive way. (AI) related technologies can facilitate our live. For example, industrial robotics, robotic medical assistants, smart games, financial forecasting software, big data analysis, algorithms in health and bioinformatics, pilotless cargo places, drone ambulances and general purpose and workplace robots and others. (Disruptors technologies: Advances that will transform life, business and the global economy).

Artificial intelligence also known as computational intelligence is defined as " the human –like intelligence exhibited by machines or software. It is theorized that intelligence of humans can be described and intelligence machines or software can simulate it. These machines software can be reasonable , learn, perceive and process information, like human mind and thus facilitate human life. They can think and act for us. So, artificial intelligence is an interdisciplinary field of study including computer science, neuroscience, psychology, linguistics and philosophy.

However, (AI) research and developments have economically impacted many industries, such as robotics, telecommunications, computer applications , health, finance, heavy manufacturing, transportation, aviation, e-service and e-commerce, military , music and movie, toys and games entertainment etc. industries.

In fact, many ideas, systems and technologies have been developing in the world of (AI) technology. However, which are net called or considered (AI) products, rather which are mentioned with their specific names, such as smart graphics, machine learning, e-commerce etc. (i.e. this is called (AI) effect).

What is influence to (AI) job nature change to change developed and countries countries digital economy

- How can (AI) technology influence digital economy?

Nowadays, (AI) related industrial applications will replace most human power in fields, including call centers, customer services and air cargo transportation. (AI) technologies also help weather forecasting based on repeated rainfall pattern (data) recognition, through robotics (i.e. floor cleaning, moving lawns etc.) transporting people and products with unmanned vehicles, sending space unmanned smart shuttles, developing robotic arms, predicting market values in stock exchanges by internet, making homes safer, helping elderly and disabled using robotic servants etc.

Among the (AI) related technologies , there are a few that significance for the impact on society and especially on digital economy . (AI) is particularly influential in machine learning. Such as robotics, transportation, finance, health and bioinformatics, e-commerce , e-games, big online data gathering and internet-of-things. For example, machine e-learning is based in bioinformatics and robots that can learn new skills for better caregiving in healthcare. What is machine e-learning? Machines can e-learn from e-data gathering, coming up generalizations and making decisions to act in certain ways from internet.

There are important applications , such as e-machine perception, electronic online natural language learning processing, online search engines, online bioinformatics, online brain –computer interface, online game playing, online robot locomotion, online advertising, online computations finances, online health monitoring, online DNA classification and decision making, online in chemistry –cheminformatics . So, online machine learning can positivcly impact productivity and it can enhance information and analytical system from (AI) online channel.

What is robotics? Robotics is one of the most strongly influenced fields in (AI). For example, heavy manufacturing industries, robots and used and man power is replaced for effectiveness, precision, and accuracy, especially in respective or dangerous tasks, including welding, assembling , picking and placing .

So, robots can acquire new skills or adapt the changing dynamic environment. Also, artificial intelligence can be applied in developing transportation. For example, automated vehicles, driver assistance systems , safety systems, collision avoidance systems and public transportation. Moreover, (AI) technology has proven to produce some of the best tools to predict stock market fluctuations from internet data gathering method. It's predictions are based on ever-evolving predictions algorithms and systems learn new models and make connections between historical data and new data to measure stock market trading more accurate from internet data gathering channel.

In health field, especially in health data processing , analysis, decision making support and medical diagnosis. So, online data can show which patients will need what treatment and what alternative drugs could be used more accurate from (AI) online data gathering method. Bioinformatics is an interdisciplinary field combining statistics, (AI) online technology can help in discovering data patterns and modeling through the application of machine learning, artificial neural networks and genetic algorithms. For example, further (AI) technology development of human genome project of online data sequences.

Online shopping can be facilitated by virtual assistants developed through (AI) technology and these assistants can offer the best advice. (AI) online purchase coming after every product image recommendations and personalization bring important revenue to shopping online sites, like Amazon . Smart computer graphics and games, artificial intelligence is useful in smarter computer, graphics, scene modeling , scene rendering processes in order to create, for example, effective human –robot interactions , online machine learning, online strategic games techniques etc. online computer related (AI) software.

So, online big data analysis and big data does have a critical need in the world of online intelligence machines and software in our future. In other words, (AI) offers online technology to enable online big data analysis to provide industrial organizations with valuable information for effective decision making in short time. For example, what IBM's Watson achieved: this machine used 200 million of structured and unstructured content with a special technology of hypothesis generation, massive evidence gathering, analysis and scoring from internet channel.

Finally, (AI) online technology another related internet invention (internet of things) (IOT) is the network of machines or objects connected through internet. These connected objects can sense their internal and external environment, communicate with each other, can send critical data and finally can make decisions to act or correct their environment from (AI) online technology. For example, factories can monitor and automatically change production processes, hospitals can monitor and regulate the health conditions of their patients , schools can collect data from facilities and cars can send data to car makers from (AI) online technology.

Partner predicts that (IOT) market will create about trillion amount value by 2020 year. Although machines collect big data from their environment, whether which gain an insight or learn from these online data largely depends on the (AI) online machine learning principals and (AI) online technology. In 2013, Mckinsey estimated that disruptive technologies closely related with potential economic impact in 2025 year between $7.1 to $13.1 trillion amount (automation of knowledge work, advanced robotics, autonomous or near-autonomous vehicles).

What is the relationship between
(AI) and global digital economy development ?

● Could work activities in China be automated
making in the nation with the world's largest automation potential?

Can (AI) technology influence China economy? Could China workers be affected and jobs made up of routine work activities and predictable? Will programmable tasks be particularly impact to China employment market ? When impact on labor market is likely to be gradual at the aggregate level, it can be sudden and dramatic at the level of specific work activities, rending some job obsolete fairly. Overall (AI) technology will raise digital skills when reducing demand for medium incomer inequality for China workers. It seems (AI) technology's effect on productivity could be crucial to China's future economic growth as the population ages are increasing.

In China, some biggest technological companies driving significant investments in research and development. Moreover, China is one of the leading global (AI) technology development county. However, China will need to focus on building its innovation capacity. For example, United States and United Kingdom are currently producing more influential (AI) technological research. However, if China planed to achieve (AI) technology success, it's traditional industries will need to develop technical know-how –to and overcoming implementation costs prepare to develop (AI) . When (AI) technology is introduced into China society, China government needs to raise concerning ethical, legal, technological security etc. business questions. Also, surrounding issues include privacy, discrimination, legal liability and regulation. It aims to encourage overseas investors to choose to invest (AI) technological industry to raise GDP growth and manufacturing industries income growth for long term in China.

If China encouraged overseas (AI) technology investment in its country. It is possible to influence China employment market to be changed. Because (AI) technology will impact to influence China people daily life. Due to (AI) technology is introduced to China society, many rich people will prefer to spend to buy any high (AI) technological products for entertainment or learning or machine man driving etc. daily necessity activities. Then it will raise GDP growth and will raise (AI) manufacturers or related-(AI) technological manufacturers profit. It is beneficial to China because it can become one high knowledgeable and (AI) technological economical society.

But it will bring bad influences to raise unemployment chance for the low skillful labor. In labor economy aspect influence , how (AI) technology can influence China low skillful labor unemployment ratio raising. The raising low skill labor unemployment reason is because China low skillful human labors are argued or are replaced by (AI) technology creating new challenges to introduce to influence China society of simply human manufacturing job nature to be changed to be high (AI) technology manufacturing job nature in any China factories. Moreover, when (AI) technology introduction to China, it will cause other related social challenges in China. The varied (AI) related challenges, including the difficulty of creating safe and reliable hardware for sensing and affecting (transportation and education), the challenges of gaining public trust, a low resource comities and public safety and security, the challenges of overcoming fears or marginalizing humans in China employment and workplace and the risk of diminishing interpersonal trust because the low skillful labors won't believe any China employers will give chance to employ them , due to (AI) technology will replace their skills and man manufacturing of productivity is much less to compare to (AI) technology manufacturing method.

- How does (AI) technology influence
the future of employment change?

Are future nature of jobs changed to computerization from (AI) technology? Where are the probability of computing occupations from (AI) technology influence? What is expected impacts of future computing on labor market from (AI) technology influence? John Maynard Keynes's frequently cited prediction of widespread technological unemployment " du to our discovery of means of economic the use of labor outrunning the pace of which we can find new used of labor" (Keynes, 1933, p.3).

In the future, (AI) technology will impact some nature of occupations to change computing. This chance will also influence some countries' economic change. For example, some factory human labors hand routine manufacturing tasks will be changed to computerization of routine manufacturing tasks by (AI) technological machine men hand manufacturing method. it will cause a structured shift in the labor market, with workers reallocating their labor supply from middle-income manufacturing to low-income service occupations.

Arguably, this is because the manual tasks of service occupations are less computerization, as who require a higher degree of flexibility and physical adaptability. So, (AI) technology will influence the human hand labor skillful occupation nature of task cheaper , such as vehicle manufacturing , ship manufacturing, computer manufacturing, steel manufacturing, television, radio etc. home electronic products of heavy machine industry change. Due to (AI) technology machine man will be proper to be used to manufacturing these electronic products when the (AI) technology innovation can develop to the mature stage. Then, any countries manufacturers will choose to use (AI) technology machine man, instead of human hand production.

Supposing the future prices of computing are fallen, seriously, problem solving skills are becoming relatively productive, explaining the substantial employment growth in manufacturing occupations, involving cognitive tasks where skilled labor has a comparative advantage, as well as the increase education needs for (AI) technology computing of machine man subject study.

Prediction of education needs for (AI) technology student numbers will increase, due to manufacturing industry needs many (AI) technology students in future employment market. Another (AI) technology influence if the future (AI) technological innovation, e.g. machine man manufacturing or machine man service industries will both increase demand, then with more sophistic software technologies will be disrupted labor markets by marketing workers redundant.

For publishing industry, what is striking about the case in paper book publishing industry will be unpopular? Due to the electronic book publishing industry will be popular, e.g. Amazon publish . (AI) technology can influence paper book manufacturing method which is replaced by machine man electronic book manufacturing method as well as it will cause the computerization is no longer confined to routine manufacturing tasks. Due to (AI) machine man manufacturing technology will be proper to be used to manufacture any products in short time efficiently and effectively , e.g. electronic book products. In the future, if it is fact to occur this case, such as (AI) technological machine man manufacturing method will be adopted (applied) to manufacture electronic books or any products in possible. (AI) technology will cause many manufacturing workers are unemployed. It is beneficial to employers, who can reduce to spend much wages expenditure to employ manufacturing workers, but it will cause many manufacturing workers loss jobs and reduce income to support whose families lives. It will cause social challenges, e.g. increasing stealing crimes if the manufacturing workers had not other skills to find other jobs to do easily. So, manufacturers need to concern over technological unemployment which will be hardly future phenomenon if who decided to dismiss all manufacturing workers, due to (AI) technology machine men replace to them.

If (AI) technology can be innovated to produce any kinds of machine man to serve any service or manufacturing industries successfully. Then, it will bring these questions: Can future that workers be influenced to be automation employment and productivity by (AI) technology influence? Does it impact to influence the (AI) technology countries' productivity and growth and natural resources development and labor markets and evolution of global financial markets and economic impact of technology and innovation and urbanization etc. issues? How will automation transform the workplace? What will be the implication for employment? What is likely to be its impact both on productivity in the global economy and on employment?

In fact, automatic of activities can enable businesses to improve performance by reducing errors chance and improving quality and speed, and same cases achieving outcomes that go beyond human capabilities. Some economists indicate (AI) technology would give a needed boost to economic growth and prosperity have of the working age population in many countries. Based on the scenario modeling, they estimate automation could raise productivity growth globally by 0.8 to 1.4 % annually. They also indicated that almost half the activities people are almost $1.6 trillion in wages to do in the global economy have the potential to be automated adapting current demonstrates technology, according to their analysis of more than 2,000 work activities across 800 occupations.

When less than 5% of all occupations can be automated entirely using demonstrated technology, about 60% of all occupations have at least 30% of worker made activities, that would be automated. More occupation will change to be automated. They also indicated for business performance benefits of automation are relatively clear, but the issues are more complicated by policy making to attract foreign investors. Beyond technical feasibility, the cost of technology, competition labor will include skills and supply and demand dynamics, performance benefits and beyond labor cost savings and social and regulatory acceptance will affect the automation. Their predictions suggest that half of today work activities could be automated by 2055 year, but this could happen 10 to 20 years earlier or latter depending on the various factors in addition to their wider economic condition.

Some scientists suggest (AI) technology is finally starting to deliver real-life business benefits. Computer power is growing significantly , algorithms are becoming more sophisticated and perhaps most important of all, the world is generating vast quantities of the fuel that powers (AI) technology data billions of gigabytes of it every day. Also, online firms are digital natives, such as Google online search service company is investing on (AI) technology. For new though most of the news if coming from the suppliers of (AI) technologies. And many new users are only in the experimental phase. Few products are on the market or are likely to arrive these soon to drive immediate and widespread adoption. As a result, analysts believe (AI) technology's potential will give true economic benefit in the future. (AI) industry will introduce to suppliers and users to raise economic potential of (AI) technology.

In the future, (AI) technology systems can solve business problems. Some scientists categorized those into five technology systems that are key areas of (AI) technology development: robotics and autonomous vehicles, computer vision language virtual agents and machine learning , which is based on algorithms that learn from data without replying on rules-based programming in order to draw conclusions or direct an action.

Such as computer vision and language includes natural language processing, analytics, speech recognition technology, some are about learning from information, such as about machine learning and others are related to acting on information, such as robotics, autonomous vehicles and virtual agents, which are computer programs that can converse with humans. Machine learning and a subfield called deep learning are artificial intelligence applications.

- Can artificial intelligence impact
global economy growth?

Artificial intelligence (AI) is a term first defined in 1956 year. It is a branch of computer science that aims to create intelligent machines that work and react like humans. In contrast today, 60 years later, (AI) is characterized by a number of applications, including computers playing games against humans and understanding human languages, virtual personal assistants, and robotics which involve computers seeing , hearing and reacting to sensory stimuli. In the future, technologists predict for (AI) technology ranging from (AI) being used as a tool to aid relatively simple processes for robots with human like mental capabilities, who expect (AI) technology can emulate human performance by learning, coming to mind its own conclusions, understanding complex content, engaging in dialog with people, enhancing human cognitive performance or replacing humans in executing both routine and non-routine tasks. In existing industry, (AI) technology is used , such as targeted advertising and virtual used personal assistant as well as the (AI) technology that my exist in the future, such as robots with human vehicle processing capabilities.

The range of (AI) technology's progress in the future will determine the economic impact future of (AI) technology on the global economy with more limited advances and applications (i.e. weak (AI) only) corresponding to more limited economic impacts and more substantial progress, i.e. strong (AI) technology is corresponding to more significant economic impact.

(AI) technology learning that automates analytical model, including predicting cause-and-effect relationship from biological data, identifying new drugs, self-driving cars and protecting against fraud etc. functions. Also (AI) learning can improve natural language processing that allows computers to continue to better analyze, understand and generate language to interface with human using the natural human language, virtual personal assistant, helps users by providing scheduling appointment, reminds organizing personal finance and finding providers of various services, machine vision allows (AI) machine man to identify object, scenes and activities in detect pedestrians and bicyclists.

We expect the economic effects of (AI) technology to include both direct GDP growth from sectors that develop or manufacture (AI) technology and indirect GDP growth through increased productivity in existing sectors that employ some from of (AI) technology. If (AI) producing sectors could grow, then it could lead to increase revenues and employment of (AI) technological professionals within these existing firms as well as the potential creation of entirely new economic activities to any countries' societies productivity improvement in existing sectors could be realized through faster and move efficient processes and decision making as well as increased (AI) technological knowledge and access to information available in societies easily.

In the future, if (AI) technology is an increasingly critical component of more products, it will become an integral part of necessary products of many people's lives. The extent of (AI)'s economy effort is also likely to vary from region to region, thought variation may be more dependent on the predominate economic activity of a region and the (AI) ability can influence economic activity, rather then the economic or developmental status of the regions. (AI) technology can move accessibility and can use source development to do international business between one country and another country.

So (AI) technology has the potential to give benefits to different income chooses and to bring significant gains to both developed and developing countries. For agricultural technology, (AI) has the potential to optimize food production around the world by analyzing agricultural regions and identifying what is necessary to improve crop yield. In total, (AI) technology gives greater economic impact to any countries agricultural regions if which implemented (AI) technology to grow crop , fruit etc. food production in the farms.

Investment in (AI) technology is such as capital investment to any countries' public or private enterprises. So, it will have large economic impact to the future . If the (AI) technology is reasonable invested to the different needs aspect by the public or private enterprises in the country. Then, it will have good economic impact to the country in the future. However, when (AI) technology is likely to affect both the productivity and employment components of economic growth in many sectors. Significant public debate has focused on projections of (AI)'s effect on the labor force. However, for instance, some researchers have argued that the rise of (AI) technology and automation will led to significant unemployment as capital is substituted for the low skillful labor. So, they point to the concern that the increasing sophistication of (AI) technology may balance skilled and semi-skilled workers and the reduce the size of the middle class. However, this is not a new argument, due to (AI) technology negatively affecting the labor force and leading to mass unemployment. Because the (AI) technology is the substitution of machinery for human labor. Although, employment in certain industries, has been reduced in the past due to technological advancement. For long term, the labor market has adapted to the introduction of new technology, giving rise to new jobs in new areas. (AI) technology may also be accomplished without a reduction to total employment in the long-term to some Asia countries, such as Hong Kong and Japan. Because Hong Kong and Japan many low skilled labor, e.g. security, cleaner who complaint that employers need them to work long time hours. (abnormal working hours) e.g. one day 12 to 15 working hour per day. Hence, if (AI) machine means invention technology success. Security or cleaning job can be worked from (AI) machine man in some hours every day in order to reduce the long time working hours cleaners or security workers, e.g. one (AI) machine man works 4 hours for cleaning or security job, one day as well as another cleaner or security labor only needs to work 8 hours one day. So total security or cleaning employers can employ 12 hours machine cleaners or security workers and human cleaners or security workers in one day. For long term benefit, Hong Kong or Japan every security or cleaning worker does not need to work 12 hours minimum working hours one day. They won't feel tried and bore and without private with whose families, so who will accept to do these cleaning or security jobs, even they can raise work efficient and performance when who feel happy and health.

So, (AI) technology of machine man invention can raise low skillful labor efficiency and it can help them to avoid abnormal working hours demand in some busy work life countries, such as Hong Kong and Japan. Before, one Japan female labor feel unhappy to work, due to who often needs to work abnormal working hours for her employer and who has less sleeping and without any private time to enjoy her life with her families every day. So this abnormal working hours factor causes her to do commit suicide behavior, then she is die unlucky. So (AI) technology of machine man invention ought avoid abnormal working hours demand for employer in any countries in the future.

The most important occurrence to any employers, some researchers had attempted to do one experiment to find that private research and development , venture capital and public research and development investment all have strong net effect or economic growth with venture capital funding further having the strongest such effect from (AI) technology. The researchers hypothesize the venture capital investment contributes to economic growth through (AI) technology innovation and by the capacity of an economy to use existing (AI) technology knowledge to increase productivity. They predict the impacts of venture capital, business-research and development and public research and development can raise multi factor productivity from (AI) technology introduction.

Can (AI) technology influence the economic development to developing countries? The developing regions of the world contain most of natural resources. If one day, (AI) technology has invent one kind of machine man which can assist any gas or oil workers to seek any new oil/gas natural resource locations easily. I believe that (AI) technology can help these natural resource exploitation countries will gain economic benefit more easily. So, (AI) driven technology can be used to change to create any new opportunities to address poor management or resources and improve human well being, such as Africa Latin America and India can use (AI) technology machine man to seek any oil/gas natural resource countries exploitation activities to attempt to gain much economic benefits.

- Why will (AI) technology grow economic development ?

Nowadays, increases in capital and labor are no longer driving the levels of economic growth, such as (AI) technology. The ability of increase in capital investment and in labor of traditional drivers of production, have no longer to be enjoyed in most developed economies ,e.g. developed country, US, UK . However, artificial intelligence has the potential to overcome the physical limitation of capital and labor to avoid missing out on this opportunity. So, policy makers and business leaders must prepare for and work toward a future with artificial intelligence. They must do with the idea that (AI) is another simply method to enhance productivity method . Rather they must see (AI) as the tool that can transform thinking about how growth is created.

Economists have always thought of new technologies are as driving growth their ability to enhancing. It can replace labor and capital factor of production. So, it brings this question: What is the factor of production (AI) technology characteristics. They key factor is to see (AI) technology as a capital-labor .

(AI) can replicate labor activities at much greater scale and speed, and to even perform some tasks began the capabilities of human. For example, by using virtual assistants , 1000 legal documents can be reviewed in a matter of days instead of taking three people six moths to complete. Some (AI) technology may be one kind of factor of production in the future. For another example, people will work in workplace digitalization environment. So, in the future, working environment and information management are automated. Such as Konica camera sale company will use workplace digitalization. So , (AI) technology can provide workplace digitalization in order to raise productivity efficiency. (AI) technology will be one kind of production which is replaced by workplace digitalization and it will grow any organization productivity efficiently. Then, (AI) technology will assist overall social economy growth , due to productivity is raised and products can be produced in short time to prepare to sell in consumption market. So, time will be shortened to increase GDP growth fast for the development of (AI) technology countries.

- How can (AI) technology impact to global economic and social and psychological changes?

What will be the development of (AI) technology and predictions concerning the future evolution? The computers and robots will develop conscious, intelligent and minds into humans, enhancing psychological and behavioral abilities and allowing for direct communication with (AI) minds. (AI) technology will be impacted human life by (AI) technology information communicative and environmental influence. A " world brain" and " world mind", this psychological system will be enhanced and enriched the capacities of both individual and collective cognition by (AI) technology of service industries.

(AI) technology with influence these human needs of service industries changes, such as , biological science, finance, entertainment, business, biological science, transportation, communication military etc. The personal computer evolution, the internet and the world wide web which exploded on the scene, linking business, homes, schools, social organizations which were a completely unpredicted phenomenon to influence human life. Kurzweil (1999) predicts that by 2029 year, most human communication will be with machines. According to Person, by 2100 year, there will be human machine convergence.

How can (AI) technology influence environmental protection to make benefits to farming economic growth? (AI) technology can be applied to predict how to solve environmental pollution challenge to avoid to damage any crop or vegetable or rice or fruit etc. food growth. Because environmental experts can gather global environmental pollution data from an environmental database to build a perform a systematic analysis from (AI) technology. The first step is this broad analysis can include understanding, statistical and data gathering techniques to obtain the relevant data, the correlation among the variables involved, and a list of possible models. The next step is to select a set of methods and models that cover all kinds of knowledge and functionalities needed for the decision making process. Once the models are selected, they must be fully implemented by means of machine learning , data mining, statistical or numerical technique. After that, those models must be integrated to build the whole EDSS. The EDSS must be tested to check its performance, accuracy, usefulness and reliability, both from the user's and (AI) technology/computer scientist's point of view. If these is any wrong feature in any development stage, such as model's integration, models' implementation, selection of models, database, problem analysis etc. the developers must come back in the update th required components. When the evaluation phase is all right, the EDSS is ready to be applied to the environment. The great contribution of artificial intelligence to EDSS the integration of several methods complementing the classical statistical models/simulation , statistical analysis, linear models, etc. and numerical models (control algorithms, optimization techniques etc.) .

This cooperation makes the resulting systems more reliable and powerful in coping with real world environment systems. Date interpretation has been a principal area of research in (AI) technology since the very beginning. The most demanding problem in the environmental assessment context. Knowledge representation permits the definition of the different types of data that the existing methods adapt to the process. There is also a lot of work to clean, repair and transform the huge available quantities of raw data. Apart from this, the availability of meta-information or background knowledge is required to guide the process. Data mining is multi-disciplinary: It covers expert systems, data based technology, statistics, data visualization and unsupervised machine learning. These techniques operate at the level of data and background information, where numerous and often incompatible new commensurate pieces of information from disparate sources have to be brought together (K, Fedra, 1994).

So, it seems that in the future, (AI) technology with the increasing maturity in particular those related to knowledge and engineering, new dimensions can be assisted to users in environmental decision making are available. For example, many environmental systems are characterized both by incomplete models and by limited data. Hence, in the future, (AI) technology will be applied to predict climate change to reduce crop or fruit etc. food agriculture challenge by climate change bad influence.

- Will (AI) technology influence digital economy change to manufacturing industry ?

To understand how the manufacturing business must adapt to prosper in the technology, we need to understand how (AI) technology will change us to shape our daily habits to satisfy our expectation of products to how we shop and even the immediate of the entire process. For example, taxi services are in the crosshairs as on demand transportation services like, available of the touch of a smart phone button expand. In fact, Yellow lab, US country , san Francisco city's largest taxi company is filing for bankruptcy as the industry starts to change faster than almost anyone expected. However, at this point, its more than an app that is changing, some our taxi passengers renting taxi transportation to catch consumption behavior.

(AI) technology will influence digital economy for taxi passenger's individual customer experience, offering a growing renting taxi to catch of service and feedback opportunities when any one taxi passenger who chooses to use mobile phone app online tool to prepaid to rent any taxi more easily.

Also in the long term, (AI) technology can influence vehicles drive themselves of behavior. Already, companies like Google and GM are working on projects to bring fleets of autonomous vehicles to cities at the path of a button. Moreover, this on-demand service model is beginning to appear across a much broader range of markets. For example , Amazon company is investing in its own fleet of trucks, planes and even drone at the same time as it pushes for same-day delivery of products. As some point, vehicles will be autonomous too. So, it seems that (AI) technique will influence any transportations choose to use digital autonomous driving technology in the future . For Amazon company case, it is not stopping of logistics. It is also aiming to automatically manage the supply of consumer home products with its recently launched Amazon replenishment service, Dash. Dash is a digital service that enables that connected derive to automatically order physical products from Amazon when supplies are running low. So, it seems (AI) technology will be applied to logistic function by digital technology method introduction in the future.

Hence autonomous vehicles will optimize industry supply chains and logistics operations through increased efficiency and flexibility. In fact, fully automated and lean supply chains will keep reduce load sizes and inventory by leveraging smart distribution technologies and smaller autonomous vehicles by machine man assistance. If Amazon continues to grow market share for online sales by reducing effort required by the consumer to place an order, when also contributing the almost immediate delivery of products to the doorstep. So, it will further fuel the trend toward on-demand derive. As Amazon company fuels the on-demand economy, consumers will expect immediacy in more parts of the digital economy. On top of speed, consumers increasing expect more personalization options.

So, (AI) technology will influence digital manufacturing, such as Amazon publishing to monitor every aspect of every process in real -time and communicating to self-optimized deep learning robotics, new methods of high volume and high customization will become possible. Then, as products merge into product platforms and even services, manufacturers have the opportunity to provide components and platforms used by smaller players. So, (AI) technology will influence manufacturing industry to choose automated SMI lines, robots installed, automation engineers.

Another future (AI) technology development can be applied to space science aspect, such as Automation engineering space in manufacturing process to achieve digital manufacturing benefits to any businesses in the future. Such as reducing cost, shortening manufacturing time, raising efficiency, shortening delivery products to client individual time. How can artificial intelligence give the need and advanced fast and evaluation methods benefits for space exploration? When US NASA (space exploration organization) achieves any space exploration missions, it will answer this question:

When is it useful to have a machine use (AI) technology to achieve a decision? After all, after millions of years of space exploration and rough 10,000 years of civilization, humans are usually quite good at making decisions in complex uncertain environments. Through, Johns Hoplains University's Applied Physical Lab. Research in (AI) technology enabled systems, which has identified three general use cases for (AI) technology to explore space mission:

First, for some tasks (AI) technology is more cost effectiveness than human. Second, (AI) technology is better suited than humans at solving some, but not all problems. Third, (AI) technology allows NASA organization's space exploration mission to develop machines that ate capable of responding faster than when a human is in the decision loop (D. Scheidt, 2012, A. Castano et. al. 2008).

So, the use of (AI) technology to enable science by observing the pace of rapidly evolving phenomena was demonstrated. It is more effectively coordinating and (AI) technology utilizing to earn economic benefits to use for space exploration mission.

However, (AI) technology also have current risk for space exploration. Today (AI) technology is immature and requires further development to reach its potential. For instance, the (AI) technology algorithms that detected the dust derive could not have identified whether the Martain weather represented a threat to the cover. Also it can not yet use instrument input to determine what, where and how to autonomously make the next space science measurement. An equally important factor limiting (AI)'s deployment is that lacks the methodology and technology to effectively test (AI) technology. So, the challenge will testing (AI) enabled system is how (AI) performance can be measured. It would be NASA organization's difficulty to find (AI) technology to develop to carry on researching any space exploration missions in the future. However, (AI) technology will be a good economic benefit choice for space exploration mission in the future.

- What is artificial intelligence potential
benefits and ethical considerations?

The ability of (AI) technology systems to transform vast amounts of complex information into insight has the potential to help solve manufacturing or service challenges for human needs. However, to reap the societal benefits of (AI) systems, humans will need to trust then and make sure that which follow the same ethical principles, moral values, professional codes and social norms that we humans would follow in the same scenario, research and educational efforts as well as carefully designed regulation in order to achieve the most effort of economic benefits goals. For example, international business machines corporation (IBM) is actively engaged both competitors , in global discussions about how to make (AI) ethical and as beneficial as possible for people as social economic benefits. (AI) is usually defined as the " capability of a computer program to perform tasks or reasoning processes " that human usually associate to intelligence in a human being. Often, it has to do with the ability to make a good decision, even when there is uncertainty, too much information to handle. As an example, play chess or complex card games of entertainment activities is believed to need some form of intelligence in a human being, as well as choosing the best medical facilities in a difficult medical case, or creating something new, such as mathematical theorem or even some form of act, or even driving automatic machine man (self driving vehicle) replacing human driving in the middle of a crowded city.

(AI) needs depends on what we consider being intelligence in the behavior of a human being act a certain point in time. If human belief about human intelligence changes and we don't believe any longer that a certain task requires intelligence, then a computer program performing that task is no longer part of (AI), it becomes just another boring computer program. So, it means that (AI) technology will replace some old computer programs, if human can invent new generation of (AI) software for any functions or activities to satisfy human needs.

As IBM, it argues intelligence. This means that we aim to build systems that enhance and scale human expertise and skills rather than replacing them. We therefore focus on practical applications of (AI) capabilities that assist people in performing well-defined tasks of needs by exploiting and wide range of (AI)-based services. We also use the term " cognitive computing" it is mean a comprehensive net of capabilities based on technology. It comprises the fields of machine learning, reasoning and decision technologies, language, speech and vision recognition and processing technologies, high performance and high efficient functions for any industries or individual consumers needs. For example, robotics, which are usually very good at doing what which are supposed to in any environment, much have public shopping center, factory etc. places which need simply services from the robot (machine man), such as cleans the floor of our houses to the robot that can work together with humans in production chains, passing through the warehouse, robots can take care of the tasks of an entire warehouse and the companion robots like Nao, Pepper, Aibo and Giraff, who can entertain use, talk to use and help elderly people to stay connected to their friends, relatives and doctors.

Google company is building automatic machine (self-driving cars) and has acquired more than 10 robotics companies. Facebook had opened whole new research facility only on (AI) research. Apply computer has developed Siri. Microsoft computer company has built a similar personalized assistant. Google has Deep mind, a UK company whose long term aim is to build general (AI) and has already great potential to win game to the world champion and IBM is investing a huge amount of resources in applying its Watson cognitive computing system to the medical domains to finance and to personalized education. In Europe, IBM is establishing new centers in Munich and Milan focused in the application of cognitive computer capabilities to the internet of things and healthcare respectively.

For example, automatic machine man (self-driving cars) are all about (AI), which used to be able to see what happens in the street (signals ,lanes, other cars, pedestrians, traffic lights, which need to able predict what other cars and pedestrians will do, and who need to be able to cope with unforeseen situations. Since, most car accidents are due to human fault, it is estimated that the adoption of self-driving cars will save about half of the lives that are usually last in car accidents.

IBM Watson company has to understand spoken language, make sense of massive amount to text , respond correctly to questions in many categories, as well as assess its own confidence in responding to such questions. In the future, (AI) technology can own question/answering capabilities that would be very useful, for example, in assisting a doctor when trying to some to the correct diagnosis for a patient and to propose the best therapy .

Intelligent machines can also rely on huge amounts of data to be used to learn how to make better decisions. This data comes from all of us over the years Facebook users have uploaded more than 250 billion pictures and every day who upload about 350 million more. Every second, we submit 40,000 google search queries. So, (AI) technology will be connected through the web from appliances to traffic lights from cars to watches. Other tasks that are very easy for humans are physical and manipulation tasks, such as walking , running, picking up an object to make its shape and location, restricted environment. But (AI) machine man technology still not able to have the general physical and manipulation capabilities even of a 6 year old.

So, it brings this question: Why do (AI) scientists need to concern ethics? Because (AI) technology is complex, information into insight has the potential to reveal long held secrets and help solve some of the world's most difficult problems. (AI) systems can potentially be used to help discover insights to treat disease, predict the whether, and manage the global economy. So, ethic issues is important to and (AI) scientists . If any one new (AI) technology research investigation could success, it will be a secret to and the (AI) scientists can not permit to their loyalty to any competitors to damage the fair (AI) technology products trading market. The country (countries) (AI) technology scientists need to concern ethic issues, who need to keep secrets for their countries economic or/and social benefits. This is moral issues to any countries/country loyalty is whose countries intangible assets. They can not sell (AI) loyalty to any their countries to assist whose economic benefits immorally.

- How can (AI) technology influence to global
health care economy development?

According to (AI) lecturer analysis, when combined key clinical health (AI) application can potentially create $150 billion in annual savings for the US healthcare economy by 2026 year. (AI) technology is re-winning modern conception of healthcare delivery. It enables machines to sense, comprehend, act and learn. So which can perform administrative and clinical healthcare functions (Accenture, 2017).

It will help health care service organizations to reduce health care cost, will improve and raise service quality and access. So, (AI) health market size will be predicted growth. (AI) applications in health care include robot-assisted surgery, virtual nursing assistant, administrative workflow assistant, fraud detection, error reduction connected machines, clinical trial participant identifier, preliminary diagnosis, automated image diagnosis and cybersecurity.

What kind of benefits (AI) technology can contribute to healthcare service? (AI) technology can deliver what many health care organizations need, such as financial and operational of labor costs, digital expectations from patient consumers how to use (AI) technology to solve interoperability challenges in any healthcare organizations. Also (AI) technology can be applied to wellness an d lifestyle management, diagnostics, delivers financially but also way of organizational and workflow improvement. So, (AI) technology will be continue to become most prevalent and adoption to healthcare organizations , which must need to enhance structure to be position to take full advantages of new (AI) technological capabilities. (AI) technology can change the nature of work and employment is rapidly changing to make the best use of both humans and (AI) talent in healthcare industry in the future. For example, (AI) technology offers a way to fill in gaps and the rising labor shortage in healthcare. According to Accenture analysis, the physicians shortage is increasing. However, (AI) technology will manufacture healthcare machine men to replace physicians in future one day(2017). Hence, (AI) technology will be invented to raise health care service staffs work efficiency and performance in any hospitals or clinics in the future.

In conclusion, (AI) technology will raise efficiency for any service or manufacturing industries in the future, although, it is possible that it will also rise low skillful workers unemployment numbers. But, the most important influence to human technological innovation will be risen and it will influence human life will be changed to be better, e.g. self drive cars, health care physician machine men, machine man cleaners etc. intelligent machine men will be manufactured to serve for our daily life. Furthermore, (AI) technological products will influence countries trading, some low technological development countries manufacturing businessmen can choose to buy any (AI) products to raise whose productivity and efficiency and reducing cost to achieve economic cost saving result. Also, GDP of trading growth income will increase to the (AI) products sale countries. Hence, it will be beneficial to economic development to both developed and developing countries both in the future as well as (AI) scientists time and money spending will be valued to continue to invest (AI) technology development for human life and economy benefits for long term.

In conclusion (AI) technology will raise macro economy growth and it can create many (AI) jobs , but it also raise the low level technological worker unemployment change. In the future, (AI) technology can be applied to digital technology to attempt to invent any new undiscovered (AI) and digital technology. So, it needs any scientists to continue to research how digital and (AI) technology can be mixed to satisfy human's future undiscovered needs.

Must Developed And Developing Countries Need Artificial Intelligent To Replace Human Job

Must developed and developing countries need artificial intelligent development? If one developed country, e.g. US, UK , Japan , Singapore it does not continue to develop artificial intelligence, robotic, then what disadvantges or weaknesses , it will encounter to compare when it chooses to continue to develop this artificial intelligent technology in society. If one developing country, e.g. China, Korea, Taiwan, it does not continue to develop artificial intelligence, robotic, then what disadantages or weaknesses, it will also encounter to compare when it chooses to continue to develop this artificial intelligent technology in in society. I shall explan the reasons why the results may cause to either the developed country, or the developing country as below:

● How AI help developing countries to communication and agriculture and learning and medical delivery development

Why can AI help developing countries ? Drones that pick inaccessible crops and mobile phones that give medical advice are two of the ways AI can transform life in the developing world. Artificial intelligence (AI) may improve the lives of the world's poor, the technology needed to revolutionise inefficient, ineffective food and healthcare systems in developing countries is well. For example, in low-income areas, agriculture and healthcare are two critical ecosystems that we can apply AI to immediately; this is not the far future, or even in five years.

Artificial intelligence (AI) has seeped into the daily lives of people in the developed world. From virtual assistants to recommendation engines, AI is in the news, our homes and offices. There is a lot of potential in terms of AI usage, especially in humanitarian areas. The impact could have a multiplier effect in developing countries, where resources are limited.

Emergency Response to developing countries' earthquake natural damage suddence occurrence predicting

AI and machine learning are still finding importance in emerging markets, but certain applications have emerged and are now widely used. For instance, predictive models for disaster relief enable first responders to automatically analyze large-scale behavior and movement through multiple sources of data including social media platforms, web forums, news sources, etc. Based on collected data, responders can scale reconstruction efforts and distribute supplies in a timely manner.

Why and how AI can assist farmers to predict when the earthquake occurs suddenly in order to avoid or reduce the natural damage to their agriculture productive number loss. For example, In 2015, when a major earthquake hit Nepal, more than 8 million people were affected. During the aftermath, drones were used to map and assess the destruction and speed up the rescue mission. The town of Sankhu, situated about 20 kilometers northeast of Kathmandu, was among the highly affected locations. In May 2018, my company Fusemachines and GeoSpatial Systems partnered with Sankhu's city officials to use drones and artificial intelligence in an effort to automatically estimate the reconstruction need. After processing data accumulated from a drone-powered aerial mapping of the region, the team fed this data to advanced machine learning algorithms. Combining drone imagery, digital mapping and machine learning, the team configured region modeling and infrastructure development with higher accuracy. Another organization known as One Concern, a California-based startup, has created a predictive AI program called Seismic Concern to accurately predict seism and is also working on solutions for wildfires, floods and hurricanes.

Smart AI Agriculture

Another application of AI in developing countries is smart agriculture. Farmers monitor crops more effectively and make better predictions on planting, weeding and harvesting using AI tools. It can also be used to analyze one plant at a time and add pesticides only to infected plants and trees instead of spraying pesticides across large swaths of crops. One California-based tech company is an example of this use of AI. So, the developing countries farmers in rural parts of India are also using AI to increase yields through better access to information about the farming season than they would normally have. Technology-enabled process automation offers the agribusiness industry the chance for remarkable growth -- not only in developed countries but around the world. There's a unique opportunity to increase yields, cut down labor costs and improve people's health.

Medicine Delivery to developing countries' patients urgent need

Companies are also leveraging AI to improve access to health care in some of the most remote areas of the world. In Rwanda, for example, Zipline is using drones to deliver medical supplies and blood to hospitals and clinics that are difficult to access by car. This has dramatically impacted people living in remote parts of the country because they are able to get medical help when needed. The drone system in Rwanda has also helped reduce waste of blood by 95%, as noted by Zipline. One Concern has created an AI program called Seismic Concern that accurately predicts seismic events and is also working on solutions for floods, wildfires and hurricanes. The medical field may actually benefit the most from emerging technologies in developing countries.

Assistance to reduce teaching work workload or psychological pressure to teachers in developing countries' schools

Another vital area benefiting from innovative technologies like AI is education. Advanced technologies can enhance how we learn, teach and perform tasks. In most developing countries, schools lack experienced teachers and resources to enhance students' knowledge. As a result, many students still have to walk long distances to get to the nearest school, which has created education gaps, especially in rural areas. AI tools such as personalized learning assistants can simplify learning by making tutoring services and learning materials accessible to all students, wherever they are. Machines can be automated to help students learn basic concepts without a tutor, which companies like Carnegie Learning are working on. This would allow students to learn at any time from anywhere. With AI, education is made easy and accessible to more people.

The initial usage of AI in developing countries has been at a micro level -- solving small, specific problems in a defined industry. As machine learning advances and there is a higher utilization of AI, we will see more complex issues being targeted and resolved. When duly adopted, AI can positively impact future developing countries people everyday lives not just in disaster intervention, education, health care and agriculture but can also help in mitigating poverty, malnutrition and pollution. Especially, in developing nations, to leverage AI's true potential and create a snowball effect. Startups are defining a holistic and humanitarian approach to building more sophisticated, AI-ready societies. Stakeholders in the AI landscape should understand the strengths and nuances of the developing world as well as the limitations of AI and create localized solutions and applications.

Why does smart phone help developing countries communication ?

Internet Seen as Positive Influence on Education but Negative on Morality in Emerging and Developing Nations. Internet access differs substantially across the 32 emerging and developing countries polled, with the lowest rates of internet use in South Asian and sub-Saharan African nations. Within countries, computer owners, young people, the well-educated, the wealthy and those with English language ability are much more likely to access the internet than their counterparts. To access the internet, people increasingly use smartphones rather than more cumbersome fixed landline connections and computers. Around the world, both smartphones and basic-feature phones alike are used for sending messages and taking pictures.

In fact, many developing countries young people, students are popular to use smart phones for internet usage aim, instead of communication. Moreover, many developing countries working people are also popular to use smart phones for any working usage in their working time , even non working time any time. So, smart phones (AI) phones will be important communication or leisure tools to developing countries people in the future. Unless, it is one day, scientists can develop another new communication tool to replace smart phones. So, artificial intelligence will be important to influence developing countries people , how to improve or bring positive learning attitudes to students in their daily learnnng lifes. as well as how to raise developing countries peoplc, how to raise working people efficiency or improve performace in their daily working lifes. So, AI may bring positive learning or working attitudes to developing countries working people and students both.

The Positive Impact of Mass Media in Developing Countries

Radio, newspapers, television, Internet, social media, etc., all of these are forms of mass media. Each of these outlets has the capability of bringing information to thousands of people with one device. While in some communities it is easy to take advantage of these communication outlets such as television and Internet access, not everyone has access to such outlets. Radio is one of the most common forms of mass media in developing countries because it's affordable and uses less electricity than many other forms of mass media, but only approximately 75 percent of people in developing countries have access to a radio, and roughly 77 percent of people in rural areas have access to electricity.

For developing countries that have implemented forms of mass media in their communities, there have been numerous positive outcomes are influenced to impact developing countries mass media by artificial intelligence as below:

When AI is participated to developing countries mass media, it can influence any radio, television audiences raise more attention to each other through social media platforms such as Facebook and Twitter and create, organize and initiate street protests and campaigns. Furthermore, having access to social media in developing countries, people are able to connect to those that they usually wouldn't have the chance to talk to. Moreover, AI Provides educational opportunities- In many countries, the division between local and national languages as well as issues of literacy can make communication difficult. With the use of mass media, a bridge can be built between these two gaps. In India, there is a radio station that provides information in local languages and respects local culture and traditions. One of the main ways is to create public awareness of what is going on with businesses and government officials. The media plays an important role in giving people the opportunity to act against injustice, oppression and misdeeds that they otherwise wouldn't know about. Information on available healthcare, a mass radio broadcast was sent out encouraging parents to seek treatment at local healthcare facilities for their sick children. With this mass outreach on healthcare, the encouragement of people to take their children to healthcare facilities saved thousands of lives. This easy way of encouraging others and bringing awareness about certain diseases was made possible through a simple radio broadcast. Finally, when AI is particiapted to media, it may bring many social issues to life that otherwise would remain unknown to many people. In developing countries and communities like Burkina Faso, when the radio broadcast was released about malaria, diarrhea and pneumonia, people were educated and moved to action and knew to take their children to healthcare facilities for preventative care. As it is seen, having access to different media outlets is vital for those in developing countries. Here are three ways that those in developing countries can implement mass media to help their people and communities.

When AI is participated to any internet radio or internet newspaper mass online listening or reading channel. It can provide online radios or newspapers in public places- By providing online radios and newspapers in public areas it gives community members to access news, information and emergency warnings. Even though radios can be on the cheaper side, there are still many people that can't afford to have a radio in their home. By providing one in a local place, not only would it better educate the community members but also it will bring the community together. So, it can make media outlets a two-way platform- Creating a two-way platform between the community and those who are behind the radio stations, newspapers or broadcasts makes the community feel involved and that their voices are being heard. An organization called Soul City in sub-Saharan Africa is showing how well two-way platforms work by engaging their listeners and having them contribute thoughts and ideas about complex issues. Because developing countries radio listening audiences or newspaper readers are popular to accept computer online radio listening channel or online newspaper reading channel to replace traditional paper newspapers or radio machines. So, AI may raise their listening news or reading news leisure feeling from online mass media channel in the future.

- Why do developed countries need to develop AI

Artificial intelligence, or AI, is driving massive shifts across the globe, and every day more questions arise. What impact will AI have on the workforce and how can we prepare for it? How can we encourage economy-boosting and job-creating technologies? How can we ensure that AI will be implemented ethically and with minimal bias? How will society benefit? For developed country, such as US example. None of the US, Israel and Russia have a formal national AI policy yet. Private sector companies such as Google, Amazon and Apple and the US department of defence are driving the bulk of AI investment in the United States. Though Israel does not have a specific policy, it is keenly focused on AI and has seen the number of AI start-ups triple since 2014.

Developed country may learn whether what weakness it is lacking when it does not continue to develop AI from one another developed country. Which countries are approaching AI most effectively, and to what degree is there opportunity for greater international collaboration? It may be too early to tell; however, when analyzing the best practices of existing national AI policies, there is much that can be learned. These are the specific areas to consider.

When one developed country continue to develop or research AI, it may bring these benefits as below:

On gathering Data aspect, from self-driving vehicles to smart cities, data is the driver behind AI. Innovation in the United States is limited without a national strategy that answers questions about protocol and ownership. France and Denmark, on the other hand, are opening government data. France is hosting troves of centrally collected public and private data that it plans to make available as part of its strategy. Conversely, by taking a restrictive position on issues of data collection (as indicated by the implementation of General Data Protection Regulation), the EU is putting manufacturers and software designers at a disadvantage while balancing the demand for privacy. On raising technologica talent aspect, the demand for AI talent far outweighs the available supply. As a result, almost every nation's strategy addresses talent development. Canada's AI strategy is distinct in that it primarily focuses on research and talent strategy. The country boasts AI degree programmes and is building a $127 million research facility in Toronto. Companies like Facebook and my own company, Uptake, are investing in Canada to access this talent pool. On AI legal technological innovation aspect, a whole host of legal questions swirl around AI. The country is developing a bill for AI liability that will be ready in March 2019. The government hopes the legal framework will attract investors by providing a simple, comprehensive guideline to enable the broad use of AI systems. So, when the developed country applied AI technology to assist any lawyers to work, then AI can help them to reduce the workload to draft any legal documents more easier. So, any developed countries lawyers' draft legal documents time must reduce if the developed countries lawyers accept to apply AI to assist their legal works. One of the great promises of AI is its potential for improving quality of life. But without the right planning and oversight, we risk exacerbating problems of inequality or marginalizing groups of people. As an example, India's AI strategy is focused on leveraging the technology not only for economic growth, but also for social inclusion.

AI may bring what benefits to developed countries

From SIRI to self-driving cars, artificial intelligence (AI) is progressing rapidly. While science fiction often portrays AI as robots with human-like characteristics, AI can encompass anything from Google's search algorithms to IBM's Watson to autonomous weapons. Artificial intelligence today is properly known as narrow AI (or weak AI), in that it is designed to perform a narrow task (e.g. only facial recognition or only internet searches or only driving a car). However, the long-term goal of many researchers is to create general AI (AGI or strong AI). While narrow AI may outperform humans at whatever its specific task is, like playing chess or solving equations, AGI would outperform humans at nearly every cognitive task.

Why research AI safety? Would AI bring war when AI is continued to develop by developed countries? In the near term, the goal of keeping AI's impact on society beneficial motivates research in many areas, from economics and law to technical topics such as verification, validity, security and control. Whereas it may be little more than a minor nuisance if your laptop crashes or gets hacked, it becomes all the more important that an AI system does what you want it to do if it controls your car, your airplane, your pacemaker, your automated trading system or your power grid. Another short-term challenge is preventing a devastating arms race in lethal autonomous weapons.

In the long term, an important question is what will happen if the quest for strong AI succeeds and an AI system becomes better than humans at all cognitive tasks. As pointed out by I.J. Good in 1965, designing smarter AI systems is itself a cognitive task. Such a system could potentially undergo recursive self-improvement, triggering an intelligence explosion leaving human intellect far behind. By inventing revolutionary new technologies, such a superintelligence might help us eradicate war, disease, and poverty, and so the creation of strong AI might be the biggest event in human history. Some experts have expressed concern, though, that it might also be the last, unless we learn to align the goals of the AI with ours before it becomes superintelligent.

There are some who question whether strong AI will ever be achieved, and others who insist that the creation of superintelligent AI is guaranteed to be beneficial. At FLI we recognize both of these possibilities, but also recognize the potential for an artificial intelligence system to intentionally or unintentionally cause great harm. We believe research today will help us better prepare for and prevent such potentially negative consequences in the future, thus enjoying the benefits of AI while avoiding pitfalls.

How can AI be dangerous when developed countries continue to develop AI to become weapon to replace soldiers?

Most researchers agree that a superintelligent AI is unlikely to exhibit human emotions like love or hate, and that there is no reason to expect AI to become intentionally benevolent or malevolent. Instead, when considering how AI might become a risk, experts think two scenarios most likely:

The AI is programmed to do something devastating: Autonomous weapons are artificial intelligence systems that are programmed to kill. In the hands of the wrong person, these weapons could easily cause mass casualties. Moreover, an AI arms race could inadvertently lead to an AI war that also results in mass casualties. To avoid being thwarted by the enemy, these weapons would be designed to be extremely difficult to simply "turn off," so humans could plausibly lose control of such a situation. This risk is one that's present even with narrow AI, but grows as levels of AI intelligence and autonomy increase.

The AI is programmed to do something beneficial, but it develops a destructive method for achieving its goal: This can happen whenever we fail to fully align the AI's goals with ours, which is strikingly difficult. If you ask an obedient intelligent car to take you to the airport as fast as possible, it might get you there chased by helicopters and covered in vomit, doing not what you wanted but literally what you asked for. If a superintelligent system is tasked with a ambitious geoengineering project, it might wreak havoc with our ecosystem as a side effect, and view human attempts to stop it as a threat to be met. So, a super-intelligent AI will be extremely good at accomplishing its goals, and if those goals aren't aligned with ours, we have a problem. You're probably not an evil ant-hater who steps on ants out of malice, but if you're in charge of a hydroelectric green energy project and there's an anthill in the region to be flooded, too bad for the ants. A key goal of AI safety research is to never place humanity in the position of those ants.

Why the recent interest in AI safety ?

Stephen Hawking, Elon Musk, Steve Wozniak, Bill Gates, and many other big names in science and technology have recently expressed concern in the media and via open letters about the risks posed by AI, joined by many leading AI researchers. The idea that the quest for strong AI would ultimately succeed was long thought of as science fiction, centuries or more away. However, thanks to recent breakthroughs, many AI milestones, which experts viewed as decades away merely five years ago, have now been reached, making many experts take seriously the possibility of superintelligence in our lifetime. While some experts still guess that human-level AI is centuries away, most AI researches at the 2015 Puerto Rico Conference guessed that it would happen before 2060. Since it may take decades to complete the required safety research, it is prudent to start it now.

Because AI has the potential to become more intelligent than any human, we have no surprise way of predicting how it will behave. We can't use past technological developments as much of a basis because we've never created anything that has the ability to, wittingly or unwittingly, outsmart us. The best example of what we could face may be our own evolution. People now control the planet, not because we're the strongest, fastest or biggest, but because we're the smartest. If we're no longer the smartest, are we assured to remain in control?

A captivating conversation is taking place about the future of artificial intelligence and what it will/should mean for humanity. There are fascinating controversies where the world's leading experts disagree, such as: AI's future impact on the job market; if/when human-level AI will be developed; whether this will lead to an intelligence explosion; and whether this is something we should welcome or fear. But there are also many examples of boring pseudo-controversies caused by people misunderstanding and talking past each other. When one developed country continue to develop AI, can itself country's all factories workers will lose jobs, due to AI can replace them to do simple works in factories, or any public transport drivers, e.g. bus drivers, ferry , tram, train drivers, they will lose jobs, when AI (non manual driving drivers) can replace all public transport drivers. So, some occupations will lose if developed countries continue to develop or research AI to replace human to do some simple jobs, such as some cooking jobs can be done by AI. So, it is possible that future cookers won't be needed, because AI cooking skills may be better than them to cook any good taste chinese or western food in restaurants. If you drive down the road, you have a subjective experience of colors, sounds, etc. But does a self-driving car have a subjective experience? Does it feel like anything at all to be a self-driving car? Although this mystery of consciousness is interesting in its own right, it's irrelevant to AI risk. If you get struck by a driverless car, it makes no difference to you whether it subjectively feels conscious. In the same way, what will affect us humans is what superintelligent AI does, not how it subjectively feels.

In fact, AI may be make any brokers jobs in financial market. the main concern of the beneficial-AI movement isn't with robots but with intelligence itself: specifically, intelligence whose goals are misaligned with ours. To cause us trouble, such misaligned superhuman intelligence needs no robotic body, merely an internet connection – this may enable outsmarting financial markets, out-inventing human researchers, out-manipulating human leaders, and developing weapons we cannot even understand. Even if building robots were physically impossible, a super-intelligent and super-wealthy AI could easily pay or manipulate many humans to unwittingly do its bidding. So, future brokers will be replaced by AI, when AI can be made to own financial brokers' analytical mind to make more accurate whether the share price will rise up or fall down to compare human financial brokers' analytical mind. The robot misconception is related to the myth that machines can't control humans. Intelligence enables control: humans control tigers not because we are stronger, but because we are smarter. This means that if we cede our position as smartest on our planet, it's possible that we might also cede control.

Not wasting time on the above-mentioned misconceptions lets us focus on true and interesting controversies where even the experts disagree. What sort of future do you want? Should we develop lethal autonomous weapons? What would you like to happen with job automation? What career advice would you give today's kids? Do you prefer new jobs replacing the old ones, or a jobless society where everyone enjoys a life of leisure and machine-produced wealth? Further down the road, would you like us to create superintelligent life and spread it through our cosmos? Will we control intelligent machines or will they control us? Will intelligent machines replace us, coexist with us, or merge with us? What will it mean to be human in the age of artificial intelligence?

Why do developed countries people need AI ?

Why do we assume that AI will require more and more physical space and more power when human intelligence continuously manages to miniaturize and reduce power consumption of its devices. How low the power needs and how small will the machines be by the time quantum computing becomes reality? Why do we assume that AI will exist as independent machines? If so, and the AI is able to improve its Intelligence by reprogramming itself, will machines driven by slower processors feel threatened, not by mere stupid humans, but by machines with faster processors? What would drive machines to reproduce themselves when there is no biological incentive, pressure or need to do so?

Who says superior AI will need or want

to have a physical existence when an immaterial AI could evolve and preserve itself better from external dangers. What will happen if AI developed by competing ideologies, liberalism vs communism, reach maturity at the same time, will they fight for hegemony by trying to destroy each other physically and/or virtually. If AI is programmed to believe in God, and competing AI emerges programmed by muslims, christians or jews, how are the different AI's going to make sense of the different religious beliefs, are we going to have AI religious wars? What if the "powers that be" greatest fear is the emergence of a super AI that police's and rationalizes the distribution of wealth and food. A friendly super AI that is programmed to help humanity by, enforcing the declaration of Human Rights (the US is the only industrialized country that to this day has not signed this declaration) ending corruption and racism and protecting the environment.Most benefits of civilization stem from intelligence, so how can we enhance these benefits with artificial intelligence without being replaced on the job market and perhaps altogether?

Key to the process of machine learning are neural networks. These are brain-inspired networks of interconnected layers of algorithms, called neurons, that feed data into each other, and which can be trained to carry out specific tasks by modifying the importance attributed to input data as it passes between the layers. During training of these neural networks, the weights attached to different inputs will continue to be varied until the output from the neural network is very close to what is desired, at which point the network will have 'learned' how to carry out a particular task. A subset of machine learning is deep learning, where neural networks are expanded into sprawling networks with a huge number of layers that are trained using massive amounts of data. It is these deep neural networks that have fuelled the current leap forward in the ability of computers to carry out task like speech recognition and computer vision.

In conclusion, when developed countries continue to develop AI, it may bring positive advantages to bring raising productivies, or efficiencies, but it may also raise unemployment ratio to any low skill or low knowledge jobs in ther societies. However, human future society will need to change to be better to raise our living standard. But AI is one kind the best choice tool to achieve this aim in our future, so I agree developed countries continue to develop or research AI to be the super -human machine.

Reference

A. Castano et. al. " Automatic detection of dust devils and clouds at Mars" Machine vision and applications, Oct. 2008, vol. 19, no 5-6, pp. 467-482.

Accenture, " Why artificial intelligence is the future of growth"(2017) <http://www.accenture.com/us-en/insight-a rtificial-intelligence-future-growth>.

D. Schedidt , Unmanned Air Vehicle Command And Control, Handbook Of Unmanned Air Vehicles, Springer-Verlag, 2014. Facebook (AI) Research Available at https://research.facebook.com/ai, research at google, machine intelligence available at
http://research.google.com/pubs/machineintellige nce.html; micro soft research-machine learning and artificial intelligence available at http://research.microsoft.com/en-us/research- areas/machine-learning-ai.aspx.

International Federation Of Robotics, 2016. IFR press release world robotics report. IFR, org . 29 Sept. Accessed Feb. 01, 2017. http://www.ifr.org/news/ifr-press-release/world-robitics report -2016-8321.

K, Fedra , "GIS and environmental modelling" in environmental modelling with GIS, edited by M.F. Goodchild.B.O. Parks and L.T. Steyaert, Oxford University press, pp. 35-50, 1994.

Keynes, J.M. (1933). Economic possibilities for our
grandchildren (1930). Essays in persuasion, pp.358-73.

Mckinsey & Company (2013, May). Disruptive
technologies: Advices that will transform life,
business and the global economy , USA.

Ministry of economy, trade and industry, Japan, 2015, Japan's robot strategy. Ministry of economy, trade and industry.

Ray Kurzweil , The age of spiritual machines (1999) is cited numerously through this chapter: Kurzweilai.net http://www.kurzweilai.net

Rich, Elaine & Knight, Kevin, Artificial Intelligence Second Edition, 1991, New York; Mc-Graw-Hill.

What Factors Can Influence Developed And Developing Countries Talent Growth By Robotic Education

IQ talent human training method factor

To discuss whether scientists can apply IQ method to create or train talent human. Firstly, we need to know what IQ means. According to the general consensus, the answer is "no." An IQ test measures a person's cognitive ability compared to the population at large. The average IQ is 100, anything above 130 is considered exceptionally smart while a score under 70 is categorized as developmental delays related to intelligence. An IQ test measures a person's cognitive ability compared to the population at large. The average IQ is 100, anything above 130 is considered exceptionally smart while a score under 70 is categorized as developmental delays related to intelligence. Intelligence is defined as general cognitive problem-solving skills.The equation used to calculate a person's IQ score is Mental Age / Chronological Age x 100. On most modern IQ tests, the average score will be 100 and the standard deviation of scores will be 15.

In organizational training working environment, managers across the organization are in touch with the employees you are grooming for their next big role. In larger organizations, talent management requires Human Resources Information Systems (HRIS) that track the career paths of employees and manage available opportunities for talented employees.

● What Is Talent Management? Is talent management be IQ training method to train talent employees in any organizations?

Talent management is just another one of those pesky Human Resources terms, right? Wrong. Talent management is an organization's commitment to recruit, hire, retain, and develop the most talented and superior employees available in the job market. So, talent management is a useful term when it describes an organization's commitment to hire, manage, develop, and retain talented employees. It comprises all of the work processes and systems that are related to retaining and developing a superior workforce.

Is talent Management as a Business Strategy to any organizations ?

Talent management is a business strategy that organizations hope will enable them to retain their topmost talented and skilled employees. Just like employee involvement or employee recognition, it is the stated business strategy that will ensure the attraction of top talent in competition with other employers. When you tell a prospective employee that you are dedicated to a talent management strategy that will ensure that he or she will have the opportunity to develop professionally, you attract the best talent. This is because studies show consistently that the opportunity to continue to grow and develop their professional and personal skills is a major motivator for why employees take and stay at a job.

Differences Depending on Stated Talent Strategy

What appears to differentiate talent management focused practitioners and organizations from organizations that use terminologies such as human capital management or performance management is their focus on the manager's role, as opposed to reliance on Human Resources, for the life cycle of an employee within an organization. For example, practitioners of the other two employee development and retention strategies would argue that, for example, performance management has the same set of best practices.

Talent management does give managers a significant role and responsibility in the recruitment process and in the ongoing development of and retention of superior employees. In some organizations, only top potential employees are included in the talent management system. In other companies, every employee is included in the process. In some companies, the talent management system is accessible via computer programs; in others, informal communication among managers and HR staff is the approach used.

What Processes Are Part of a Talent Management System? You can include the following systems when you approach talent management as your overall business strategy to recruit and retain talented employees.

Recruitment planning meeting
Job description development
Job post writing and recruiting location placement for the posting
Application materials review
Phone or online screening interview
In-house interviews that can involve multiple meetings with many of your current employees
Credential review and background checking
Making the job offer to the selected person
Agreeing on the amount of the offer
Employee starting day and onboarding process
New employee welcome information and introductions
On-the-job training
Goal setting and feedback
Coaching and relationship building by the manager
Formal feedback systems such as performance management or an appraisal process
Ongoing employee development
Career planning and pathing
Promotions, lateral moves, transfers
Employment termination by choice of the employee or cause by the employer

Manager's Key Role in Talent Management

As stated, the majority of these work systems are squarely in the hands of the employee's manager. HR can provide support, training, and backup, but the day-to-day interactions that ensure the new employee's success comes from the manager. Developing and coaching the employee comes from his or her active, daily interaction with the manager.

HR can take the lead in some of the activities you see on this list, especially in recruiting and selecting new employees, and in the case of employment termination. HR is also deeply involved in the performance management system, career planning, and so forth leading the development of the systems. Hence, talent management is a business strategy and you must fully integrate it within all of the employee-related processes of the organization. Attracting and retaining talented employees in a talent management system is the job of every member of the organization, but especially managers who have reporting staff (talent). An effective strategy also involves the sharing of information about talented employees and their potential career paths across the organization. This enables various departments to identify available talent when opportunities are made or arise. On conclusion, An organization that does this kind of effective succession planning makes sure that the best talent is trained and ready to assume the next position in their career path. Succession planning benefits the employees and it benefits the organization. Managers across the organization are in touch with the employees you are grooming for their next big role. Thus, organizational talent mangement strategy may be one kind of IQ training method to raise or improve employee individual work efficiency or productive effort or creating talent skillful abilities for any organizations.

● Is An IQ Test An Accurate Way To Measure Intelligence Or Are Mental Abilities ?

The IQ test is an exam most of us are familiar with, regardless of whether we have taken it or not. The test was originally designed by the French psychologist Alfred Binet in the early 1900s. But in the new millennium, is the IQ test still an effective means of measuring general intelligence? According to the general consensus, the answer is "no." An IQ test measures a person's cognitive ability compared to the population at large. The average IQ is 100, anything above 130 is considered exceptionally smart while a score under 70 is categorized as developmental delays related to intelligence.

However, some scientists believe that intelligence is defined as general cognitive problem-solving skills. Since the days of Binet, psychologists have agreed that intelligence is much more complex than a single number and may be in fact divided into many subcategories. This is where the IQ test falls short. A Canadian study published online in the journal Neuron concluded that the IQ test is "fundamentally flawed," seeing that its questions "grossly oversimplify the abilities of the human brain." The report identified three indications of human intelligence: short term memory, reasoning skills, and verbal ability. None of these skills are at all accurately measured in the traditional IQ test. So what does the IQ test accurately measure? While the IQ test may give an indication of general intelligence, it can't measure the entire complexity of the human thought process. Creativity, emotional sensitivity, social understanding, and various acquired skills such as music or art, are excluded from test's measurements of intelligence. If you'd like to get an idea of your IQ take this test, but just remember that whatever your score be, it doesn't necessarily define how smart you really are. So, it seems that IQ test training can not guarantee to raise or improve any foolish person to be talent person in possible. So, IQ test training may not be the most suitable talent human creative method to any one. It depends on whether the suitation how to provide IQ training , e.g. organizations may attempt to apply IQ skill test method to raise or improve employee efficiency or skill to serve their organizations effectively. But, schools will be difficult to apply IQ test training method to raise student learning effort effectively.

However, I believe that IQ test training can create these kinds of talent human as below:

Your answer to that question probably depends pretty heavily on your grades in school or, if you've ever taken one, the results of an IQ test. But are those a fair basis to assess a person's mental capabilities? Everyday language suggests maybe not. We speak of street smarts and EQ, for instance. Both terms suggest that there are abilities that deserve to be considered as forms of intelligence but that fall well beyond the scope of traditional academic measures. Does science go along with common sense in seeing that intelligence and IQ are far from the same thing?

More than you probably imagine. According to Harvard's Howard Gardner, intelligence actually comes in an incredible eight flavors. I shall indiate these most popular six kinds of IQ test training talent human as below:

1. Musical intelligence.

"People say, well, music is a talent. It's not an intelligence. And I say why, if you're good with words, is that an intelligence, but if you're good with tones and rhythms and timbres, it's not. And nobody's ever given me a good answer, which is why it makes sense to talk about musical intelligence.

2. Spatial intelligence.

This is the easy grasp of how things lay in space that allows a chess master to win or a surgeon to perform near miracles. It's also "what an airplane pilot or a sea captain would have. How do you find your way around large territory and large space," Gardner notes.

3. Bodily kinesthetic Intelligence.

Forget the cliché of the dumb jock. Coordinating your body actually takes a great deal of intelligence -- just not the kind measured by IQ tests. This type of smarts "comes in two flavors. One flavor is the ability to use your whole body to solve problems or to make things, and athletes and dancers would have that kind of bodily kinesthetic intelligence. But another variety is being able to use your hands or other parts of your body to solve problems or make things. A craftsperson would have bodily kinesthetic intelligence" too, according to Gardner.

4. Interpersonal intelligence.

This one seems a bit similar to the popular concept of EQ. "Interpersonal intelligence is how you understand other people, how you motivate them, how you lead them, how you work with them, how you cooperate with them," says Gardner, who adds that it's a particularly important type of intelligence for leaders to have.

5. Intrapersonal intelligence.

Intrapersonal intelligence, or self-knowledge, is both very hard to assess and very important, Gardner says, particularly in today's fast-changing world. "Nowadays, especially in developed society, people lead their own lives. We follow our own careers. We often switch careers. We don't necessarily live at home as we get older. And if you don't have a good understanding of yourself, you are in big trouble," he explains.

6. Naturalist intelligence.

This one is "the capacity to make important, relevant discriminations in the world of nature between one plant and another, between one animal and another. It's the intelligence of the naturalist, the intelligence of Charles Darwin," Gardner says. And before you argue that you live in Detroit or Manhattan and so have no need for this type of smarts, he adds that "everything we do in the commercial world uses our naturalist intelligence. Why do I buy this jacket rather than another one? This sweater rather than another one?" Those fine distinctions are made by the part of the brain that used to discern a tasty small animal from a poisonous one. "When an old use of a brain center no longer is relevant, it gets hijacked for something new. So we're all using our naturalist intelligence even if we never walk out into the woods," Gardner concludes.

The bottom line? If you've been thinking of your mental horsepower solely in terms of high school report cards or a single number from an IQ test, you're probably selling yourself short. And if you're focusing your energies only on boosting your book learning, you may be wasting efforts that could be better focused on other strengths. Hence, above six kinds of talent human ought be trained to raise skill effort on music, calculation, interpersonal communication , athletes and dancers etc. sport skill aspects. IQ test training method may be one kind of effective talent human training method choice.

● Can IQ test train talent human ?

What does science say? Is innate talent a myth? This question is the focus of the new book Peak: Secrets from the New Science of Expertise by Florida State University psychologist Anders Ericsson and science writer Robert Pool. Ericsson and Pool argue that, with the exception of height and body size, the idea that we are limited by genetic factors—innate talent—is a pernicious myth. "The belief that one's abilities are limited by one's genetically prescribed characteristics....manifests itself in all sorts of 'I can't' or 'I'm not' statements," Ericsson and Pool write. The key to extraordinary performance, they argue, is "thousands and thousands of hours of hard, focused work."

To make their case, Ericsson and Pool review evidence from a wide range of studies demonstrating the effects of training on performance. In one study, Ericsson and his late colleague William Chase found that, through over 230 hours of practice, a college student was able to increase his digit span—the number of random digits he could recall—from a normal 7 to nearly 80. In another study, the Japanese psychologist Ayako Sakakibara enrolled 24 children from a private Tokyo music school in a training program designed to train "perfect pitch"—the ability to name the pitch of a tone without hearing another tone for reference. With a trainer playing a piano, the children learned to identify chords using colored flags—for example, a red flag for CEG and a green flag for DGH. Then, the children were tested on their ability to identify the pitches of individual notes until they reached a criterion level of proficiency. By the end of the study, the children had seemed to acquire perfect pitch. Based on these findings, Ericsson and Pool conclude that the "clear implication is that perfect pitch, far from being a gift bestowed upon only a lucky few, is an ability that pretty much anyone can develop with the right exposure and training." This sort of evidence makes a compelling case for the importance of training in becoming an expert. No one becomes an expert overnight, and the effects of extended training on performance can be larger than might seem possible. But does the fact that training leads to improvements—even massive improvements—in skill level mean that innate talent is a myth? This is a much harder scientific argument to make, and is where Peak runs into trouble. Ericsson and Pool gloss over or omit critical details of research they review that undermine the anti-talent argument. As one example, although they claim that the results of Sakakibara's training study imply that "pretty much anyone" can acquire perfect pitch, the sample in that study did not include pretty much anyone. It included children who had been enrolled in a private music school from a very young age (the average age at which training began was 4). It does not seem likely that this non-random sample was representative of the general population in music aptitude or interest—factors that are known to be influenced by genetic factors. It's also not clear whether the children had acquired true perfect pitch, because there was no comparison of the children after training to people who possess this rare ability—for example, in terms of speed of identifying notes or neural correlates of performance.

As another example, describing the results of a study of ballet dancers by Ericsson and colleagues, Ericsson and Pool claim that "the only significant factor determining an individual ballet dancer's ultimate skill level was the total number of hours devoted to practice" and that there was "no sign of anyone born with the sort of talent that would make it possible to reach the upper levels of ballet without working as hard or harder than anyone else." Not mentioned is the exact magnitude of the correlation—a value of .42, where 1.0 is perfect. The fact that the correlation was modest in magnitude means that factors not measured in the study—including heritable aptitudes—could have actually accounted for more of the differences in ballet skill than deliberate practice did. As it always is in scientific debates, the devil is in the details in the debate over the origins of expertise.

Ericsson and Pool also leave out a good deal of evidence that runs counter to the anti-talent argument. For example, they claim that professional baseball players have "no better eyesight than an average person," but there is evidence to suggest otherwise. In a study published in the American Journal of Ophthalmology, Daniel Laby and colleagues assessed the vision of major and minor league baseball players in the Los Angeles Dodgers organization over the course of four spring training seasons. As David Epstein recounts in his book The Sports Gene, in the first year of the study the researchers used a standard test of visual acuity, and it turned out to be too easy. Over 80% of the players got a perfect score of 20/15, meaning that they could see at 20 feet what an average person can see at 15 feet. In the following seasons, using a custom test, Laby and colleagues found that 77% of the 600 eyes tested had visual acuity of 20/15 or better, with a median of about 20/13. Even for young adults, this is excellent vision. Overall, Laby and colleagues concluded that "professional baseball players have excellent visual skills. Mean visual acuity, distance stereoacuity, and contrast sensitivity are significantly better than those of the general population."

Based on scientists own evaluation of the evidence, we argue in a recent Psychological Bulletin article that training is necessary to become an expert, but that genetic factors may play an important role at all levels of expertise, from beginner to elite. In other words, some people take much more training than other people to acquire a given level of skill. As it happens, Sakakibara's pitch training study provides some of the most compelling evidence of this type. There was a large amount of variability in how long it took the children to pass the test for perfect pitch—from around 2 years to 8 years. As Sakakibara notes in her article, this evidence implies that factors other than training may be involved in acquiring perfect pitch, including genetic factors. This finding is consistent with the results of recent reviews of the relationship between deliberate practice and skill, which include numerous studies Ericsson and colleagues have used to argue for the importance of deliberate practice. Regardless of domain, deliberate practice leaves a large amount of individual differences in skill unexplained, indicating that other factors contribute to expertise.

The more direct evidence for the multifactorial view of expertise comes from "genetically informative" research on skill—studies that estimate the contribution of genetic factors to variation across people in factors that may influence expert performance. In a study of over 10,000 twins, two of us found that music aptitude was substantially heritable, with genes accounting for around half of the differences across people on a test of music aptitude. As another example, in a pioneering series of studies, the Australian geneticist Kathryn North and her colleagues found a significant association between a variant of a gene (called ACTN3) expressed in fast-twitch muscle fibers and elite performance in sprinting events such as the 100 meter dash. There is no denying the importance of training for becoming an elite athlete, but this evidence (which is not discussed in Peak) provides compelling evidence that genetic factors matter, too.

On conclusion, based on this sort of evidence, many human behavioral scientists have argued that the experts are "born versus made" debate is over—or at least that it should be. There is no doubt that training is required to become an expert. Notwithstanding a report by North Korea's state-run news agency that Kim Jong-il made five holes-in-one his first time playing golf and rolled a perfect 300 his first time bowling, no one is literally born an expert. Expertise is acquired gradually, often over many years. However, as science is making increasingly clear, there is more to becoming an expert than training. Moving ahead, the goal for scientific research on expertise is to identify all of the remaining factors that matter. Hence, it seems that human behavioral scientists feel that IQ test may be one good skills to create talent skill to some kinds of humans if IQ test may prove the person has the kind of talent skill, e.g. music, sport, reading, writing, draw picture etc. art creative skills for future new creative art talent human method.

Businessmen expect to improve better economic environment, they will prefer to recruit the most sought after skills of intelligent employees to bring positive beneficial impact to organizations. However, technology and digization has had a significant influence on workers. Future globalization will trend digital economic development. Hence, it will influence workers' skills to be changed also. In fact, not all changes are positive because some workers will possible lose jobs, either due to new technology replaces their jobs or they lack enough effort to improve their skills in global digital economic labour market environment.

It brings this question: What are regional dynamic skills need with digital business environment is growing. In fact, organizations will continue to deal with skills shortages, labour markets across the global are continually changing. so, more employers and workers will need to adopt innovate working pattern, e.g. on call jobs, freelance jobs will grow popularly. The greater flexibility afforded to employ regard.

Finally, digitalisation includes artificial intelligence, big data , online platforms. All these new technology will influence future employees how to worker. For example, they can apply online platform to work at home conveniently. So, they do not need to go to offices. They can finish their jobs and send to their employers by email easily. This kinds of job pattern can raise efficiencies and employers do not need go to offices often.

An important implication of innovating working which needs the employees who own digital skills in order to serve organizations more efficiently. So, employers are increasingly able to access demographics that were hitherto less active in labour markets. For example, future more women are joining the labour market because part time and self employment opportunities make it easier. This kinds of job pattern can raise efficiencies and employees do not need go to offices often.

An important implication of innovating working which needs the employees who own digital skills in order to serve organizations more efficiently. So, employers are increasingly able to access demographic that were hitherto less active in labour markets. For example, future more women are joining the labour market because part time and self employment opportunities make it easier to manage family with work life. So, digital skilling needs will cause many women lose jobs in possible. If the women lack digital job skills. Because high digital skill occupations need, like those requiring research, medical treatment and architectural design occupational digital skills are more common in the services sector, more women who own digital skill who can compete to win.

High digital skill occupations more easier than men because employers usually select female to do high skill occupations easier than make. However, if those professional service female employees can not learn how to apply digital skills to do these researchs medical treatmentm architectural design professional service jobs. Then, it is also different for these professional service femal employees to raise competition in global labour professional service market. So, these professional service female employees need to learn how to apply digital to do themselves jobs in future global professional service labour market. Otherwise, if the male professional service employees can attempt to learn how to apply digital skill to do themselves jobs in order to improve efficiencies and service performance to satisfy patients, such as medical service needs, school search service needs, construction firms' building needs. Then, the owning high digital technology skillful female employees will be more easier to find the professional service jobs which need digital skill more easier than the lacking digital skill female service professionals in future global digital service professional labour market.

On the other robotic communication skill need aspect, future employers expect workers to know how to communicate with robots to work efficiently in any working environment if the employers need robotc to serve their organizations. For example, communication between the robots on factory floors, and between people and robots could allow robots to start and stopr processes based on real-time conditions around them and alert people when there is a problem, so robots could increase their own efficiency if the workers could monitor themselves and determine when they needed maintenance; efficiency would also be improved if machines and robots could make production decisions on their own by. For example, ordering new suppliers when existing inputs into a production process run low. The increase in productivity of industrial robots will likely reduce the number of manual jobs on the shop floor.

At the same time, the increased output made possible by such robots will mean that manufacturers need more people in accounting, finance, sales, advertising and other roles. The increase in putput may also drive increased employment on manufacturers' supply chains. Hence, future employers expect to employ the workers who can know how to communicate with robots to work efficiently in order to raise productivity in any working environment. It means that it the worker can know how to control and communicate with the robots to work together in the team. Then, his/her communication and controlling robotic skill will help the organization's team to work efficiently and raise productivity in order to reduce time waste and human waste and resource waste considerately. So, future shortage of communication and controlling robotic skillful workers number will increase. It has much beneficial to workers who choose to attempt to learn how to communicate and control robots to work together in any working environment team efficiently. Because future employers will like to use robots to assist manual workers to attempt to raise productive efficiency in any working environment. So, the need of employees who know how to cooperate or communicate with robots whose talent skills will be useful to any future employers.

Future global business leaders will need human machine cooperation skill. This technological skill includes artificial intelligence (AI and internet of things (IOT), will reshape our working change. These machines will participate to our daily working environment. For instance, many business leaders agree that automated systems will free-up their time as well as they also believe they'll have more job satisfaction by offloading the tasks that they don't want to do to intelligent machines.

Therefore, future leaders will expect humans and machines can work as integrated teams within their organization in order to their workforce and machines are already successfully working this way. So, they need to expect future employees can know or learn how to work with automated systems more easily, because many jobs will be participated by automated systems, e..g simple accounting tasks, legal administration tasks etc. clerical tasks. They will be participated with (AI) technology, it learns how to cooperate with (AI) technology to finish simple clerical tasks efficiently.

Future workers will need have automated system operational skills: They include that how to operate automated systems to free -up workers' time. Workers will need to learn how to operate automated system to better with healthcare tracking devices workers will need to learn how to operate automated systems to absorb and manage information in completely different ways. Workers will need to learn how to operate automated systems of smart machines to work as admin. in any working environments. Workers need be needed to learn how to operate (AI) automated machines to make more accurate clerical tasks or efficiencies. So, the automated system (robotic) operational skillful workers' demand and number will increase.

In the future, employers need automated machine manufacturing and service with workers cooperation reasons include that clear protocols, will need to be established if autonomous machines fail. So, they need their workers to learn how to control and manage and communicate with autonomous machines skillfully. They believe move they depend upon technology, the more they'll have to lose in the event of a cyber attack. So, skillful workers are real required to let them to know how to cooperate with autonomous machines more efficiently and easily. Computers will need to be able to decipher between good and bad commands, so future employers have much chance to need the owning automated machines operating workers to assist any robots to make more accurate good or bad decision when robots and workers have need to make immediate judgement in their any related job responsibilities aspect.

Therefore, future owning automated machines operating workers' skillful level will be high. It bases on automated machine manufacturing environment trend factor. Finally, future technology will connect the right employee to the high task at the right time. It implies that when future global employers began to accept to apply robots to help them to raise any productivities efficiently. It will influence many manufacturing positions which need to employ any proficient skillful workers who own automated machines operational skills to know how to communicate or manage or control , even supervise any robots to work in teams in any organizational manufacturing environment efficiently. In the future, employers also expect employees to own sufficient digital vision and strategic skills, manifest among other things. They can know how to apply data to demonstrate any senior support and sponsorship digital technological skill. They expect to reduce a skill gap and avoid a lack of employee buying and a workforce culture to change in their digital technological manufacturing organizations. Future employers also believe outdated technology that can't work fast enough, data overload, privacy and security concerns. So, it explains why it is possible that future employers also need digital working environment and automated robots machines to attempt to achieve raising productive efficient aim.

Moreover, it also explains why digital transformation need will be raised. The reasons include: They feel digital technology can gain employees' buying in , making customer experience a boardroom concern, achieving fair compensation , training and goals and strategy achievement more easily, tasking senior leaders with digital working environment change putting policies and technology to support a fully remote, flexible workforce , empowering lines of team work more efficient, teaching all employees how to code/understanding how to adopt to work with automatic machines or rots in any team efficiently. So, automate machine can raise efficiency in manufacturing society.

In conclusion, in the future business society, employees need to be stronger human machine partnerships. So , future manufacturing or service industries will have digital technology and automated machine robotic technology to assist workers to work in any working environment efficiently. They expect digital technology and automated machine robotic technology anticipation to workers' daily jobs in order to bring positive impacting to the customer experience from business owners to decision makers in marketing, customer service, research and developmnt and finance etc. They also expect technological productivity can bring positive relationship between technology and workers emerging technologies' impact on business and the way workers and automated machine work together.

In the future whether in general organizations need what kinds of employees' skills, they expect employee individual own. It is one interesting question. The common skills that employees need to own in order to any duties to any organizational departments efficiently, e.g. human resource, marketing, administrative, logistic etc. different departments. For hospital, school, business, professional occupations etc. different organizations. Whether future school ought implement one system educational method to teach different common skills to students in order to let them to leave schools to jobs more easier.

Future employers need to create new technologies including automation and algorithms, in order to create new high quality jobs and improve the job quality and productivity of the existing work of human employees in any organizations, e.g. accounting department will need intelligence (AI) to assist account clerks to do simple repeating accounting job tasks in order to share their work load and raise performance efficiency or legal organizations will need (AI) to assist law clerks to do simple repeating legal draft or legal document revising job tasks . All future general clerical jobs will apply (AI) technological tools to assist human to job, it will produce a comprehensive platform for managing workforce change.

Hence, human manual(employees) need to learn how to adopt (AI) job participation to assist them to do different kinds of simple clerical jobs in any organizational administrative departments . They , clerical employees or white color workers need to learn how manage or dominate (AI) tool to improve job performance to be better. However, (AI) administrative workforce change, it is not only one kind of job automation change role in any physical offices. It influences future administrative clerks need change a more flexible manner, utilizing remote staffing beyond physical offices and decentralization of operations organizational workforce change.

Instead of (AI) participation to administrative job aspect, (AI) will also participate to manufacturing industry environment aspect, a new human-machine manufacturing workforce change will exist to any factories, warehouses working environment. Scientists predict that in present an average of 71% of total task hours across the industries are performed by humans, compared a 29% by machines. In this average is expected to have shifted to 58% task hours performed by humans and 42% by machines. In fact, nowadays, in terms of total working hours, no work task was yet estimated to be predominantly performed by a machine or an algorithm (AI). But, this picture is predicted to have somewhat changed with machines and algorithms (AI) on average increasing their contribution to specific tasks by 57% . For example, in the future, 62% of organization's information and data processing and information search and transmission tasks will be performed by machines compared to 46% today.

Therefore, these high technological skillful job change will bring negative influence to some demotive-skillful or low skillful labors to be dismissed, if they can not upgrade or raise or re-skillful their skill level to improve their analytical thinking , technology design and programming skills to cooperate with (AI) tools to work efficiently together in any organizational manufacturing or office work environment. Because it will have many employers apply (AI) automation tools to participate with blue -color or white -color workers' tasks in order to raise efficiencies or improve performance in any working environment. So, it is right time to young or mid age employees need to upskill and/or reskill their rihgt type of skills to prepare future technology rich work environment changing needs.

Future technological advances will permit an increasing number of tasks traditionally performed by humans to become automated. It seems that , such automation focused primarily on routine tasks, e.g. clerical work, bookkeeping, basic paralegal work and reporting etc. However, with the advent of big data, artificial intelligence (AI), the internet of things and ever-increasing computing power , i.e. the digital revolutions, non-routine tasks are also increasingly likely to become automated. For example, the recent development in robotics and 3D printing allow firms in advanced economies to locate production closer to domestic markets in fully automated factories. As a result, the future strongest incentive to automate because of their relatively higher labour costs will be reduced, when production automated will bring the negative influence to dismiss some foolish or low productive or low skill workers , the owning high automated productive skillful workers will replace the low productive skillful workers in any factories' manufacturing environments. So, technological progress participates to raise quantity of jobs will cause result in significant job losses to low skillful workers. Because future employers will need many high automated productive employees to help them to cooperate with (AI) automated machine to work together efficiently. For example, many proportion of occupations at high risk is greatest in Germany and lowest in Korea, these countries organizations will accept to spend technology investments and education of workers to prepare future automatability manufacturing development successfully.

However, future automat ability manufacturing development will bring technological unemployment in possible, due to workers need to adjust to the challenge of automation by switching tasks. Thus, preventing technological unemployment, also technological change does not just destroy jobs, but also generates new roles through its effect on productivity and the demand for new technologies. For example, it has been estimated that, for each high tech-job created in the industries , such as computing equipment or electrical machinery, some 4.9 % additional jobs are created for lawyers, taxi, drivers and waiter in the local economy (Moretti, 2011).

Therefore, automated will also influence service industries‘ job nature change, e.g. taxi drivers need to apply (AI) automated machines to assist them to drive their taxis. When the passenger tells the taxi driver where he/she wants to go. Then, the (AI automated machine will follow the GPS road direction map to be indicated how to drive the taxi to go to the destination automatically . So, future taxi driver is one assistance role to assist the (AI) automated driving tool to dominate the (AI) tool to drive the taxi to catch the passenger to arrive the destination safety in the short time in possible. For another example, future restaurant waiters will need (AI) automated machines's assistance to help them to deliver or dispatch any foods and soft drinks to send to the identified eater's table carefully in accurate and efficient service performance way from the kitchen, in especially in the busy time and many people are sitting in the large size restaurant environment. So, future, waiter roles will be the leader , they need to manage or control or supervise the (AI) robotics how to make decisions to arrange to dispatch which foods or soft drinks to the different tables in preference immediately. Also, future law clerks need to supervise or manage the law robotics how to help them to make decisions to do revision or draft or filing legal tasks in preference in order to avoid any typing words are mistaken to type on computers or revised draft in wrong way to assist manual legal clerks' mistaken words are appearance on any legal documents. So, the law clerk future role will be the trainer role , he/she needs to teacher the robots how to check any words, e.g. grammars to correct them to be right grammars, or giving the accurate revision legal documents‘ instruction to let the legal robots to know how to revise each legal draft to prove whether which part of the legal draft will have wrong to be needed to revise.

In conclusion, future many manual workers' service or manufacturing job natures will become automated assistance to robotics. So, employees need to upgrade their skills in order to adopt new technological work nature change.

Reference

Moretti, E. (2011) local labor market in O, Ashentelter and D. Card (eds.) handbook of labor economics, Elsevier, North Halland.

Becoming robotic talent human how influences social changes

Nowadays we tend to think about social and digital technology more from a personal or consumer perspective than their business or professional applications, but as the Digital Era continues to progress, many of technology's most profound impacts are likely to be in the world of work. In addition to changes in product and business development, knowledge management, data analysis, and other operational processes, transforming talent management will be a key priority for organizations striving to be employers of choice.

● Digital technology encourages to create talent human

Why does digitized technology encourage to create talent human or excite human to learn new things ? The human capital implications of social and digital technologies impact virtually everyone, regardless of the type of organization they work for, their profession, their functional area, or their career stage. That means that the talent management functions in all organizations, as well as the professionals who staff and lead them, have a critical role to play in ensuring the efficient and effective transition and transformation from Industrial Era models and processes to their Digital Era upgrades.

It's no surprise that talent management has already become more "high tech." Many employment related activities have been digitized, and there has been a corresponding increase in employee self-service. It's important to remember, however, that digitization is not the same thing as digital engagement, and that the rise of "high tech" solutions doesn't necessitate the loss of a "high touch" approach to managing an organization's human assets. Transforming talent management requires digitization, to be sure, but it also involves leveraging social and digital technologies in ways that promote and enhance communication, collaboration, and engagement - not just between an employee and the organization, but between and among employees themselves.

Talent Acquisition

The logical place to start when talking about the impact of social and digital technologies on talent management is talent acquisition, where the greatest advances have been made. Anyone who has searched and applied for jobs in the past 10 years is very familiar with how technology has transformed the application process, which in most organizations (and virtually all large ones) is now almost completely digitized and automated. However, there are other ways in which social and digital technologies are impacting talent acquisition that may not be as well-known or commonly understood. Social media sites in particular (such as Facebook, YouTube, and Pinterest) are a great way to promote an employer's brand and offer realistic previews of work life, people and culture in organizations. Online games and simulations can also be used to get a sense of what working for an organization would be like, and give organizations themselves an opportunity to determine if a prospective candidate would be a good cultural fit and potentially successful.

On organizational working environment aspect, some employers are recognizing the value of digital alumni networks or communities to maintain strong relationships with former employees. One of the primary motivations for doing this is that the employees may return one day and/or make referrals to or from their personal and professional networks. Similarly, talent networks enable organizations to establish and maintain relationships with professionals in key areas like IT and engineering, even when there isn't a current opportunity to have those folks be a part of the organization. Moreover, social media can bring positive influence to impact organizations to encourage employees to attempt or feel needs to learning new things for their tasks needs. Due to social media has actually transformed every stage of the recruiting process in significant ways - so much so that the traditional recruiting funnel can be recast in "social" terms. At the top of the funnel are activities like social advertising (i.e., placing job ads on social networks like Facebook), social sourcing (i.e., searching for candidates who meet certain criteria on networks like LinkedIn), and social referrals (i.e., having current employees share position openings with their online personal and professional networks). And at the bottom of the funnel is social screening (i.e., reviewing a candidate's public activity in social networks to identify potential hiring risks).

● Learning management is needed to feel needs to any organizations

Learning management is probably the another most advanced area when it comes to adopting and adapting to new technologies. As with recruiting and other processes, the initial advances are in the area of digitization, with social software applications evolving next. One of the obvious digital impacts is the increased use of el-earning and online learning platforms with self-paced study. There are also countless instructional videos on the web, both free and fee-based, that address a virtually unlimited range of topics. And we can't forget MOOCs - massive, open, online courses - which have proliferated in the past couple of years. Finally, many organizations have also started to leverage tablets and other mobile devices for learning, as well as using simulations and games to help employees develop specific skills. In addition to offering training through a variety of multimedia channels, organizations are increasingly using a range of digital tools for assessing employees' skills. They're also allowing employees to play an enhanced role in identifying their key skill sets and training needs, and can even have them create their own learning and development plans. Allowing employees to take a more active role in their own learning and skills management enables organizations to develop and maintain a more complete and accurate knowledge and skills database, which in turn enables them to maximize the value of the workforce in which they've already invested.

1. Formal learning management systems and platforms are also beginning to incorporate social technologies in a variety of ways. Promoting connections and interactions among participants, as well as with the instructor, can enhance the learning experience both during and after a course. Creating course-based cohorts that allow people to continue to interact with each other via a digital community - even when their shared learning experience is face-to-face - can promote both knowledge transfer and retention, in addition to increasing commitment and engagement through interpersonal connections.

2. Informal learning - which is now also referred to as social learning - is greatly enhanced by social technologies as well. In fact, this is probably the greatest opportunity and area of growth for organizations of all types and sizes. Through private social networks, intranets and other internal platforms that have incorporated social technology elements, organizations are better able to facilitate employee learning as they perform their job duties and complete work activities. Along with the networks themselves, features like advanced search, identified subject matter experts, digital communities of practice, wikis and more enable employees to access and learn from colleagues who are not just next door or down the hall, but even in another city, state or country!

As organizations move forward with leveraging technology to enhance learning initiatives, it will become increasingly important for them to address issues related to digital literacy and digital competencies. For the past several decades we've generally taken what I refer to as an LIY, or Learn It Yourself, approach to digital knowledge and skills. Although organizations may invest in teaching someone how to use a specific application related to their job, they make virtually no investment in helping individuals learn how to use general digital tools like Microsoft Office and even email. Left to their own devices, most people - and I include myself in this group- are much less efficient and effective at using these tools than they could or should be. As our tools get even more sophisticated, we need the foundational knowledge and skills to be able to use them well - and this foundation should probably be provided via more formal training. In other words, many people need to be "taught how to learn" in the Digital Era. If organizations aren't going to provide the formal training workers need to do that, it's probably in an individual's best interests to pursue those kinds of development opportunities on their own.

- What are social influences on human behavior when talent human number increases?

Social Influences on Human Behavior Because human beings are social and learn from observation rather than depending entirely on instinct, almost all aspects of human psychology and behavior are socially influenced. Languages, modes of dress, gender roles and avoided taboos are all agreed upon at a group level and form the basis of culture. What are the characteristics of social change? Small-scale and short-term changes are characteristic of human societies, because customs and norms change, new techniques and technologies are invented, environmental changes spur new adaptations, and conflicts result in redistributions of power. This universal human potential for social change has a biological basis.

This universal human potential for social change has a biological basis. It is rooted in the flexibility and adaptability of the human species—the near absence of biologically fixed action patterns (instincts) on the one hand and the enormous capacity for learning, symbolizing, and creating on the other hand. Because human beings are social and learn from observation rather than depending entirely on instinct, almost all aspects of human psychology and behaviours are socially influenced. Languages, modes of dress, gender roles and avoided taboos are all agreed upon at a group level and form the basis of culture.

On conclusion, when one country can create many talent human, e.g. students, workers. Then they can bring more positive attribution to help or assist themselves country to develop rapidly. Consequently, the country's economic growth speed will be rapid. So, I believe that it has close relationship between economy growth and talent human number to any countries in nowadays societies.

Robotics can improve internet technology development to raise economic growth factors

Why do we improve to improve internet technology? What long term social benefits will benefits if scientists can improve internet speed and research any information function ? I shall research these questions to give suggestion as below:

Why does internet improvement make life better? Internet of Things Benefits In short, the scale of change that IoT technology offers can be scary. At the same time, the benefits of a well-executed IoT strategy can be more need for an organization: Safety, Comfort, Efficiency. Also, the Internet offers teens the ability to make friends with peers with whom they would not otherwise connect. With pop culture deteriorating into many distinct subcultures, teens' interests are more variable than they have ever been. With internet communication, employees can effortlessly communicate with one another at anytime from anywhere in the world. This allows employees situated in different parts of the world to give their opinion and voice their concerns. Through internet access, individuals in developing countries are able to gain access to more of the modern economy. With internet connectivity, those living in remote areas can now easily take out microloans, participate in e-banking and more. A large share of respondents predict enormous potential for improved quality of life over the next 50 years for most individuals thanks to internet connectivity, although many said the benefits of a wired world are not likely to be evenly distributed.

● How internet can excite young to learn?

Internet can learn younger to learn much different new knowledge when they research any questions and find answers from internet channel.

As one major aspect of teen life is social environment, changes in how teens connect impact the ways in which teens develop social skills. ** Luckily, the Internet offers many social-skill enhancement opportunities for teens of all different personalities . One advantage the Internet brings that the standard school environment cannot is the ability for teens to adjust their amount of social interaction. Teens who are extremely outgoing can spend their free time in social environments both offline and online, making new connections and catching up with friends.For example, a teen who finds large amounts of face-to-face interaction to be intimidating can use the Internet to engage in conversations while reducing the potential for social anxiety. In a way, this trains less social teens to be more social . In the past, these types of teens did not have the advantage of this social training provided by the Internet.

● Internet can encourage Social Network Growth

The Internet offers teens the ability to make friends with peers with whom they would not otherwise connect. With pop culture deteriorating into many distinct subcultures, teens' interests are more variable than they have ever been. Whereas in the past, children at school might have discussed the current top 40 when discussing music, today's kids define their musical tastes as specific genres, such as post-industrial, dubstep or jpop. Today, it's harder for teens to find peers who share the same interests in their schools. But online, not so. The Internet's social networks help teens find communities of peers who share similar interests, allowing a teen to grow his social network in a way that is specific to him 2. Today's teens are increasingly willing to make friends with different groups of people due to the ability to actually meet them, and this can be useful when they reach adulthood, a time in which accepting people of different backgrounds and demographics is crucial to career and academic growth. The Internet offers teens the ability to make friends with peers with whom they would not otherwise connect.

Today's teens are increasingly willing to make friends with different groups of people due to the ability to actually meet them, and this can be useful when they reach adulthood, a time in which accepting people of different backgrounds and demographics is crucial to career and academic growth.But the Internet can help teens foster self identity through exposure to new people, communities, hobbies and concepts. As teens go through more experiences, they learn more about themselves. And as the Internet can offer teens a wealth of experience, it can play the role of hastening the development of self identity.For many teens, the hardest part of life is figuring out identity.But the Internet can help teens foster self identity through exposure to new people, communities, hobbies and concepts.

● What Are Main Benefits of Internet Communication speed improvement ?

It may include as below:

1 Makes communication easier

Doing business through phone or mail doesn't work well ? Before the internet came into existence, the only way to communicate was through a phone. Or if you needed to send a note you had to send letters via mail. With the arrival of the Internet, staff and team managers can connect instantaneously without leaving their work place. ezTalks Meetings, a one-stop internet communication provider, is a perfect example. With this platform, participants can communicate as if they were right next to one another thanks to its quality video and audio. The tool comes with a rich set of features like screen sharing, cross platform chat, innovative whiteboard, and more.

2 Enhances collaboration

Internet communication brings teams together across the globe. Staff can collaborate easily without limitations and make more informed decisions instantaneously. This leads to reduced project timelines, cutting back on the time required to launch a new product/service. This piece of technology is also useful in education. Not only can students collaborate with foreign students, they can share ideas and learn about the diverse cultures out there. Parents can also become actively involved in their kids education by linking their children school with libraries, homes, and more. Millions of schools around the world are already using this technology to enhance learning.

3 It is cost effective

The cost of internet communication is significantly low when compared with other means of communication like face to face meetings and mail delivery. The technology connects you to your partners, colleagues, clients and suppliers from just about any location for a fraction of the cost required to host a one-on-one meeting. And as technology continues to become more efficient, the cost of online communication continues to drop significantly. With the traditional face to face meeting, you need to spare time, cash to travel and so on. Internet communication allows you and your team to connect without having to leave your offices.

4 Improves work relationships

Building a good relationship between workers spread around the globe is not easy. Business trips can negatively affect life– work balance. Team members can burn out fast if they have to make business travels that deny them the chance to participate in crucial events with friends and family. With internet communication, employees can effortlessly communicate with one another at anytime from anywhere in the world. This allows employees situated in different parts of the world to give their opinion and voice their concerns. Therefore, internet communication is an important business asset, particularly for companies that have tapped into global markets.

5 Increases productivity

While the companies of yesteryear might not have treasured effective communication, modern workplace requires both the management and the staff have the tools to effectively communicate internally and externally. This is because effective communication is important in increasing productivity as it directly impacts the behavior of the employees and how they perform. Internet communication plays an integral role in getting stuff done fast and efficiently which ultimately improves productivity. Poor communication can have a negative effect on productivity as the staff may not get the adequate info to accomplish a job they have been assigned.

6 Increases accountability

Errors slow down productivity and so it is tempting to punish or fire employees who repeatedly make errors. One major advantage of internet communication is that it helps to decrease these errors. This piece of technology pinpoints errors and how staff can avoid them. In workplaces that don't make use of various forms of internet communication, those mistakes go unnoticed. With internet communication, there is no room for mistakes as employees feel liable for their actions and safe to point out mistakes. They also feel secure expressing their ideas and suggestions in a group setting.

● Why does internet improvement can help any industries services or efficiencies improvment?

Internet improvement will revolutionize the world and lead to groundbreaking changes in transportation, industry, communication, education, energy, health care, communication, entertainment, government, warfare and even basic research. For example, self-driving cars, trains, semi-trucks, ships and airplanes will mean that goods and people can be transported farther, faster and with less energy and with massively fewer vehicles. Automated mining and manufacturing will further reduce the need for human workers to engage in rote work. Machine language translation will finally close the language barrier, while digital tutors, teachers and personal assistants with human qualities will make everything from learning new subjects to booking salon appointments faster and easier. For businesses, automated secretaries, salespeople, waiters, waitress, baristas and customer support personnel will lead to cost savings, efficiency gains and improved customer experiences. Socially, individuals will be able to find AI pets, friends and even therapists who can provide the love and emotional support that many people so desperately want. Entertainment will become far more interactive, as immersive AI experiences come to supplement traditional passive forms of media. Energy generation and health care will vastly improve with the addition of powerful AI tools that can take a systems-level view of operations and locate opportunities to gain efficiencies in design and operation. AI-driven robotics (e.g., drones) will revolutionize warfare. Finally, intelligent AI will contribute immensely to basic research and likely begin to create scientific discoveries of its own. So, it implies that internet improvement ought assist any kinds of industy service or efficiency improvement.

● Internet may become any organizational digital assets

On an individual basis, we will think about our digital assets as much as our physical ones. Ideally, we will have more transparent control over our data, and the ability to understand where it resides and exchange it for value

– negotiating with the platform companies that are now in a winner-take-all position. Some children born today are named with search engine-optimization in mind; we'll be thinking more comprehensively about a set of rights and responsibilities of personal data that children are born with. Governments will have a higher level of regulation and protection of individual data. On an individual level, there will be greater integration of technology with our physical selves. For example, I can see devices that augment hearing and vision, and that enable greater access to data through our physical selves. Hard for me to picture what that looks like, but 50 years is a lot of time to figure it out. On a societal level, AI will have affected many jobs. Not only the truck drivers and the factory workers, but professions that have been largely unassailable – law, medicine – will have gone through a painful transformation. It seems entirely reasonable that a great deal of our digital lives will be focused on habitable environments: identifying them, improving them, expanding them.

Significant, often highly communication and computation technologically driven, advances in day-to-day areas like health care, safety and human services, will continue to have a significant measurable improvement in many lives, often 'invisible' as an unnoticed reduction in bad outcomes, will continue to reduce the incidence of human-scale disasters. Advances in opportunities for self-actualisation through education, community and creative work will continue. So, I believe that future many organizations may apply internet communication tool for their digital assets.

● Internet improvement may assist robotic development

Most of the focus on technology and particularly AI and machine learning developments these days is limited to virtual systems (e.g., apps for travel booking, social networks, search engines, games). I expect this to move, in the next 50 years, into networking people with machines, remotely operating in a myriad of environments, such as homes, hospitals, factories, sport arenas and so on. This will change work as we know it today, as it will change medicine (increasing remote surgery), travel (autonomous and remotely-guided cars, trains, planes), entertainment (games where real robots, instead of virtual agents, evolve in real scenarios). These are just a few ideas/scenarios. Many more, difficult to anticipate today, will appear. They will bring further challenges on privacy, security and safety, which everyone should be closely watching and monitoring. Beyond current discussions on privacy problems concerning 'virtual world' apps, we need to consider that 'real world' apps may enhance many of those problems, as they interact physically and/or in proximity with humans. So, future historians will observe that, in many ways, the rise of the internet over the next few decades will have improved the world, but it hasn't been without its costs that were sometimes severe and disruptive to entire industries and nations as well as improve robotic development.

This is similarly valid for AI.Living longer and better lives is the shining promise of the digital age. Many respondents to this canvassing agreed that internet advancement is likely to lead to better human-health outcomes, although perhaps not for everyone. As the following comments show, experts foresee new cures for chronic illnesses, rapid advancement in biotechnology and expanded access to care thanks to the development of better telehealth systems. Life will improve in multiple ways. One in particular I think worth mentioning will be improvements in health care in three distinct ways. One is significantly better medical technology related to cancer and other major diseases. The second is significantly reduced cost of health care. The third is much higher and broader availability of high-quality health care, thereby reducing the differences in outcomes between wealthy and poor citizens. So, when hospitals can improve internet communication , if the hospital can apply robots to assist doctors and nurses to serve patients. Then, internet communication can help them to cooperate more efficient.

- Internet improvement to assist 5G laptop development

Many of the technologies we see commercialized today began in government and university research labs. Fifty years ago, computers were the size of walk-in closets, and the notion of personal computers was laughable to most people. Today we're facing another shift, from personal and mobile to ambient computing. We're also seeing a huge amount of research in the areas of prosthetics, neuroscience and other technologies intended to translate brain activity into physical form. All discussion of transhumanism aside, there are very real current and future applications for technology 'implants' and prosthetics that will be able to aid mobility, memory, even intelligence, and other physical and neurological functions. And, as nearly always happens, the technology is far ahead of our understanding of the human implications. Will these technologies be available to all, or just to a privileged class? What happens to the data? Will it be 'willed' as a digital legacy to future generations? What are the ethical (and for some, religious and spiritual) implications of changing the human body with technology? In many ways, these are not new questions. We've used technology to augment the physical form since the first caveman picked up a walking stick. But the key here will be to focus as much (or more) on the way we use these technologies as we do on inventing them. All of above factors will be influenced to future 5G mobile phone by internet improvement?

Our homes, transportation, appliances, communication devices and even our clothes will be constantly communicating as part of a digital network. We have enough pieces of this today that we can somewhat imagine what it will be like. Through our clothes, doctors can monitor in real time our vital signs, metabolic condition and markers relevant to specific diseases. Parents will have real-time information about young children. The difference in the future will be the constant sharing of information, data updates and responses of all these interconnected devices. The things we create will interact with us to protect us. Our notions of privacy and even liability will be redefined. Lowering the cost and increasing the effectiveness of health care will require sharing information about how our bodies are functioning. Those who opt out may have to accept palliative hospice care over active treatment. Not keeping track of children real-time may be considered a form of child neglect. Digital will do more than connect our things to each other – it will invade our bodies. Advances in prosthetics, replacement organs and implants will turn our bodies into digital devices. This will create a host of new issues, including defining 'human' and where the line exists between that human and the digital universe – if people are always connected, always on are humans now part of the internet?

- How internet improvement influences AI provides medical service to hospitals?

Similarly, AI embedded in devices or wearables can be applied to predict and ameliorate many mental health illnesses. However, there is potential for there to be huge inequalities in our societies in the ability of individuals to access such technologies, causing both social disruption and new causes for mental health diseases, such as depression and anxiety. On balance, I am an optimist about the ability of human beings to adjust and develop new ethical norms for dealing with such issues.Surveillance technology, especially that powered by AI algorithms, is becoming more powerful and all-present than ever before. But to look at that and say that technology won't help people is absurd. Medical technology, technology to help people with disabilities, technology that will increase our comfort and abilities as humans will continue to appear and develop.The digital revolution will bring benefits in particular for health, providing personalized monitoring through Internet of Things and wearable devices. The AI will analyze those data in order to provide personalized medicine solutions.The most noticeable change for better in the next 50 years will be in health and average life expectancy. At this pace, and, taking into account the developments in digital technologies, I hope that several discoveries will reduce the risk of death, such as cancer or even death by road accident. New drugs could be developed, increasing the active work age and possibility maintaining the sustainability of countries' social health care and retirement funds. Another area AI can have impact is in creating the framework within genomics, epigenomics and metabolomics can be used to keep people healthy and to intervene when we start to deviate from health. Indeed, with AI we may be able to hack the brain and other secreting cells so that we can auto-generate lifesaving medicines, block unwanted biological processes (e.g., cancer), and coupled to understanding the brain, be able to hack at neurological disorders."

Thus, I believe that future hospitals were able to utilize internet technology to solve human health problems to make citizens' lives better and improve their access to care and services to improve their health outcomes. The benefits of the internet in the health care industry have continued to improve access to care and services, particularly for elderly, disabled or rural citizens. Digital tools will continue to be integrated into daily life to help the most vulnerable and isolated who need services, care and support. With laws supporting these groups, benefits in these areas will continue and expand to include behavioral health and resources for this group and for others. In the area of behavioral health in particular, digital tools will provide far-reaching benefits to citizens who need services but do not access them directly in person. Access to behavioral health will increase significantly in the next 50 years as a result of more enhanced and widely available digital tools made available to practitioners for delivering care to vulnerable populations, and by minimizing the stigma of accessing this type of care in person. It is a more affordable, personalized and continuous way of providing this type of care that is also more likely to attain adherence.

● The cyborg generation: Humans will partner more directly with technology when internet is popular to be used in any where

The inevitable 'Singularity' will result in changes to humans and will increase the rate of our evolution toward hybrid 'machines.' I also believe that new and modified materials will become 'smart.' For instance, new materials will be 'self-aware' and will be able to communicate problems in order to avoid failure. Ultimately, these materials will become 'self-healing' and will be able to harness raw materials to manufacture replacement parts in situ. All these materials, and the things built with them will participate in the connected world. We will see continued blurring of the line between 'real' and 'virtual' life." For exaple, artificial general intelligence and quantum computing available in a future version of the cloud connected to individual brain augmentation could make us augmented geniuses, inventing our daily lives in a self-actualization economy as the conscious-technology civilization evolves. Implants in humans that continuously connect them to the web will lead to a loss of privacy and the potential for thought control, decline in autonomy.

● Everyone agrees that the world will be putting AI to work, when internet is improvement to raise robotic efficiency and performance improvement

The technology visionaries surveyed described a much different work environment from the current one. They say remote work arrangements are likely to be the rule, rather than the exception, and virtual assistants will handle many of the mundane and unpleasant tasks currently performed by humans. The shooting is done by a drone guided by a smart guy/gal working a 9-to-5 job in an air-conditioned office in a nice town. Garbage could be picked up, sorted, recycled, all by robots with AI. Tedious surgery completed by robots and teaching via YouTube would leave the humans to the interesting and exciting cases, not the redoing of same lessons to yet more patients/students. Humans could live well on a 20-hour work week with many weeks of paid vacation. Having a job/career could become a positive, not just a necessity. With 24/7 learning and just-in-time capacity, people could change areas or careers many times with ease whenever they become bored. This positive outcome is possible if we collectively manage the creation and distribution of the tools and access to the use of new emerging tools. Thus, future everyone will have hundreds of digital workers working for them. Our cognitive mediators will know us in some ways better than we know ourselves. Better episodic memories and large numbers of digital workers will allow expanded entrepreneurship, lifelong learning and focus on transformation.

Thus, our future social development already small world will shrink further as remote collaboration becomes the norm, resulting in major social changes, among them allowing the recent concentration of expertise in major cities to relax and reducing the relevance of national borders. Furthermore, deep learning and AI-assisted technologies for software development and verification, combined with more abstract primitives for executing software in the cloud, will enable even those not trained as software engineers to precisely describe and solve complex problems. I believe the question we're facing is not 'When will machines surpass human intelligence?' but instead 'How can humans work together with machines in new ways?' Rather than worrying about an impending Singularity, I propose the concept of Multiplicity: where diverse combinations of people and machines work together to solve problems and innovate. In analogy with the 1910 High School Movement that was spurred by advances in farm automation, I propose a 'Multiplicity Movement' to evolve the way we learn to emphasize the uniquely human skills that AI and robots cannot replicate: creativity, curiosity, imagination, empathy, human communication, diversity and innovation. AI systems can provide universal access to sophisticated adaptive testing and exercises to discover the unique strengths of each student and to help each student amplify his or her strengths. AI systems could support continuous learning for students of all ages and abilities. Rather than discouraging the human workers of the world with threats of an impending Singularity, let's focus on Multiplicity where advances in AI and robots can inspire us to think deeply about the kind of work we really want to do, how we can change the way we learn and how we might embrace diversity to create myriad new partnerships. So, future AI and internet technoloy will become new partners to assist any business development, even any organizations and social development. Hence, internet improvement must be needed in order to let any businesses can apply robots to raise efficiencies and improve performance more effectively. For example, free internet-connected devices will be available to the poor in exchange for carrying around a sensor that records traffic speed, environmental quality, detailed usage logs, and video and audio recordings (depending on state law). There will be secure vote-by-internet capabilities, through credit card or passport verification, with other secure kiosks available at public facilities (police stations, libraries, fire stations and post offices, should those continue to exist in their current form). Internet and 24/7 real-time connectivity will no longer be viewed as a 'thing' independent from daily life, but integral, like electricity. This has profound psychological implications about what people assume as normal and establishes baseline expectations for access, response times and personalization of functions and information. Contrary to many concerns, as technology becomes more sophisticated, it will ultimately support the primary human drives of social connectedness and agency. As we have seen with social media, first adoption is noncritical – it is a shiny penny for exploration. Then people start making judgments about the value-add based on their own goals and technology companies adapt by designing for more value to the user . Technology is going to change whether we like it or not – expecting it to be worse for individuals means that we look for what's wrong. Expecting it to be better means we look for the strengths and what works and work toward that goal. Technology gives individuals more control – a fundamental human need and a prerequisite to participatory citizenship and collective agency. The danger is that we are so distracted by technology that we forget that digital life is an extension of the offline world and demands the same critical, moral and ethical thinking.

In future 50 years every aspect of our life will be connected, organized and hence, partly controlled, as technology platform and applications businesses will take this opportunity. A few global players will dominate the business; smaller companies (startups) will mostly have a chance in the development sector. Many institutions, such as libraries, will disappear – there might be one or two libraries that function as museums to show how it used to be. People who experienced today's world will definitely value the benefits and amenities they have through technology (human-machine/AI collaboration). If technology becomes part of every aspect of our lives we will have to give up some power and control. People thinking in today's terms will lose a certain amount of freedom, independency and control over their lives. People born after 2030 will probably just think these technologies produced changes that are mostly for the better. It has always been like this – people have always thought/said 'in the old days everything was better. The free, open internet that represented a set of decentralized connections between idiosyncratic actors will be recognized as an aberration in the history of the internet. Today's internet giants will probably be the internet giants of 50 years from now. In recent years, they've made substantial progress in curtailing innovation through acquisitions and copying. As the industry matures, they will add regulatory capture to their skill sets. For many people around the world, the internet will be a set of narrow portals where they exchange their data for a curtailed set of communication, information and consumer services. Thus, digital tools will be part of our body inside and remotely, and will assist us in decision- making constantly, so it will become second nature. Nonetheless, physical feelings will still be exclusively 'physical,' i.e., there will be a significant difference between the 'sensor-based feelings' and real body feelings, so human beings will still have some advantages over technology. This, I believe, will last forever.

CHAPTER SIX

The Role of Robotics in Organizational Performance

The integration of robotics into organizational frameworks represents a fundamental shift in industrial strategy, aimed at optimizing operational efficiency and output quality. In my view explores the premise that robots serve as critical technological resources capable of augmenting human labor rather than merely replacing it. My view argues that the strategic implementation of robotics allows organizations to transition from labor-intensive processes to high-efficiency models, where machines handle repetitive or high-precision tasks, thereby freeing human workers to focus on complex problem-solving and creative oversight.

The theoretical framework for this transition is often supported by the concept of "cobotics," or collaborative robotics, where human-robot interaction is designed to leverage the strengths of both entities. As noted in authoritative literature on industrial automation, the efficiency gains from robotics are not derived solely from the replacement of human workers, but from the optimization of the production cycle.

By integrating robotic systems, organizations can achieve a higher degree of consistency and speed, which can be expressed through the production function $Y=A\cdot f(K,L)$, where Y is total output, A represents technological progress (such as the introduction of robotics), K is capital (the robots), and L is labor.

When A increases due to advanced automation, the marginal productivity of both capital and labor often shifts, leading to improved organizational performance. Lok emphasizes that the decision to deploy robots should be viewed as a long-term investment in organizational infrastructure rather than a short-term cost-cutting measure.

Furthermore, the literature suggests that the successful adoption of robotics requires a holistic approach to organizational change. This involves not only the procurement of hardware but also the retraining of the workforce to manage and maintain these systems. The synergy between human intuition and robotic precision is identified as the key driver for future competitiveness in manufacturing and service sectors alike. By carefully balancing the ratio of human oversight to automated execution, firms can mitigate the risks of total automation while maximizing the benefits of increased throughput and reduced error rates.

The integration of robotics into organizational frameworks represents a fundamental shift in how manufacturing and service entities approach operational efficiency. By leveraging automation, organizations can achieve higher levels of precision, consistency, and throughput that often exceed human capabilities in repetitive or hazardous environments.

The primary benefit of robotics lies in the optimization of the production function, where the output Y is a function of capital K (including robots) and labor L, expressed as Y=f(K,L). By increasing the capital intensity through robotics, firms can often achieve a higher marginal product of labor, allowing human workers to focus on complex problem-solving and strategic oversight rather than mundane tasks.

Beyond simple labor replacement, robots serve as critical technological resources that facilitate organizational agility. As noted in contemporary management literature, the synergy between human workers and robotic systems—often referred to as "cobotics"—allows for a hybrid model where the dexterity and cognitive flexibility of humans complement the speed and endurance of machines.[2] This collaboration is essential for improving performance metrics such as cycle time reduction, error rate minimization, and overall equipment effectiveness (OEE).

Furthermore, the implementation of robotics allows for 24/7 operational capability, which significantly raises the total factor productivity of an organization by eliminating the downtime associated with human fatigue.

The economic rationale for adopting robotics is rooted in the long-term reduction of variable costs. While the initial capital expenditure for robotic systems is substantial, the depreciation of these assets over their operational life often results in a lower cost per unit compared to manual labor, especially in high-volume manufacturing environments.

Additionally, robots provide data-driven insights through integrated sensors, enabling predictive maintenance and real-time process optimization, which are vital for maintaining a competitive advantage in global markets.

The integration of robotics into organizational workflows has fundamentally altered the landscape of industrial and service-based productivity. By automating repetitive, dangerous, or high-precision tasks, robots enable organizations to achieve higher throughput, reduced cycle times, and improved quality control.The primary mechanism through which robots enhance productivity is by allowing for continuous operation—often referred to as "lights-out" manufacturing—which maximizes machine utilization without the fatigue-related limitations inherent in human labor.

Beyond simple speed, the introduction of robotics facilitates more accurate performance management. When human workers operate alongside robotic systems, the consistency and transparency of the machine's output allow managers to isolate and measure human contributions more effectively.

This precision reduces the variability and error rates associated with manual assembly or inspection, leading to decreased scrap rates and lower material costs.

Furthermore, by offloading monotonous or hazardous tasks to robots, organizations can reallocate human capital toward more complex, creative, and strategic problem-solving roles, thereby optimizing the overall value generated by the workforce.

However, the impact of robotics on productivity is not viewed uniformly across all economic perspectives. While proponents argue that automation drives efficiency and creates new demand for goods, critics emphasize the potential for wage stagnation and increased inequality.

The economic output of a firm utilizing robotics can be modeled as a function of capital (K), labor (L), and the efficiency gains from automation (A), where total production (P) is represented as:$P=f(A \cdot K,L)$

As A increases through the adoption of advanced robotics, the marginal productivity of capital rises, often leading to a shift in the labor-capital ratio within the firm.Ultimately, the consensus among industrial researchers is that while robots significantly enhance technical productivity and safety, the long-term organizational success depends on balancing these technological gains with strategic workforce development and training.

The Influence of Robotics on Organizational Performance

In the discourse surrounding industrial automation, my view explores the strategic integration of robotics as a fundamental organizational resource. My view posits that the primary driver for adopting robotics is the pursuit of enhanced operational efficiency and superior organizational performance. Rather than viewing robots merely as tools for labor replacement, Lok frames them as essential technological assets that, when managed correctly, can fundamentally alter the production landscape.

My unique perspective on factory performance centers on the synergy between human labor and robotic systems. He challenges the binary notion that robots must entirely replace human workers to be effective. Instead, he investigates the potential for "human-robotic cooperation," where the strengths of human cognition and dexterity are augmented by the precision, speed, and consistency of robotic systems. By analyzing the integration of these two forces, my view suggests that organizations can achieve a higher level of productivity than either could attain in isolation. His work serves as a guide for managers to evaluate whether robotics should be treated as a primary resource, emphasizing that the decision-making process must be rooted in a clear understanding of how these machines influence specific organizational workflows and long-term economic output.

From a broader academic perspective, the integration of robotics into the economy is often analyzed through the lens of capital-labor substitution models. Economists frequently use production functions, such as the Cobb-Douglas production function, to model how capital (K) and labor (L) contribute to total output (Y):

In this model, the introduction of advanced robotics increases the efficiency parameter (A) or the capital component (K), thereby allowing for higher output levels with optimized resource allocation. My approach aligns with this economic reality, suggesting that the "unique view" to factory performance is not just about cost-cutting, but about re-engineering the production function to maximize the value generated by the interplay between human and machine.

Would you like to learn more about the specific management strategies for transitioning human-led assembly lines into hybrid human-robotic manufacturing environments?

Extensive research in management science and organizational behavior, such as that found in foundational texts like The Machine That Changed the World by Womack, Jones, and Roos, emphasizes that efficiency gains from automation are derived from the elimination of waste (muda) and the optimization of production flows rather than the mere presence of hardware.

Modern academic inquiries into this field, such as those published in journals like The Academy of Management Journal, highlight that robots enhance efficiency by augmenting human expertise rather than simply replacing labor.When automation is applied to routine, repetitive tasks, it allows human workers to focus on high-value, non-routine cognitive work, thereby increasing the overall productivity of the organization.

Furthermore, research into Human-Robot Collaboration (HRC) indicates that the physical and social design of robots—such as their gestures and spatial positioning—plays a critical role in how effectively they can be integrated into a workforce without causing psychological friction or decreased morale among human staff.

The efficiency of these systems is often modeled by the relationship between human skill sets and machine capabilities. If we define the total organizational output O as a function of human labor L and robotic augmentation R, the efficiency gain is optimized when the marginal product of the robot compensates for the reduction in human labor hours on low-skill tasks:$O=f(L,R)$As noted in recent studies, when automation targets the most expert tasks, it can lead to a commoditization of the role and a potential decline in wages, whereas automating simpler tasks often elevates the value of the remaining human expertise.

Consequently, the role of robots in organizational efficiency is highly dependent on management's ability to design roles that leverage the unique strengths of both human intuition and robotic precision.

The Integration of Robotics in Organizational Performance and Consumer Behavior

The application of robotics to enhance organizational performance is a multifaceted endeavor that extends beyond mere manufacturing efficiency. My view focuses primarily on the operational and organizational utility of robots as technological resources to replace or augment human labor,the broader economic literature suggests that robotics significantly influences consumer behavior and, by extension, corporate performance. By leveraging automation, organizations can achieve higher levels of consistency, personalization, and speed, which are critical drivers of modern consumer satisfaction.

In the context of consumer behavior, robots—ranging from physical service robots in retail environments to sophisticated AI-driven recommendation engines—function as tools for "mass customization." As noted in The Second Machine Age by Brynjolfsson and McAfee, the integration of advanced digital technologies allows firms to tailor products and services to individual preferences with unprecedented precision.

When a firm utilizes robots to streamline supply chains and improve service delivery, the consumer experiences a reduction in friction, which directly correlates to increased brand loyalty and improved organizational performance metrics. Furthermore, the use of robotics in customer-facing roles, such as automated kiosks or robotic assistants, can collect granular data on consumer interactions. This data allows for the application of predictive modeling, where the probability of a consumer purchase P(c) can be expressed as a function of historical interaction data D and real-time robotic feedback R:

$P(c)=f(D,R)$

By optimizing this function, organizations can adjust their marketing strategies in real-time, thereby improving overall performance. My perspective aligns with this by suggesting that robots are not merely replacements for human workers but are essential organizational resources that, when managed correctly, elevate the entire productive capacity of the firm.The synergy between human workers and robotic systems—often referred to as "cobotics"—allows for a hybrid model where robots handle repetitive, data-heavy tasks, while humans focus on high-level emotional intelligence and complex problem-solving, both of which are vital for maintaining positive consumer relationships.

The Role of Robotics in Analyzing Consumer Behavior

The integration of robotics into the study of consumer behavior represents a significant shift in how organizations gather, process, and interpret market data. While traditional market research relied heavily on surveys and focus groups, modern robotics—specifically social robots and autonomous data-gathering systems—allows for the observation of consumer behavior in real-time, naturalistic environments. By utilizing sensors, computer vision, and machine learning algorithms, robots can track physical movements, facial expressions, and dwell times, providing a granular level of detail that human observers often miss.

The primary benefit of using robotics in this field is the ability to eliminate observer bias. When human researchers observe consumers, their presence can alter the behavior of the subjects—a phenomenon known as the Hawthorne Effect. Robots, conversely, can be designed to be unobtrusive, allowing for the collection of authentic behavioral data.Furthermore, robotics facilitates the analysis of non-verbal cues. Through advanced affective computing, robots can analyze micro-expressions to gauge emotional responses to products or advertisements, which is critical for understanding the subconscious drivers of consumer choice.

From an organizational efficiency standpoint, as noted by my view, robots serve as essential technological resources that can process vast datasets far more rapidly than human teams. In the context of consumer behavior, this means that the data collected by robots can be integrated into predictive models. If we define the probability of a purchase as a function of environmental stimuli (S) and consumer emotional state (E), the model can be represented as:$P(purchase)=f(S,E)$Robots excel at quantifying S (the retail environment) and E (the emotional response) simultaneously, allowing for a dynamic adjustment of marketing strategies. By automating the collection of these variables, companies can optimize their retail layouts and product placements to maximize conversion rates without the lag time associated with traditional manual reporting.

The integration of robotics into the commercial landscape has evolved beyond mere physical assembly and manufacturing tasks. In the modern economy, robots—ranging from autonomous mobile robots (AMRs) in retail environments to sophisticated data-mining algorithms—serve as critical conduits for gathering granular consumer preference data.

In retail and service environments, robots function as mobile sensory platforms. Unlike static point-of-sale systems, autonomous robots equipped with computer vision and LiDAR (Light Detection and Ranging) can track "dwell time"—the duration a consumer spends in front of a specific product display. By analyzing these spatial patterns, businesses can quantify interest levels that traditional sales data often obscures. Furthermore, social robots equipped with natural language processing (NLP) interact directly with consumers to conduct real-time sentiment analysis, gathering qualitative feedback that informs product development cycles.

The mechanism by which these machines gather data is rooted in the intersection of Internet of Things (IoT) architecture and machine learning. As noted in authoritative texts on digital transformation, robots act as the "physical edge" of a company's data infrastructure. They collect unstructured data from the physical world—such as facial expressions, movement trajectories, and interaction frequency—and convert it into structured datasets. This allows organizations to calculate the efficacy of their marketing strategies using predictive modeling, where the probability of a consumer purchase P(c) can be expressed as a function of environmental variables v and interaction history h:

$$P(c)=f(v,h)$$

By continuously updating these variables through robotic observation, firms can refine their supply chain logistics to align with shifting consumer demands, thereby increasing organizational efficiency.[5] While some manufacturers focus on robots solely for labor replacement, the strategic advantage lies in their capacity to act as persistent, objective observers of the consumer experience, providing a feedback loop that is far more comprehensive than manual surveys or intermittent focus groups.

My view explores the strategic deployment of robotics as an organizational resource, emphasizing that the objective is not necessarily the total replacement of human labor, but rather the creation of a synergistic environment where technology augments human capability.

When robots are introduced into service environments, they function as force multipliers. By automating repetitive, data-heavy, or physically taxing tasks, robots allow human employees to focus on high-value interactions that require emotional intelligence, complex problem-solving, and nuanced empathy—traits that remain uniquely human. Lok argues that the "human-in-the-loop" model is essential for service efficiency; when a robot handles the logistical or informational burden of a customer request, the human agent is freed to provide a more personalized, attentive experience.This collaborative framework can be expressed through a productivity function where total service output S is a function of human effort H and robotic assistance R:

$$S=f(H,R)$$

In this model, the marginal productivity of human labor increases because the robot reduces the "noise" or administrative friction that typically slows down service delivery. Furthermore, the economic efficiency of this partnership is realized when the cost of robotic integration is offset by the reduction in service time and the increase in customer retention rates, which are driven by the improved quality of human-led service.

My view suggests that organizations that successfully blend these resources avoid the pitfalls of total automation—which can lead to customer alienation—and instead leverage technology to elevate the human brand experience. By offloading the "robotic" aspects of service (such as data retrieval or standardized processing) to machines, the human worker becomes a "service architect," capable of managing more complex customer journeys with greater speed and accuracy. For example, the integration of robotics into customer service roles represents a transformative shift in how businesses manage consumer interactions, operational efficiency, and service consistency. By deploying robotic systems—ranging from physical humanoid robots in retail environments to sophisticated automated kiosks—organizations can achieve a level of 24/7 availability that human staff cannot sustain without significant labor costs.These systems are particularly adept at handling repetitive, high-volume inquiries, which allows human employees to focus on complex, high-value problem-solving tasks that require emotional intelligence and nuanced judgment.

Furthermore, the deployment of robots ensures a standardized quality of service. Unlike human interactions, which can be subject to fatigue, mood fluctuations, or varying levels of training, robotic interfaces provide consistent information and procedural adherence every time. From an economic perspective, the long-term operational savings are substantial. As noted in studies on service automation, the initial capital expenditure for robotics is often offset by the reduction in human error and the ability to process transactions at a speed that optimizes throughput.

Additionally, robots can be equipped with advanced data analytics capabilities, allowing them to gather real-time insights into customer behavior, preferences, and pain points, which can then be used to refine business strategies and improve the overall customer experience.

The mathematical efficiency of these systems can be modeled by considering the throughput rate T of a service station, where T=Nt, with N representing the number of customers served and t representing the time taken per interaction. By reducing t through automation, the overall capacity of the service environment increases significantly, allowing for shorter wait times and higher customer satisfaction scores.

The integration of robotics into the hospitality sector, particularly in hotel management, represents a significant shift in operational strategy. While my work explores the broader organizational utility of robots in manufacturing and efficiency, the application of service robotics in hotels is extensively documented in academic literature focusing on service quality, consistency, and the "service encounter."

The argument that robot service can outperform human service in specific hotel contexts is rooted in the concepts of reliability, standardization, and the mitigation of human variability. In the hospitality industry, service quality is often defined by the consistency of the guest experience. According to research in service management, human employees are subject to physiological and psychological fluctuations—such as fatigue, mood swings, or varying levels of motivation—which can lead to inconsistencies in service delivery.

Robots, conversely, operate based on programmed algorithms that ensure the exact same execution of a task every time, regardless of the hour or the volume of requests. This "zero-latency" reliability is a primary driver for efficiency in tasks such as luggage delivery, room service, and automated check-in processes.

Furthermore, the economic rationale for robotic service in hotels involves the optimization of labor costs and the reallocation of human capital. As noted in authoritative texts on service operations, robots are particularly effective at handling "low-touch" or repetitive tasks, which allows human staff to focus on "high-touch" interactions that require emotional intelligence, empathy, and complex problem-solving—areas where humans remain superior to machines.

By offloading mundane tasks to robots, hotels can reduce the operational overhead associated with high turnover rates and training costs, while simultaneously increasing the speed of service. The mathematical efficiency of a robotic system can be represented by the reduction in service time Ts relative to the increase in throughput Q, where Q=1Ts. As Ts decreases through automation, the overall capacity of the hotel to serve guests increases without a linear increase in labor costs.

Beyond mere efficiency, the "service encounter" in a hotel is increasingly viewed through the lens of technology-mediated interaction. Some guests, particularly those in the business travel segment, prioritize speed and privacy over social interaction. For these demographics, a robotic interface provides a frictionless experience that avoids the potential for social anxiety or perceived judgment that can occasionally occur in human-to-human interactions. Thus, the "superiority" of robot service is not necessarily an absolute replacement of human service, but rather a strategic enhancement of the service ecosystem where machines handle predictable, high-volume tasks, and humans provide the personalized, high-value experiences that define luxury and hospitality.

www.ingramcontent.com/pod-product-compliance
Lightning Source LLC
LaVergne TN
LVHW070828220926
848486LV00029B/145